Reusable and Sustainable Building Materials in Modern Architecture

Gülşah Koç
Yildiz Technical University, Turkey

Bryan Christiansen
Global Research Society, LLC, USA

A volume in the Advances in Civil and Industrial Engineering (ACIE) Book Series

IGI Global
DISSEMINATOR OF KNOWLEDGE

Published in the United States of America by
IGI Global
Engineering Science Reference (an imprint of IGI Global)
701 E. Chocolate Avenue
Hershey PA, USA 17033
Tel: 717-533-8845
Fax: 717-533-8661
E-mail: cust@igi-global.com
Web site: http://www.igi-global.com

Library of Congress Cataloging-in-Publication Data

Names: Koc, Gulsah, 1987- editor. | Christiansen, Bryan, 1960- editor.
Title: Reusable and sustainable building materials in modern architecture / Gulsah Koc and Bryan Christiansen, editors.
Description: Hershey, PA : Engineering Science Reference, [2019] | Includes bibliographical references.
Identifiers: LCCN 2018011700| ISBN 9781522569954 (hardcover) | ISBN 9781522569961 (ebook)
Subjects: LCSH: Building materials--Recycling. | Building materials--Environmental aspects. | Sustainable buildings. | Sustainable architecture--Materials. | Architecture, Modern--21st century.
Classification: LCC TA403.6 .R48 2019 | DDC 691.028/6--dc23 LC record available at https://lccn.loc.gov/2018011700

This book is published in the IGI Global book series Advances in Civil and Industrial Engineering (ACIE) (ISSN: 2326-6139; eISSN: 2326-6155)

British Cataloguing in Publication Data
A Cataloguing in Publication record for this book is available from the British Library.

All work contributed to this book is new, previously-unpublished material.
The views expressed in this book are those of the authors, but not necessarily of the publisher.

For electronic access to this publication, please contact: eresources@igi-global.com.

Advances in Civil and Industrial Engineering (ACIE) Book Series

ISSN:2326-6139
EISSN:2326-6155

Editor-in-Chief: Ioan Constantin Dima, University Valahia of Târgovişte, Romania

MISSION

Private and public sector infrastructures begin to age, or require change in the face of developing technologies, the fields of civil and industrial engineering have become increasingly important as a method to mitigate and manage these changes. As governments and the public at large begin to grapple with climate change and growing populations, civil engineering has become more interdisciplinary and the need for publications that discuss the rapid changes and advancements in the field have become more in-demand. Additionally, private corporations and companies are facing similar changes and challenges, with the pressure for new and innovative methods being placed on those involved in industrial engineering.

The **Advances in Civil and Industrial Engineering (ACIE) Book Series** aims to present research and methodology that will provide solutions and discussions to meet such needs. The latest methodologies, applications, tools, and analysis will be published through the books included in **ACIE** in order to keep the available research in civil and industrial engineering as current and timely as possible.

COVERAGE

- Materials Management
- Engineering Economics
- Quality Engineering
- Production Planning and Control
- Coastal Engineering
- Structural Engineering
- Ergonomics
- Productivity
- Construction Engineering
- Earthquake engineering

IGI Global is currently accepting manuscripts for publication within this series. To submit a proposal for a volume in this series, please contact our Acquisition Editors at Acquisitions@igi-global.com or visit: http://www.igi-global.com/publish/.

The Advances in Civil and Industrial Engineering (ACIE) Book Series (ISSN 2326-6139) is published by IGI Global, 701 E. Chocolate Avenue, Hershey, PA 17033-1240, USA, www.igi-global.com. This series is composed of titles available for purchase individually; each title is edited to be contextually exclusive from any other title within the series. For pricing and ordering information please visit http://www.igi-global.com/book-series/advances-civil-industrial-engineering/73673. Postmaster: Send all address changes to above address.

Titles in this Series

For a list of additional titles in this series, please visit:
https://www.igi-global.com/book-series/advances-civil-industrial-engineering/73673

Contemporary Strategies and Approaches in 3-D Information Modeling
Bimal Kumar (Glasgow Caledonian Univerity, UK)
Engineering Science Reference • ©2018 • 313pp • H/C (ISBN: 9781522556251) • US $205.00

New Approaches, Methods, and Tools in Urban E-Planning
Carlos Nunes Silva (University of Lisbon, Portugal)
Engineering Science Reference • ©2018 • 407pp • H/C (ISBN: 9781522559993) • US $205.00

Designing Grid Cities for Optimized Urban Development and Planning
Guiseppe Carlone (Independent Researcher, Italy) Nicola Martinelli (Politecnico di Bari, Italy) and Francesco Rotondo (Politecnico di Bari, Italy)
Engineering Science Reference • ©2018 • 305pp • H/C (ISBN: 9781522536130) • US $210.00

Design Solutions for nZEB Retrofit Buildings
Elzbieta Rynska (Warsaw University of Technology, Poland) Urszula Kozminska (Warsaw University of Technology, Poland) Kinga Zinowiec-Cieplik (Warsaw University of Technology, Poland) Joanna Rucinska (Warsaw University of Technology, Poland) and Barbara Szybinska-Matusiak (Norwegian University of Science and Technology, Norway)
Engineering Science Reference • ©2018 • 362pp • H/C (ISBN: 9781522541059) • US $225.00

Handbook of Research on Perception-Driven Approaches to Urban Assessment and Design
Francesco Aletta (University of Sheffield, UK) and Jieling Xiao (Birmingham City University, UK)
Engineering Science Reference • ©2018 • 641pp • H/C (ISBN: 9781522536376) • US $295.00

Convective Heat and Mass Transfer in the Free Atmosphere Emerging Research and ...
Kaliyeva Kulyash (Lorraine University, France)
Engineering Science Reference • ©2018 • 127pp • H/C (ISBN: 9781522530503) • US $135.00

For an entire list of titles in this series, please visit:
https://www.igi-global.com/book-series/advances-civil-industrial-engineering/73673

IGI Global
DISSEMINATOR OF KNOWLEDGE

701 East Chocolate Avenue, Hershey, PA 17033, USA
Tel: 717-533-8845 x100 • Fax: 717-533-8661
E-Mail: cust@igi-global.com • www.igi-global.com

Editorial Advisory Board

Table of Contents

Detailed Table of Contents

This chapter focuses on how sand, the second most used natural resource on earth after water, is facing one of the greatest environmental challenges of the new millennium. Sand is a crucial material used in all sorts of building projects, from asphalt, concrete, and glass. Globally, construction accounts for the largest portion of the 15 billion tons of sand consumed annually. Yet, sand is a finite resource and the depletion of alluvial sand used in construction is destroying the ecosystem of riverbeds, sea beds, and coastal beaches, and is contributing seriously to climate change. This chapter will discuss how these threats have developed, including coastal construction and erosion, river dredging, and sand "mafias" whereby illegal sand miners strip beaches and use sand in inferior concrete that has led to building collapses and deaths. The authors then discuss potential solutions to this crisis, including regulation and enforcement of environmental and construction standards, as well as materials substitution such as desert sand, sand created from sandstone, and recycled glass.

Current approaches to designing sustainable buildings are inadequate for meeting environmental goals. Buildings continue to consume nearly half of all resources, and architects, engineers, and contractors remain complicit in their deficient environmental performance—as well as the consequential global overshoot of resource consumption. It is imperative that the AEC industry pursue an alternative approach to green rating systems with the intent to determine measurable, absolute outcomes. The most appropriate existing model is the ecological footprint (EF) method devised by Mathis Wackernagel and William Rees at the University of British Columbia in the early 1990s. EF quantifies the human demand on the environment in terms of both resources and waste, translating these impacts into land area equivalents. This chapter aims to evaluate EF methodology for buildings by analyzing existing models and proposing new approaches while identifying their respective opportunities and limitations.

Urbanization is creating enormous pressure for the effective utilization of the existing land with demolition of old structures for new and modern structures. The debris produced in demolition of these structures are in large amount and disposal of this waste in sustainable manner is the biggest challenge being faced today and should be considered as a resource. With the increasing waste production and public concerns regarding the environment, it is desirable to recycle these materials. If suitably processed in appropriate industrial plants, these materials can be profitably used in concrete. This chapter highlights the composition of construction and demolition waste, the necessity for its recycling, and possibilities that can be implemented for its resourceful use, further focusing on current trends in this field by elaborating various ways to use these waste from laboratory research scale to commercially available technologies around the globe. The chapter concludes with future research directions and guidelines for sustainable use of these wastes.

Increase in urbanization leads to more construction of houses, dams, and streets. Reduction of the global warming effects can be carried out by recycling of construction material and searching for eco-friendly construction material. Greenhouse gas emissions can be reduced with the help of construction material which requires

This chapter describes the approach adopted within the framework of a multi-destination development project; the goal of which is to promote innovative technologies and methods to evaluate the environmental quality of an urban district under construction. This method of analysis has been tested on an area located in the former historic district of the Fiera di Milano, where a series of typical urban functions are inserted within a large public park. The success of the work is represented by indicators (air quality, acoustic, microclimate) that relate to the finished district and that can be compared with average values in the same city. The system may constitute a protocol capable of bringing benefits to local authorities. This type of assessment could be requested of developers/builders for complex projects, resulting in changes to the initial plan if the assessment identifies critical issues related to the design choices (orientation of buildings, green areas, traffic emissions, etc.) with the ultimate goal of creating neighborhoods with better environmental conditions.

Preface

While structures are necessary to meet housing demands, supplying, processing, and transporting of the construction materials to the construction sites reduces a significant amount of energy and resources and creates a global problem. With increasing urbanization, land use changes and this creates great pressures on both natural resources and the ecosystem. To reduce waste, preserve resources, and increase energy efficiency, it is necessary to consider reusable and sustainable building materials and ecological design. In this context, energy and resource use and material selection are important criteria. Selected building materials should be suitable for the ecosystem and should not produce waste. Consequently, reusable and sustainable building materials and ecological design understanding can be effective in ensuring energy and resource efficiency. This book is a reflection of that goal via the following nine chapters that are outlined here.

Chapter 1 focuses on how sand, the second most used natural resource on earth after water, is facing one of the greatest environmental challenges of the new millennium. Sand is a crucial material used in all sorts of building projects from asphalt to concrete and glass. Globally, construction accounts for the largest portion of the 15 billion tons of sand consumed annually. Sand is a finite resource and the depletion of alluvial sand used in construction is destroying the ecosystem of riverbeds, sea beds, and coastal beaches, and is contributing seriously to climate change. This chapter will discuss how these threats have developed, including coastal construction and erosion, river dredging, and sand "mafias" whereby illegal sand miners strip beaches and use sand in inferior concrete that has led to building collapses and deaths. We then discuss potential solutions to this crisis, including regulation and enforcement of environmental and construction standards, as well as materials substitution such as desert sand, sand created from sandstone, and recycled glass.

Chapter 2 investigates preliminary methods for measuring the true environmental impact of buildings. Current approaches to designing sustainable buildings are inadequate for meeting environmental goals. Buildings continue to consume nearly half of all resources, and architects, engineers, and contractors remain complicit in their deficient environmental performance—as well as the consequential global overshoot of resource consumption. It is imperative that the AEC industry pursue an alternative approach to green rating systems with the intent to determine measurable, absolute outcomes. The most appropriate existing model is the Ecological Footprint (EF) method devised by Mathis Wackernagel and William Rees at the University of British Columbia in the early 1990s. EF quantifies the human demand on the environment in terms of both resources and waste, translating these impacts into land area equivalents. This chapter aims to evaluate EF methodology for buildings by analyzing existing models and proposing new approaches while identifying their respective opportunities and limitations.

Chapter 3 explores the recycling and reuse of building materials for construction and demolition for sustainable growth. Urbanisation is creating enormous pressure for the effective utilisation of the existing land with demolition of old structures for new and modern structures. The debris produced in demolition of these structures are in large amount and disposal of this waste in sustainable manner is the biggest challenge being faced today and should be considered as a resource. The increasing waste production and public concerns regarding the environment, it is desirable to recycle these materials. If suitably processed in appropriate industrial plants, these materials can be profitably used in concrete. This chapter highlights the composition of Construction and Demolition waste, the necessity for its recycling and possibilities that can be implemented for its resourceful use, further focusing on current trends in this field by elaborating various ways to use these waste from laboratory research scale to commercially available technologies around the globe and finally the chapter concludes with future research direction and guidelines for sustainable use of these waste.

Chapter 4 covers eco-friendly construction. The increase in urbanization leads to increasing numbers of constructions of houses, dams and streets. Reduction of the global warming effects can be carried out by recycling of construction material and searching eco-friendly construction material. Green house gas emissions can be reduced with the help of construction material which requires less energy for their production. Concept of eco-friendly construction is based on biomimetic (i.e., finding natural material with potential of endurance and self cleaning property). Construction materials like portland cement and concrete can be replaced by eco-friendly biocement and bioconcrete. Production of biocement and bioconcrete can be done by using plants, algae and bacteria. The use of less cement in concrete leads to less pollution. Concrete is the mixture of cement, sand, gravel and water. By addition of pozzolan in concrete, requirement of cement will be reduced. In the current review, major emphasis is given on eco-friendly construction material.

Chapter 5 studies the proliferation of architectural follies that use recycled or recyclable materials in a move to promote better practices in waste and recycling. Given the slow uptake of this impetus in the architectural world proper, the text will investigate the obstacles in engaging in materially sustainable practices in the construction industry as well as case studies for rethinking currently problematic materials. However, while some improvements have been made in the construction industry's use of recycled materials, the industry often dismisses the after-life of materials used throughout the process. What are the motivations of the industry and how can we incentivize circular thinking in an industry that produces hundreds of millions of tons of waste per year in the USA.

Chapter 6 provides a comprehensive evaluation for mortars and concretes incorporating wastes. Sustainability is concerned with the most efficient use of resources where the residues play an essential role. Trends in concrete technology include natural or artificial additions and additives in order to reduce the consumption of cement. The characterization of the wastes is of great importance with respect to the amount that must be incorporated into the matrices of construction materials both for its economic and engineering impacts (strength and durability). The authors study the impact in strength, durability and sustainability, of the use of finely ground waste of ferroalloys in concrete. The behaviour of durability of sustainable concrete also is evaluated. The proportioning between traditional materials and these additions involves preliminary tests on pastes and mortars. They also study the impact of the use of different plastic wastes (polyethylene) in different percentages. They evaluated consistency, compressive strength, suction capability and leaching.

Chapter 7 addresses energetic forms of matter. One of the challenges that architects and designers are confronted with in contemporary contexts is the need to address an ethical responsibility towards the health of the environment through understanding the energetic processes embedded in materials and their compositions. A scientific explanation of material fundamentals, including chemistry, physical structure, and embodied energy, provides the greatest insight to material property performance values and relative environmental impacts. This information aids architects in making informed decisions about building materials in the design process. This chapter addresses the book topic of reusable and sustainable building materials through the position that all matter is a form of energy, just as living systems are the transmutation of matter and energy. The seven major material groups, which include natural materials, non-technical ceramics, technical ceramics, metals, polymers, foams and elastomers, and composites, are presented with examples and applications discussed.

Chapter 8 describes how digital immersion, changing social values and environmental and economic pressures have the potential to create a paradigm shift in relationships between people and their built environment with the growing sustainability imperative. It responds to emerging opportunities provided by digital technologies for the construction, maintenance and heritage curation of the life of buildings, and draws on aligned changes in thinking apparent in manufacturing, healthcare, business and education in the twenty-first century. The ideas that shape this chapter are relevant to architects and educators, but also to scholars and practitioners across disciplines because they provide an innovative approach in responding to the types of changes currently impacting societies worldwide.

Chapter 9 describes the approach adopted within the framework of a multi-destination development project, the goal of which is to promote innovative technologies and methods to evaluate the environmental quality of an urban district under construction. This method of analysis has been tested on an area located in the former historic district of the Fiera di Milano, where a series of typical urban functions are inserted within a large public park. The success of the work is represented by indicators (air quality, acoustic, microclimate) that relate to the finished district and that can be compared with average values in the same city. The system may constitute a protocol capable of bringing benefits to local authorities. This type of assessment could be requested of developers/builders for complex projects, resulting in changes to the initial plan if the assessment identifies critical issues related to the design choices (orientation of buildings, green areas, traffic emissions, etc.), with the ultimate goal of creating neighbourhoods with better environmental conditions.

We trust this book will spur further, applicable research into reusable and sustainable building materials in modern architecture. It was a great pleasure to develop this publication and the authors contemplate additional work in the exciting field of architecture.

Gülşah Koç
Yildiz Technical University, Turkey

Bryan Christiansen
Global Research Society, LLC, USA

Chapter 1
The End of Sand:
Confronting One of the Greatest Environmental Challenges of the New Millennium

David T. A. Wesley
Northeastern University, USA

Sheila M. Puffer
Northeastern University, USA

ABSTRACT

This chapter focuses on how sand, the second most used natural resource on earth after water, is facing one of the greatest environmental challenges of the new millennium. Sand is a crucial material used in all sorts of building projects, from asphalt, concrete, and glass. Globally, construction accounts for the largest portion of the 15 billion tons of sand consumed annually. Yet, sand is a finite resource and the depletion of alluvial sand used in construction is destroying the ecosystem of riverbeds, sea beds, and coastal beaches, and is contributing seriously to climate change. This chapter will discuss how these threats have developed, including coastal construction and erosion, river dredging, and sand "mafias" whereby illegal sand miners strip beaches and use sand in inferior concrete that has led to building collapses and deaths. The authors then discuss potential solutions to this crisis, including regulation and enforcement of environmental and construction standards, as well as materials substitution such as desert sand, sand created from sandstone, and recycled glass.

DOI: 10.4018/978-1-5225-6995-4.ch001

A WORLD BUILT ON SAND

[A] foolish man... built his house on sand. The rain came down, the streams rose, and the winds blew and beat against that house, and it fell with a great crash. – Matthew 7:26-27, Holy Bible, New International Version

After water, sand is the second most used natural resource on earth. A United Nations report estimated that in 2012, the world used as much as 29.6 billion tons of sand, or, "enough concrete to build a wall 27 meters high by 27 meters wide around the equator" (Peduzzi, 2014). Sand is a major internationally traded commodity. According to the Observatory of Economic Complexity, the world's largest importer of sand is Singapore, which imports 13% of the world's sand, followed by Canada (11%), Belgium-Luxembourg (9%), The Netherlands (6.2%), Germany (4.4%), and Mexico (4.4%). The top ten largest exporters of sand in 2015 were the United States, at $385 billion being more than twice The Netherlands, the next country. Germany, Belgium, Australia, Vietnam, Cambodia, France, China, and Egypt rounded out the top ten (World Atlas, 2017).

Asphalt, concrete, and glass all contain copious amounts of sand. The silicon that goes into computer chips that power our iPhones and computers is made from it. Sand even finds its way into toothpaste, cosmetics, and fast food.[1] Without sand, we would be without bridges, skyscrapers, and paved roads.

Sand is the primary raw ingredient in the manufacture of silicon wafers used in computer chips and other electronics, a process that requires a highly purified form that can only be mined in specific locations. Turning sand into silicon involves a 300-step process (Marshall, 2016). Once the process is complete, the resulting crystals are the purist on earth, surpassing even the purist diamonds (McWhan, 2012).

Silicon is also a key ingredient in other forms of electronics. For instance, it is essential to the manufacture of solar panels. Despite its importance, silicon production accounts for a tiny fraction of overall sand production. In the United States, 173,000 metric tons of the nearly 30 million metric tons of industrial sand and gravel consumed annually goes toward silicon manufacturing (Dolley, 2012).

By far, the biggest consumer of industrial sand in the United States is the petroleum industry, which uses sand for hydraulic fracturing, commonly known as fracking. Sand forms part of a slurry that is injected into wells at high pressure to release oil and gas. In 2010, fracking consumed 3.8 million metric tons of industrial sand (12.7 percent of total production for that year) (Dolley, 2012). Fracking is considered a godsend to energy security advocates, as it has greatly reduced America's dependence on imported fuel. The use of sand is also costly for fracking companies. As a result, in 2017, sand use per foot of drilling declined as oil companies deployed new methods that require lower quantities of the substance (Nair & Bhattacharjee, 2017).

Aside from silicon and frac sand, industrial sand is also used in large quantities in glassmaking, in creating molds for steel foundries, and in the manufacture of water filtration systems and abrasives.

Despite these impressive numbers, industrial sand use is dwarfed by construction sand. In the United States, the construction industry consumes 443 million metric tons of sand, which is nearly 15 times the amount used for industrial purposes. In 2016, the US Geological survey estimated, that about 44% of construction sand and gravel was used as concrete aggregates; 25% for road base and coverings and road stabilization; 13% as asphaltic concrete aggregates and other bituminous mixtures; 12% as construction fill; 1% each for concrete products, such as blocks, bricks, and pipes; plaster and gunite sands; and snow and ice control; and the remaining 3% for filtration, golf courses, railroad ballast, roofing granules, and other miscellaneous uses *(US Department of the Interior, 2017).*

Globally, construction accounts for the largest portion of the 15 billion tons of sand consumed each year, with 10 billion tons of sand destined for cement production (Allen, Thallon, & Schreyer, 2017).

Today, entire cities are rising from the oceans, built on little more than sand. Singapore, for instance, has added 52 square miles to its territory since 1965, primarily by importing sand from neighboring countries (Comaroff, 2014). In Dubai, the famous "Palm" housing project was constructed on imported coastal sand reshaped into the form of a massive palm, with each branch forming a neighborhood of residential properties. A more recent and ambitious project known as the "World" involved the construction of an artificial archipelago of 300 islands within eyeshot of Dubai. Each island represents a country, allowing owners to relax in their own private "Germany" or "Australia."

Not only are these cities consuming vast quantities of sand to form new land, they are giving rise to some of the tallest skyscrapers in the world. At 828 meters, the Burj Khalifa towers over Dubai. To support a structure this size, high quality sand was imported from Australia, more than 12,000 km away (Peduzzi, 2014). Dubai has grown at such a pace that it continues to break world records in concrete use. The latest occurred in May 2017 with a concrete pour of 21,580m^3 in just over 35 hours, surpassing a 2016 record of 19,793m^3 that occurred in another part of the city (Bambridge, 2017).

Construction of all forms consumes massive quantities of sand. An average hospital requires 3,000 tons of sand, which is a fraction of the 12 million tons used to build a nuclear power plant. Still, almost nothing compares to highways, which require 30,000 tons of sand per kilometer (Rappeneau, Mini, & Delestrac, 2012). That's the equivalent of one nuclear plant per 400 km of roadway. Such massive projects make smaller scale construction seem unimportant. For instance, an average home consumes 200 tons of sand. However, if one considers that the United States has built an average of 1.44 million new homes each year since 1959, the amount of consumed sand is staggering - nearly 300 million tons per year in residential home construction alone (Taborda, 2017).

FOCUS OF THE CHAPTER

This chapter focuses on how sand, the second most used natural resource on earth after water, is facing one of the greatest environmental challenges of the new millennium. Sand is a crucial material used in all sorts of building projects, from asphalt, concrete, and glass. Without sand, we would be without bridges, skyscrapers, and paved roads. Globally, construction accounts for the largest portion of the 15 billion tons of sand consumed annually. Yet, sand is a finite resource and the depletion of alluvial sand used in construction is destroying the ecosystem of riverbeds, sea beds, and coastal beaches, and is contributing seriously to climate change. This chapter will discuss how these threats have developed, including coastal construction and erosion, river dredging, and sand "mafias" whereby illegal sand miners strip beaches and use sand in inferior concrete that has led to building collapses and deaths. We then discuss potential solutions to this crisis, including regulation and enforcement of environmental and construction standards, as well as materials substitution such as desert sand, sand created from sandstone, and recycled glass.

AN INVISIBLE CRISIS

Even as new landmasses and structures rise in one part of the world, they are disappearing elsewhere. Like all mineral resources, the supply of sand is limited. And like water and air, we often take it for granted. Walking along a sandy beach, we can be forgiven for thinking that the supply of sand is endless. Behind this facade, a hidden crisis is looming as the world's supply fast disappears. The consequences could prove to be as devastating as climate change, ocean acidification, and ozone depletion.

Most sand originates from granite and sandstone high in the mountains. Individual grains are released from stone through erosion and begin their way to the sea. Over a period extending many thousands of years, they are deposited in small streams and eventually make their way to the ocean. Today, this natural process is being arrested by hydroelectric dams and irrigation reservoirs (Welland, 2009).

Alluvial sand, the form used in construction, is primarily found in riverbeds, sea beds, and coastal beaches. Their violent journey through fire and ice imbibes them with unique bonding properties that give products like cement their essential strength. Desert sands, on the other hand, have had their sharp edges worn away by constant winds. This makes it generally unsuitable for construction (Guettala & Mezghiche, 2011), and, as a result, desert cities like Dubai must use sand pulled from the ocean or import sand from other countries. And Dubai is not unique. An insatiable global demand for alluvial sand is having devastating impacts on coastal communities.

Although more attention has been given to the role of climate change and rising sea levels in beach erosion, the primary culprit is reduced supply. Seawalls, harbors, and other coastal structures cut off the supply of new sand from nearby cliffs and other inland sources, while river and sea dredging remove existing sand (Bird & Lewis, 2015). Even sand dredged from sea beds far from shore can have an impact on coastlines as ocean currents seek to replenish the seabed with sand pulled from nearby beaches. Efforts to arrest coastal erosion with dykes, breakwaters, and seawalls are a zero-sum game, as the problem is just pushed downstream to neighboring beaches. In many cases, coastal hardening has only produced more problems by further disrupting natural processes (Pilkey & Wright III, 1988).

Reservoirs, Dams, and Levees

Coastal erosion is exacerbated by the construction of reservoirs and hydroelectric dams. California's 500 dams prevent most of that state's beaches from being properly replenished (Griggs, 2017). The combined impact of coastal construction, sand mining, and river damming is causing nearly 90 percent of America's beaches to recede (Gopalakrishnan, Smith, Slott, & Murray, 2011). Similar impacts can be seen in other parts of the world, where an estimated 30 percent of alluvial sand is trapped behind dams (McCarney-Castle, Voulgaris, & Kettner, 2010). No country has been more ambitious than China. Since 1949, it has constructed 86,000 dams, arresting the natural flow of nearly all its rivers and forcing 24 million people to relocate (Piesse, 2016).

Dams are beneficial in several ways, providing irrigation and renewable energy. For instance, in the Canadian province of Quebec, electricity contributed only 0.4 percent of all greenhouse gas emissions, largely because of the province's massive investment in hydropower (Houle, Chhem, & Bougie, 2002). "But the consequences of stopping a sand grain on its journey are huge," notes geologist Michael Welland (2009):

The effects of dams...include major modifications to the delivery of sediment to the river system downstream and to the ocean; this, in turn alters entire landscapes. Changing with the seasons, a river system left to its own devices irrigates and drains, excavates and backfills the terrain through which it flows. Disturb that delicate but massive balance by disrupting the water flow and impounding sediment, and you change landscapes and ecosystems.

Rivers also feed into wetlands that protect against hurricanes and other natural disasters. The American state of Louisiana, which was devastated by Hurricane Katrina in 2005 loses 100 square kilometers of wetlands each year due to human activities, such as the construction of dams, levees, and canals (Day, Boesch, Clairain, Kemp, Laska, Mitsch, ... Shabman, 2007).

River Dredging

The sand that makes it past dams is being dredged from river beds for use in cement and other construction products. Only a small fraction ever reaches the ocean. By depriving beaches of replenishment, river dredging not only diminishes the natural beauty of ocean shores, it increases the severity of natural disasters.

Long before Hurricane Katrina, geologists at the University of New Orleans warned that dredging would have catastrophic consequences if the region were hit by a major storm. Louisiana's southernmost port, Port Fourchon,

loses 40 to 50 feet a year - the fastest rate in the country. The network of canals also gives saltwater easy access to interior marshes, raising their salinity and killing the grasses and bottomwood forests from the roots up. No vegetation is left to prevent wind and water from wearing the marshes away. (Fischetti, 2001).

The depletion of New Orleans' natural barriers is believed to have greatly contributed to flooding that engulfed 80 percent of the city when Katrina struck in 2005, resulting in significant loss of life and property.

Similar consequences can be seen in other parts of the world, particularly in low-lying areas, such as Bangladesh, which is periodically devastated by cyclones and monsoons. Although coastal flooding can have many causes, including deforestation, coastal hardening, and climate change-driven sea level rise, dredging has certainly exacerbated the severity (Hyndman & Hyndman, 2016). A study of seven unregulated dredging sites on the River Kivou in Kenya found evidence of redirected land use, river bank degradation, habitat degradation, and water use conflicts between sand transporters and the locals (Wambua, 2015).

Sand Mafias

Currently, the world consumes twice as much sand per year, more than 40 billion tons, than the combined flow of the entire world's rivers (Calas, 2017). The rapid depletion of high quality sand has led to illicit trade, particularly in less developed countries. In Morocco, illegal sand miners are rapidly depleting the country's beaches, leaving nothing behind but exposed rocks and a devastated tourism industry (Pilkey & Cooper, 2014). The questionable quality of cement produced from illegal mining has contributed to numerous building collapses (Umesh & Murthy, 2014). In India, illegal mining has given rise to one of India's most powerful organized crime syndicates - the "sand mafia" (Torres, Brandt, Lear, & Liu, 2017). Government officials and environmentalists there who oppose the criminal gangs have been murdered, including an 81-year-old teacher and a 22-year-old activist (Beiser, 2017b), while uncooperative land owners have been beaten. In many cases, authorities are bribed to turn a blind eye (Muthomi, Okoth, Were, & Vundi, 2015).

In Makueni County in eastern Kenya, from 2015 to 2017 alone, ten people who fought against the powerful sand cartels were killed, including farmers defending their land and a police officer who advocated sand conservation (Koigi, 2017). Truckers transporting illegal sand from Makueni to Nairobi openly told Kenyan researchers that they gave bribes to police along the road (Beiser, 2017b). Yet extreme joblessness in eastern Kenya has drawn young people to work for the illegal sand cartels. According to 23-year-old Simon Kativo:

We have families we have to take care of. The sand business pays very well. I can sell a 90 kg bag at $30 or earn $40 scavenging for sand for a day. No other casual jobs pay that high. That is why the majority of the youth in this area have found solace in this business. (Koigi, 2017)

While sand mining has been banned in much of Makueni County, it continues under cover of darkness. According to Timothy Maneno, a member of the local legislature: "Between tonight and 7am tomorrow morning you can stand on the highway and count 100 lorries heading to Nairobi [full of sand]" (Beiser, 2017b). This semi-arid region, whose economy consists of livestock and subsistence farming, has suffered a major drought due to lowered water tables and rerouted rivers. The Kenyan government issued an official bulletin claiming that sand harvesting "has wrought wanton destruction on our rivers, farms and land to the point where it is now a human catastrophe" (Beiser, 2017b).

Farmers, including many in Nigeria, have turned away from agriculture to mine sand and gravel on their land. Some tribal chiefs and land owners have reputedly given out land for sand mining for monetary gain without regard for the negative environmental effects. This practice has caused problems for soil structure, vegetation, and local wildlife, as well as air pollution due to sand dust production that has negative effects on plants as well as on human health (Nthame, 2015). A study in

Kenya warned that such pollution can cause respiratory and behavioral problems in children, leading to absenteeism from school (Nthame, 2015).

Environmental degradation can lead to waves of environmental refugees who cross borders and create destabilizing effects domestically and in international relations (Homer-Dixon, 1991). The devastation is perhaps most evident in island communities displaced by erosion. In the Maldives, each day six tons of sand is extracted per boat (Padmalal & Maya, 2014). The combined effect of rising sea levels due to climate change and sand erosion are forcing "climate refugees" to relocate, often far from their homelands. "[T]he prognosis has become sufficiently dire that the residents of Tuvalu and other low-lying atoll islands 'are beginning to envision the wholesale abandonment of their nations.' Around one-fifth of the 12,000-some inhabitants have already left, most bound for New Zealand, where the Tuvaluan community has nearly tripled since 1996" (Morris, 2017). It also is causing international boundaries to shift in places like Indonesia, where numerous islands have dissolved into the ocean (Peduzzi, 2014). Although Vietnam, Malaysia, and Indonesia have banned the export of sand, illegal mining continues to severely impact these nations, as beaches and rivers are stripped bare (Beiser, 2017a).

Less visible is the impact on marine life. Dredging river beds and sea beds is highly disruptive to the ecosystem, as coral reefs are smashed and seabed habitats are destroyed. The US Gulf Coast, which has suffered intense dredging to satiate the region's rapid economic growth, has lost 70 percent of its coral reefs (Van Lavieren, Burt, Feary, Cavalcante, Marquis, Benedetti, ... Sale, 2011). Those that survive are smothered by debris and sediment, leading to disease and death (Lee, 2014). Elsewhere, the disruption of the seabed is leading to biotic crises. For instance, in South Carolina, a Department of Natural Resources study found a marked reduction in biodiversity as the number of species in the area dwindled from 28 before dredging to only 8 (Fretwell, 2016). Even low frequency dredging can have a significant long-term impact on the food chain, potentially disrupting fisheries (Hale, Godbold, Sciberras, Dwight, Wood, Hiddink, & Solan, 2017). The disruption has already been felt in Indonesia and Kenya, where dredging has impaired the livelihoods of subsistence fishers and led to protests against mining companies (Rahim & Rahim, 2010; *Environmental Justice Atlas, 2017*). And while extraction is devastating to ocean life, sand replenishment, a process by which sand is returned to eroding beaches, can be similarly destructive, as life forms are smothered under the new sand (Wooldridge, Henter, & Kohn, 2016).

Yet some signs of stakeholders standing up to the sand mafia are emerging. In China, a dozen members of rival gangs were imprisoned in 2015. In Kenya, in 2016 the national environmental tribunal responded to an appeal by local citizens and stopped China Roads and Bridge Corporation from harvesting sea sand in Kwale Country for construction of the Standard Gauge Railway (*Environmental*

Justice Atlas, 2017). Elsewhere in Kenya in 2015, a group of tourism stakeholders, the Kenya Wildlife Service, environmentalists, marine scientists, government representatives, members of the fishing community, and local residents organized a meeting to eventually stop the impending harvesting of sand from Diani Beach south of Mombasa. At risk was the destruction of an award-winning tourist beach, the death of reefs, and the demise of fish and other marine life (*Environmental Justice Atlas*, 2017). Officials declared:

The ruling was unprecedented. It has spelt doom to those who think they can please the government by quickening development projects without following the due process. (Environmental Justice Atlas, 2017)

POTENTIAL SOLUTIONS

Nothing is built on stone; all is built on sand, but we must build as if the sand were stone. – Jorge Luis Borges, Ficciones (1944)

Given the ubiquity of sand in essential products, any solution will need to focus on finding substitutes. The most obvious fix is better resource utilization. Vanity projects, like Dubai's "The World" are a poor use of this limited resource.

"The World" was the brain child of Dubai's Sheik Mohammed, who in 2003, saw the project as ideal representation of the city's tourism development, with its excesses and extravagance. It consumed 321 million cubic meters of dredged sand and more than 60 percent of the islands were sold prior to project completion (Gupta, 2015). However, the appetite for mega real estate projects evaporated following the global economic crisis of 2008. The following year, the Dubai government bailed out the developer to the tune of $25 billion (Spencer, 2011). Shortly afterward, John O'Dolan, a property developer who purchased "Ireland" and "England" for more than $60 million, committed suicide (Mclean & McDonald, 2012).

After dredging vast quantities of sand for the project, it is now in danger of sinking back into the sea. As noted earlier, one of the causes of beach erosion is the natural flow of sea currents which pull sand back to previously dredged areas. The erosion of the islands has created new hazards as it backfills nearby shipping lanes. Despite the environmental problems, investment losses, and shipping hazards, developers are hoping to further expand the project in 2020 (Berman, 2017).

A geopolitical arms race is developing, as other Gulf states follow Dubai's lead to construct their own artificial islands. Kuwait, Qatar, and Bahrain have recently completed their own projects to compete with those in Dubai, with more in the planning phases.

Regulation and Enforcement

Some policy analysts have focused on the need for better enforcement of existing regulations. Ghaffari, Habibzadeh, Asfad, and Mousazadeh (2017) cite the United Nations Convention on the Law of the Sea (UNCLOS), Convention on Biological Diversity (CBD), and United Nations Framework Convention on Climate Change (UNFCCC) as existing frameworks that require member nations to protect marine environments and clean up contaminants. "These principles should be put into action to protect the environment," they insist.

The important identified principles with legal value are as follows:

- Prevention
- Cooperation and common but differentiated responsibility
- Precautionary principle
- Environmental impact
- Assessment
- Sustainable development
- Polluter pays principle
- Inter-generational and intra-generational equity

Some of these general principles or rules reflect customary law, others may reflect emerging legal obligations, and yet others might have a less developed legal status. In each case, however, the principle or rule has broad support and is reflected in extensive state practice through repetitive use or reference in an international legal context. (Ghaffari, Habibzadeh, Asfad, & Mousazadeh, 2017)

However, existing regulatory frameworks are too general and do not directly address issues such as dredging and construction of artificial islands. To be effective, political leaders must agree on new frameworks and the strengthening of enforcement mechanisms (Van Lavieren, Burt, Feary, Cavalcante, Marquis, Benedetti, ... Sale, 2011).

Still, even if new frameworks can be agreed upon, the barriers to an effective regulatory approach are significant and will be difficult to overcome. First, no enforcement mechanism exists to ensure that member states comply with international law. And even if there were, military or police forces are under the direction of national and regional authorities and will only enforce international law when directed to do so by their own governments. Nations are more likely to enforce international law when it is in their interests to do so. For instance, Southeast Asian countries that have seen their fisheries and tourism industries impacted by illegal shipments of their sand to Singapore are the most aggressive in enforcing environmental regulations in the region.

Despite increased efforts to intercept sand dredging ships and barges, law enforcement is more often circumvented. The vastness of the ocean and its many archipelagos make enforcement difficult. A good example is Indonesia. Although its navy has actively intercepted illegal ships (Albrecht-Saavedra & Lippelt, 2016), the decimation of Indonesia's islands and fisheries continues.

One promising area is the development of satellite vessel monitoring systems, which can help to locate and monitor traffic in international waters (Ewell, Cullis-Suzuki, Ediger, Hocevar, Miller, & Jacquet, 2017). In 2015, human rights activists used satellite tracking to monitor illegal fishing in Papua New Guinea and to uncover a hidden slave base in Indonesia (McDowell, 2015). Although satellite tracking has only a minor impact on enforcement, it will likely improve the success rate of international intercepts as more governments deploy tracking technologies.

Inland rivers face similar challenges as illegal miners strip river beds, destroy irrigation, and disrupt navigation. The Yangtze River in China is instructive. Criminal gangs mined tons of sand from the river on a daily basis until police intervened. However, instead of stopping the practice, the gangs simply moved to Poyang Lake, causing similar disruptions (Dyer, 2016).

In China and India, the rise of criminal mining gangs is being driven by government policies that encourage investment in the real estate sector. This necessitates the continual demolition and rebuilding of cities. In China, the average building lifespan is only 25 years, as that country constructs new cities at an average rate of 10 cities per year, more than any other country (Shepard, 2015).

Under existing policy, regional governors in China are rewarded for achieving economic growth targets (Jia, 2013). This has led to perverse incentives, such as encouraging smoking because tobacco production and sales contribute to GDP targets and provide 10 percent of the central government's tax revenue (Dreyer, 2015). It has also given rise to what have come to be known as "ghost cities." Investors buy properties, not to live in or commercialize, but for speculation. Some cities in China are nearly vacant.

Similar speculative practices have occurred in Egypt, Spain, Ireland, and India. In Spain, nearly 85 percent of housing built in the last decade remains vacant, compared to a European average of 5 percent. In contrast, Sweden's vacancy rate is below 2 percent (Moreno & Blanco, 2014). The enormous waste of resources has not only fueled the illegal sand trade, but it has also displaced existing residents, contributed to global warming, and destroyed the natural ecosystems where the cities are being constructed.

Governments can take steps to reduce the demand for illegal sand by first eliminating incentives for investment that are not in the broader interests of society. For instance, although cigarette smoking has stimulated the Chinese economy, alternative industries could provide a similar benefit without longer-term health costs. Similarly, construction projects need to be strategic. Possibilities include building a solar farm instead of an apartment building, and following stricter environmental assessments prior to approving new projects.

China is currently taking steps to mitigate the ghost city problem by encouraging smaller, more dispersed, urban centers that better fit the needs of its citizens. It also plans to move 100 million rural citizens to vacant properties over the next five years and has refocused its construction on projects better suited to its development needs (Rapoza, 2015). For instance, after facing pressure to address air pollution, China more than tripled its annual installation of solar energy between 2015 and 2017, from 15 megawatts per year to 47 megawatts. In doing so, it has reduced its dependence on imported fossil fuels, trimmed its greenhouse gas emissions, and curtailed the demand for cement. In 2017, China became the world leader in new solar installations and continues to develop other renewable resources (Urban, 2018).

Other countries with less centralized policymaking may find it more difficult to address real estate speculation. As mentioned above, India's sand mafia wields enormous power, and corruption continues to be a serious barrier to effective policy implementation. It is no wonder that countries with ghost city problems, like China, Egypt, and India, also tend to rank low on Transparency International's corruption index.

Corruption is not the only barrier. Regulation is likely to be more successful in developed countries that have the resources to enforce compliance. Lucas, Wheeler, and Hettige (1992) explain that national wealth has a net positive impact on regulation enforcement for three reasons:

- **Income Equality:** Citizens in wealthier countries demand a cleaner environment over economic growth.
- **Absorptive Capacity:** An ability to locate pollution away from population centers.
- **Regulatory Capability:** A well-funded and effective enforcement mechanism.

Absorptive capacity should also be considered across national boundaries. Wealthier nations can simply externalize environmental harms by exporting them to less developed neighbors.

Even if there were no environmental regulation in wealthier economies, free trade under these conditions might well lead to disproportionately rapid growth of industrial pollution in developing countries. (Lucas, Wheeler, & Hettige, 1992)

Therefore, developed nations cannot simply shift the blame to countries like China and Indonesia.

Materials Substitution

Given the challenges described above, regulation is likely to have a limited impact on the sand crisis. How can industry, whose only duty is to provide a return for shareholders within the legal system, fill the regulatory gap? As with any non-renewable resource, sand will eventually run out. Yet, demand for new buildings, roads, and electronics will continue to be strong. The only viable solution is substitution.

Substitution has already proven effective at mitigating other harms. For instance, solar, wind, hydroelectric, and nuclear power have significantly limited growth in greenhouse gas emissions, albeit by introducing different environmental challenges, such as nuclear waste. And styrofoam packaging has largely been replaced by recyclable bubble wrap and cardboard. Industrial and construction sand also has potential substitutes, including aeolian (desert) sand and crushed glass and sandstone, each of which come with their own benefits and limitations. The need for sand substitutes will create economic opportunities for innovators. As a result, private R&D efforts designed to create viable sand substitutes have bloomed in recent years. Several potential substitutes for alluvial sand are discussed below: desert sand, sand from sandstone, and recycled glass.

Desert Highways

The first and most obvious substitute is sand itself. The world's deserts contain vast quantities of sand, most of which is currently unsuitable for construction. As Elipe and Lo´pez-Querol (2014) note,

their fine grain size and poor grading, with negligible plasticity, and their rounded shape, indicate these sands cannot be utilized for construction purposes, like for example, backfill materials of granular soils for embankments, in their natural condition.

Yet, because necessity is the mother of innovation, researchers are uncovering innovative ways to make desert sand suitable. For instance, in the late 1990s researchers began experimenting with desert sand mixed with kiln dust as a stabilizer for making pavement (Freer-Hewish, Ghataora, & Niazi, 1999). Later efforts revealed that with the right stabilizers, some desert sands could be used in mortar and cement products (Zhang, Song, Yang, & Liu, 2006). Other types of stabilizers like hydrated lime and bentonite have shown promise, but kiln dust is the most promising due to its relative low cost. More recently, municipal waste ash has shown promise as a potential stabilizer (Elipe & Lo´pez-Querol, 2014). Still, practical applications have primarily been limited to road base and asphalt. When desert sand is used as an alluvium substitute, the resulting cement shows "a possible degradation in compressive strength" (Luo, He, Pan, Duan, Zhao, & Collins, 2013). Therefore, its use in construction cement is unlikely.

Sandstone Towers

As noted earlier, alluvial sand begins its life trapped in granite and sandstone. Over millennia, erosion releases the sand and carries it downstream toward the sea, only to be trapped behind dams or dredged from riverbeds. Turning sandstone into sand could be a way to bypass this natural process. However, crushed sandstone, much like desert sand, has the wrong characteristics for use in cement, although it is commonly used as a substrate for playgrounds or equestrian tracks.

Consider Scotbar (Wesley, Puffer, & Wade, 2018), a sandstone mining company in eastern Australia that developed a superior road base made from crushed sandstone. The company invested heavily in R&D to develop a non-dusting road base that was less expensive and lasted longer than traditional road base, which normally had to be dug up after years of use.

Despite its superior characteristics, the Australian government prohibited its use in new projects, after other suppliers provided inferior quality crushed sandstone that caused multiple road failures. Scotbar CEO Alan Payne explained that creating the correct formula is

like putting medicine together. You've got prescribed amounts of a whole lot of different products. When you blend that and you put the right amount of water in and it's laid properly, you get those benefits.

To reduce costs, other companies had taken short cuts that undermined the legitimacy of the entire category.

Despite the road base setback, Scotbar continued to focus on innovation, with the aim of achieving 100 percent resource utilization. Most mining operations use only a small portion of the extracted minerals, discarding the rest. Sandstone covers a large portion of the world's surface, but sandstone quarries, like other mining operations, only extract stones with aesthetically desirable characteristics for use in luxury homes and commercial buildings. Turning the waste into sand could help alleviate the demand for river and sea sand.

Through years of trial and error, the Australian quarry developed a crushing process that results in a manufactured sand that preliminary testing has shown to be superior in strength to traditional sand. Early prototype testing in partnership with a global cement company has produced a premium cement that can be manufactured at a lower temperature, thereby reducing greenhouse gas emissions, and has superior strength. It is not sold as a cement replacement, but as an additive. Builders are instructed to mix the formula with lower quality cement when extra strength is needed. The product is currently at an early stage, but feedback from contractors has been positive and the company is moving from prototype to production.

The next challenge is to reduce production costs, which are currently 50 percent higher than traditional sand mining. The company hopes that full-scale production will result in economies of scale that could eventually reach cost parity with river sand. Scotbar is also working with the Australian government and academia to win approval for its product as a sand replacement in conventional cement production.

Finally, Scotbar decided to not protect its process innovations through patents. Instead, the CEO hopes that it can be adapted for use at other mines, both to reduce waste and to help solve the sand crisis. However, he was also skeptical that other mining companies would embrace new methods, as it would require a new mindset on how minerals are extracted. "Unfortunately, unless there is a trend, you're left dragging people kicking and screaming to achieve something that should be a common practice," he complained.

Early results suggest that Scotbar's technique for creating mined construction sand will produce stronger cement that requires less energy to produce. Despite its higher up-front cost, the superiority of manufactured sands can reduce overall project costs due to lower longer-term maintenance and repair expenses. Moreover, cement companies can improve their reputation for supplying quality products.

Yet, even if mining companies can perfect the process of turning sandstone into construction sand, a few bad actors could undermine widescale adoption. Scotbar's own experience developing a road substrate is instructive. If sand producers take the same approach, without strict oversight and testing, catastrophic building collapses could ensue. Would regulators then ban the product completely, as they did with road base in Australia? Or would they turn a blind eye, as they often do in less developed countries?

Current practice suggests that both scenarios are likely. For example, Nigeria has experienced frequent building failures, including a church that collapsed in 2016, killing more than 200 worshipers (Biobaku, 2017). And in 2013, a factory collapse in Bangladesh killed more than a thousand workers (Malkin, 2013). Although similar tragedies are commonplace across Africa and Asia, they also occur in developed countries. For instance, the deaths of 200 Italians in a 2009 earthquake was largely blamed on "the use of low quality cement and inadequate supporting iron rods" (Aloisi, 2009).

In most cases, builders and cement producers seek to cut costs by purchasing cement from illegal miners who supply product mixed with organic material and other impurities or by mixing it with cheaper aggregates that are not suitable for commercial construction. In Morocco, for instance, nearly 50 percent of the sand used in construction comes from illegal sources. Some experts believe that Morocco's recent building boom will eventually lead to structural failures like those seen in Nigeria (Rappeneau, Mini, & Delestrac, 2012).

These risks are mitigated to some extent by the fact that sandstone processing requires highly specialized and expensive equipment. Scotbar expects to spend approximately $20 million on a new sand plant in partnership with CDE Global, an Irish manufacturer of wet sand processing equipment. Although this creates a high barrier to entry that may limit adoption, particularly in less developed countries, it could also prevent bad actors from entering the industry, particularly if regulators certify the plant and process and conduct regular audits. Once the process is perfected, manufactured sand may also eventually drive smaller suppliers out of business if economies of scale become large enough to reduce the cost below that of dredged sand. Even at prices that are 50 percent higher than natural sand, Scotbar has been unable to meet demand for its manufactured sand, which is used by a global cement company for its "premium" cement category.

Glass Beaches

In the early 20th century, before the widespread use of plastics, the community of Fort Bragg in northern California created an open-air waste dump on the local beach. For decades, the community dumped appliances, vehicles, scrap metal, and glass on the beach. When the dump was closed in 1967, the metal waste was cleared out and sold for scrap, leaving only a large quantity of broken glass. Over the years, ocean waves broke down the glass into sand-like particles (Makowski, Thomson, Foye, & Higgins, 2007). In time, it became known as "glass beach," and began attracting tourists who are drawn to the beach's multi-colored "gems." Collectors have even begun to scour the beach for rare "ruby reds" created from the tail lights of automobiles and "sapphire gems from apothecary bottles" (Danger, 2016).

Recycled glass as a beach replenishment substrate was the brainchild of Encinitas, California, resident Dan Dalager after a visit to Glass Beach. Although glass can be recycled into new glass, mixed color glass is unsuitable and must be discarded (Taha & Nounu, 2008). Dalager thought that crushed glass might be suitable to shore up his community's eroding beaches. Encinitas town planners and engineers thought the idea was "really stupid," but eventually agreed to conduct tests. "It looks almost identical to beach sand, unless you look at it under a microscope," observed geologist Dave Schug at the time (Masciola, 1992).

Replenished beaches are renourished by offshore dredging, but the dredged sand eventually washes back into the ocean. After seeing the success in Encinitas, Florida Atlantic University graduate student Loisa Kerwin undertook a study to see if crushed glass could be used in Broward County, Florida, where beach replenishment was costing $45 million each year (Williams & Micallef, 2011). The key to the study was determining whether recycled glass could be refined to have similar qualities to the sand grains found on Broward County beaches, particularly in strength, durability, compaction, and workability. Not only was crushed glass compatible, the study found that it "could help stabilize [the] beaches and extend the life span of artificial beaches between full-scale renourishments" (Kerwin, 1997).

One of the problems with traditional beach replenishment projects is that it often kills existing organisms. They are either ground up by the augers, pumps, and suction pipes used to extract sand from the ocean floor, or smothered by new sand deposits. Having determined that crushed glass can have similar physical qualities to beach sand, the next step was to determine its biological viability. Makowski and Rusenko (2007) found that beach organisms not only survived, but thrived in an artificial environment composed of little more than glass. Beach erosion has been particularly hard on sea turtles that nest in the sand, but glass beaches can provide a suitable alternative nesting area for these endangered animals (Makowski, Rusenko, & Kruempel, 2008).

Despite its promise, Broward County, as well as other parts of Florida, continue to favor costly and environmentally harmful dredging to replenish eroded beaches (Alvarez, 2013). To date, recycled glass has seen only limited use on beaches in New Zealand, Hawaii, and the Caribbean (Williams & Micallef, 2011).

This begs the question, if crushed glass can have the same physical characteristics as beach sand, why not use recycled glass in cement production? Taha and Nounu (2008) considered this very question, but unfortunately discovered that glass particles and powder "reduce the consistency of the concrete mix and adhesive bond of the ingredients inside the concrete mix," primarily because glass absorbs more water than silica sand. Nevertheless, a widescale adoption of crushed glass in beach replenishment could allow more silica sand to be used for construction instead of renourishment.

CONCLUSION

Every year, stunning new buildings all over the world tower higher above the landscape. And part of what makes it possible is the low cost of sand and the low barriers to entry for sand miners. Global economic and population growth will only accelerate the unsustainable consumption of the world's remaining alluvial sand. Reducing waste, recycling, and developing substitutes are the best available solutions to the problem. However, embracing these alternatives will require a political will that has to date been lacking.

Regulation and better enforcement are key to providing solutions. Otherwise, sand will continue to be dredged more cheaply by illegal miners. In parts of Southeast Asia, it took denuded beaches and collapsed fisheries before governments acted, and yet illegal dredgers continue to evade enforcement in Indonesia, Vietnam, and elsewhere. With 75 percent of the world's cities situated along coastlines, the types of devastation seen in New Orleans, Bangladesh, and elsewhere will likely continue at an accelerated pace. Many factors contribute to coastal flooding, while the role of erosion and dredging remain largely hidden from view. Yet, despite the risks, sand mining remains low on the global environmental agenda. As such, sand depletion is likely to remain the ugly stepchild of climate change.

Governments also need to address real estate vacancy rates in major cities, where speculators and poor policy planning produce ghost cities that waste precious resources like sand. Many major cities have large numbers of vacant properties that were purchased by real estate speculators. Sydney, Australia, for example, has 200,000 such empty homes, which some blame on the city's higher rents (White, 2017). In many cases, these "ghost homes" fall into disrepair, creating hazards for nearby residents (Alvarez, 2015). Policymakers need to disincentivize real estate investment, while encouraging occupancy of existing sites. China, in particular, needs to break the cycle of tearing down and rebuilding.

Substitutes for alluvial sand are also becoming more viable including desert sand, crushed sandstone, and crushed glass. Yet most substitutes currently cost significantly more than dredging sand from public lands and waterways. Over time, process innovations can reduce the costs of manufactured sand and substitutes.

Even if the regulatory and cost problems can be overcome, commercial acceptance of alluvial sand substitutes will remain a major hurdle. Unconventional substitutes in construction projects are often viewed as potential liabilities to builders. If a project fails, builders and cement companies are the ones held accountable. This is particularly true in developed countries where regulatory standards are strictly enforced.

In the longer term, the private sector will likely drive improvements. As sand becomes scarcer and different actors compete over limited resources, the price of alluvial sand will continue to rise, perhaps significantly. Meanwhile, innovations such as those described in this chapter will continue to reduce the cost of viable substitutes. Once resources such as crushed sandstone and crushed glass reach parity with their natural sand counterparts, new business opportunities will drive the market in much the same way as declining costs for photovoltaic cells have driven the recent global boom in solar energy installations (Aanesen, Heck, & Pinner, 2017).

In conclusion, a combination of solutions is needed in order to solve the "invisible" sand crisis. Governments need to acknowledge the crisis and become serious about regulating and enforcing sound practices, while at the same time multiple stakeholders from governments to builders need to accept sand substitutes. All the while, more product and process R&D needs to be funded and conducted to make sand alternatives viable economically and from a materials and engineering perspective. Additionally, when proposing changes in the use of sand and its potential alternatives in construction, it is important to keep in mind the cultural traditions of the locations in which the materials are used in order to determine their suitability and likelihood of acceptance in specific parts of the world (Vatan, 2017). Without action to reduce waste, curb illegal mining, and develop sand alternatives, our world built on sand may soon fall with a great crash.

REFERENCES

Aanesen, K., Heck, S., & Pinner, D. (2017). *Solar power: Darkest before dawn.* McKinsey on Sustainability & Resource Productivity.

Albrecht-Saavedra, J., & Lippelt, J. (2016). Climate notes: Is sand running between our fingers? *CESifo Forum, 17*(1), 92.

Allen, E., Thallon, R., & Schreyer, A. (2017). *Fundamentals of residential construction*. Retrieved from https://books.google.com/books?id=CU4aDgAAQBAJ

Aloisi, S. (2009). Italy earthquake exposes poor building standards. *Thomson Reuters*. Retrieved from https://uk.reuters.com/article/us-italyquake-buildings/italy-earthquake-exposes-poor-building-standardsidUKTRE5364T720090407

Alvarez, J. (2015). Are 'ghost homes' becoming a problem? *Embarcadero Media.* Retrieved from https://www.paloaltoonline.com/news/2015/04/24/are-ghost-homes-becoming-a-problem

Alvarez, L. (2013, August 24). Where sand is gold, the reserves are running dry. *The New York Times*.

Bambridge, J. (2017). Dubai concrete pour breaks Guinness world record. *Construction Week Online*. Retrieved from http://www.constructionweekonline.com/article-46299dubai-concrete-pour-breaks-guinness-world-record/

Beiser, V. (2017a). Sand mining: The global environmental crisis you've never heard of. *Guardian News and Media*. Retrieved from https://www.theguardian.com/cities/2017/feb/27/sand-mining-global-environmentalcrisis-never-heard

Beiser, V. (2017b). He who controls the sand: Killing each other to control cities. *The Guardian*. Retrieved from https://www.theguardian.com/cities/2017/feb/28/sand-mafias-killing-each-other-build-cities

Berman, D. (2017). Everything you need to know about Dubai's man made islands. *Time*. Retrieved from http://www.travelandleisure.com/tripideas/island-vacations/dubai-man-made-islands-facts

Biobaku, S. (2017). Building collapse in Nigeria: 6 early warning signs. *Private Property*. Retrieved from https://www.privateproperty.com.ng/news/buildingcollapse-in-nigeria-6-early-warning-signs/

Bird, E., & Lewis, N. (2015). Causes of beach erosion. In E. Bird & N. Lewis (Eds.), Beach Renourishment (pp. 7–28). Cham, Switzerland: Springer International Publishing. 2 doi:10.1007/978-3-319-09728-2_2

Calas, G. (2017). Mineral resources and sustainable development. *Elements: A Magazine of Mineralogy. Geochemistry, and Petrology*, *13*(5), 301–306.

Comaroff, J. (2014). Built on sand: Singapore and the new state of risk. *Harvard Design Magazine: Architecture, landscape architecture, urban design and planning*, (39), 138.

Constable, H. (2017). *Kenya's sand wars: Communities are pitted against sand harvesters, powerful cartels and one another as demand for sand in Kenya grows*. Retrieved from https://interactive.aljazeera.com/aje/2017/kenya-sand-wars/index.html

Danger, T. (2016). The history of Fort Bragg glass beach. *Roadtrippers, Inc*. Retrieved from https://roadtrippers.com/trips/14332753

Day, J. W., Boesch, D. F., Clairain, E. J., Kemp, G. P., Laska, S. B., Mitsch, W. J., ... Shabman, L. (2007). Restoration of the Mississippi delta: Lessons from hurricanes Katrina and Rita. *Science, 315*(5819), 1679–1684. doi:10.1126cience.1137030 PMID:17379799

Dolley, T. P. (2012). Silica. *U.S. Department of the Interior, U.S. Geological Survey*. Retrieved from https://minerals.usgs.gov/minerals/pubs/ commodity/silica/myb1-2010-silic.pdf

Dreyer, J. (2015). *China's political system*. Retrieved from https://books.google.se/ books?id=j%5CooCgAAQBAJ

Dyer, J. (2016). Sand: It's as precious as oil - and its extraction is destroying the world's coasts and rivers. *Vice Media LLC*. Retrieved from https:// news.vice.com/ article/sand-its-as-precious-as-oil-and-its-extractionis-destroying-the-worlds-coasts-and-rivers

Elipe, M. G., & Lo'pez-Querol, S. (2014). Aeolian sands: Characterization, options of improvement and possible employment in construction – the state-of-the-art. *Construction & Building Materials, 73*(Supplement C), 728–739. doi:10.1016/j.conbuildmat.2014.10.008

Environmental Justice Atlas. (2017). *Offshore sea sand harvest in Kwale district, south of Mombasa, Kenya*. Retrieved from https://ejatlas.org/conflict/offshore-sea-sand-harvest-from-in-south-coast-of-the-indian-ocean-kenya, December 28.

Ewell, C., Cullis-Suzuki, S., Ediger, M., Hocevar, J., Miller, D., & Jacquet, J. (2017). Potential ecological and social benefits of a moratorium on transshipment on the high seas. *Marine Policy, 81*, 293–300. doi:10.1016/j.marpol.2017.04.004

Fischetti, M. (2001). Drowning New Orleans. *Scientific American, 285*(4), 76–85. doi:10.1038cientificamerican1001-76 PMID:11570045

Freer-Hewish, R., Ghataora, G., & Niazi, Y. (1999). Stabilization of desert sand with cement kiln dust plus chemical additives in desert road construction. *Proceedings of the Institution of Civil Engineers – Transport*. 10.1680/itran.1999.31285

Fretwell, S. (2016). Marine life dwindles after beach renourishment at folly, report says. *The State*. Retrieved from http://www.thestate.com/ news/local/article96794212.html

Ghaffari, K., Habibzadeh, T., Asfad, M. N., & Mousazadeh, R. (2017). Construction of artificial island in southern coast of the Persian Gulf from the viewpoint of international environmental law. *Journal of Politics and Law, 10*(2), 264. doi:10.5539/jpl.v10n2p264

Gopalakrishnan, S., Smith, M. D., Slott, J. M., & Murray, A. B. (2011). The value of disappearing beaches: A hedonic pricing model with endogenous beach width. *Journal of Environmental Economics and Management, 61*(3), 297–310. doi:10.1016/j.jeem.2010.09.003

Griggs, G. (2017). *Coasts in crisis: A global challenge*. Berkeley, CA: University of California Press.

Guettala, S., & Mezghiche, B. (2011). Compressive strength and hydration with age of cement pastes containing dune sand powder. *Construction & Building Materials, 25*(3), 1263–1269. doi:10.1016/j.conbuildmat.2010.09.026

Gupta, P. (2015). Futures, fakes and discourses of the gigantic and miniature in 'the world' islands, Dubai. *Island Studies Journal, 10*(2).

Hale, R., Godbold, J. A., Sciberras, M., Dwight, J., Wood, C., Hiddink, J. G., & Solan, M. (2017). Mediation of macronutrients and carbon by post disturbance shelf sea sediment communities. *Biogeochemistry, 135*(1), 121–133. doi:10.100710533-017-0350-9

Homer-Dixon, T. (1991). On the threshold: Environmental changes as causes of acute conflict. *International Security, 6*(2), 76–116. doi:10.2307/2539061

Houle, G., Chhem, C., & Bougie, R. (2002). *Inventaire qu´eb´ecois des gaz `a effet de serre 1990-2000*. Quebec, Canada: Minist`ere de l'environnement du Qu´ebec.

Hyndman, D., & Hyndman, D. (2016). *Natural hazards and disasters*. Retrieved from https://books.google.com/books?id=mChTCwAAQBAJ

Jia, R. (2013). *Essays on the political economics of China's development* (Unpublished doctoral dissertation). Department of Economics, Stockholm University, Stockholm, Sweden.

Kerwin, L. (1997). *Potential applications for recycled glass in beach management: Emergency stabilization of erosional "hot spots" in Broward County, Florida* (Unpublished doctoral dissertation). Miami, FL: Florida Atlantic University.

Kiogi, B. (2017). Deadly and destructive: Kenya cartels fan sand wars. *Fair Planet*. Retrieved from https://www.fairplanet.org/story/deadly-and-destructive-kenya-cartels-fan-sand-wars/

Lee, J. (2014). Digging up the seafloor makes coral reefs sick. *National Geographic Society*. Retrieved from https://news.nationalgeographic.com/ news/2014/07/140716-australia-coral-reef-dredging-sediment-diseaseenvironment-ocean/

Lucas, R., Wheeler, D., & Hettige, H. (1992). *Economic development, environmental regulation, and the international migration of toxic industrial pollution, 1960-88 (WPS1062)*. Washington, DC: World Bank. Retrieved from https://books.google.se/books?id=3mJ4UHd0CjgC

Luo, F. J., He, L., Pan, Z., Duan, W. H., Zhao, X. L., & Collins, F. (2013). Effect of very fine particles on workability and strength of concrete made with dune sand. *Construction & Building Materials*, *47*, 131–137. doi:10.1016/j.conbuildmat.2013.05.005

Makowski, C. & Rusenko, K. (2007). Recycled glass cullet as an alternative beach fill material: Results of biological and chemical analyses. *Journal of Coastal Research*, 545–552. doi:10.2112/06-0715.1

Makowski, C., Rusenko, K., & Kruempel, C. J. (2008). Abiotic suitability of recycled glass cullet as an alternative sea turtle nesting substrate. *Journal of Coastal Research*, *243*, 771–779. doi:10.2112/07-0864.1

Makowski, C., Thomson, G., Foye, P., & Higgins, S. (2007). Broward county beach demonstration project: From beers to beaches. In N. C. Kraus & J. D. Rosati (Eds.), *Coastal sediments' 07* (pp. 2265–2278). American Society of Civil Engineers. doi:10.1061/40926(239)178

Malkin, B. (2013). 'Poor quality construction materials to blame' for deadly Bangladesh factory collapse. *Telegraph Media Group*. Retrieved from http://www.telegraph.co.uk/news/worldnews/asia/bangladesh/10075098/Poor-quality-construction-materials-to-blame-for-deadlyBangladesh-factory-collapse.html

Marshall, G. (2016). From sandy beach to Kaby Lake: How sand becomes silicon. *TechRadar pro IT insights for business*. Retrieved from http://www.techradar.com/news/computing-components/processors/howsand-is-transformed-into-silicon-chips-599785

Masciola, C. (1992, November 30). He dreams of a beach–all covered with glass. *Los Angeles Times*. Retrieved from http://articles.latimes.com/1992-11-30/news/mn-9881crushed-glass

McCarney-Castle, K., Voulgaris, G., & Kettner, A. J. (2010). Analysis of fluvial suspended sediment load contribution through anthropocene history to the south Atlantic bight coastal zone, U.S.A. *The Journal of Geology, 118*(4), 399–416. doi:10.1086/652658

McDowell, R. (2015). AP tracks slave boats to Papua New Guinea. *Associated Press*. Retrieved from https://www.ap.org/explore/seafood-fromslaves/ap-tracks-slave-boats-to-papua-new-guinea.html

Mclean, J., & McDonald, B. (2012). Extravagant Dubai island project sinks under weight of the credit crunch. *Independent.ie*. Retrieved from https: //www.independent.ie/world-news/middle-east/extravagant-dubaiisland-project-sinks-under-weight-of-the-credit-crunch-26565663.html

McWhan, D. (2012). *Sand and silicon: Science that changed the world*. Oxford, UK: Oxford University Press. Retrieved from https://books.google.com/books?id=UZl4LVarWycC

Moreno, E. L., & Blanco, Z. G. (2014). Ghost cities and empty houses: Wasted prosperity. *American International Journal of Social Science*, *3*(2), 207–216.

Morris, R. (2017). What happens when your country drowns? *Mother Jones*. Retrieved from http://www.motherjones.com/environment/2009/11/tuvalu-climaterefugees/

Muthomi, S., Okoth, P., Were, E., & Vundi, S. (2015). An examination of the nature of sand harvesting conflicts and their influence on poverty alleviation initiatives in Makueni county, Kenya. *Journal of Education and Practice*, *6*(27), 28–36.

Nair, A. S., & Bhattacharjee, N. (2017). Pullback in U.S. fracking sand use pressures producers. *Thomson Reuters*. Retrieved from https://www. reuters.com/article/us-usa-oil-sand/pullback-in-u-s-fracking-sanduse-pressures-producers-idUSKCN1AW07F

Nthambi, M. V., & Orodho, J. A. (2015). Effects of sand harvesting on environment and educational outcomes in public primary schools in Kathiani Sub-County, Machakos County, Kenya. *Journal of Education and Practice*, *6*(24), 88–97. Retrieved from https://files.eric.ed.gov/fulltext/EJ1078871.pdf

Padmalal, D., & Maya, K. (2014). Sand mining: The world scenario. In D. Padmalal & K. Maya (Eds.), Sand mining: Environmental impacts and selected case studies (pp. 57–80). Dordrecht: Springer Netherlands. doi:5 doi:10.1007/978-94-017-9144-1

Peduzzi, P. (2014). Sand, rarer than one thinks. *Environmental Development, 11*, 208–218. doi:10.1016/j.envdev.2014.04.001

Piesse, M. (2016). *Livelihoods and food security on the Mekong River*. Future Direction International, Strategic Analysis Paper.

Pilkey, O. H., & Cooper, J. A. G. (2014). Are natural beaches facing extinction? *Journal of Coastal Research, 70*(sp1), 431–436. doi:10.2112/SI70-073.1

Pilkey, O. H., & Wright, H. L. III. (1988). Seawalls versus beaches. *Journal of Coastal Research*, 41–64.

Rahim, L., & Rahim, L. (2010). *Singapore in the Malay world: Building and breaching regional bridges*. Routledge. Retrieved from https:// books.google.com/ books?id=1utb8ZYyUeQC

Rapoza, K. (2015). What will become of China's ghost cities? *Forbes Magazine*. Retrieved from https://www.forbes.com/sites/kenrapoza/2015/ 07/20/what-will-become-of-chinas-ghost-cities/#43d551cb2e7b

Rappeneau, G., & Mini, L. (Producers), & Delestrac, D. (Director). (2012). *Sand Wars* [Motion Picture]. France: PBS International.

Shepard, W. (2015). *Ghost cities of China: The story of cities without people in the world's most populated country*. London: Zed Books Ltd.

Spencer, R. (2011). The world is sinking: Dubai islands 'falling into the sea'. *Telegraph Media Group*. Retrieved from http://www.telegraph.co.uk/ news/ worldnews/middleeast/dubai/8271643/The-World-is-sinkingDubai-islands-falling-into-the-sea.html

Taborda, J. (2017). United states housing starts 1959-2018. *Trading Economics*. Retrieved from https://tradingeconomics. com/united-states/housing-starts

Taha, B. & Nounu, G. (2008). Properties of concrete contains mixed colour waste recycled glass as sand and cement replacement. *Construction and Building Materials, 22*(5), 713–720. doi:.conbuildmat.2007.01.019 doi:10.1016/j

Torres, A., Brandt, J., Lear, K., & Liu, J. (2017). A looming tragedy of the sand commons. *Science*, *357*(6355), 970–971. doi:10.1126cience.aao0503 PMID:28883058

Umesh, M., & Murthy, A. (2014). Sand mining: Curbing the evil to the environment through sustainable substitution and legislative action. *Ontario International Development Agency International Journal of Sustainable Development*, *7*(3), 17–26.

Urban, F. (2018). China's rise: Challenging the north-south technology transfer paradigm for climate change mitigation and low carbon energy. *Energy Policy*, *113*, 320–330. doi:10.1016/j.enpol.2017.11.007

U.S. Department of the Interior. (2017). *Mineral Commodity Summaries 2017*. Reston, VA: U.S. Geological Survey.

Van Lavieren, H., Burt, J., Feary, D., Cavalcante, G., Marquis, E., Benedetti, L., & Sale, P. (2011). *Managing the growing impacts of development on fragile coastal and marine ecosystems: Lessons from the gulf*. Hamilton, Ontario, Canada: United Nations University Institute for Water, Environment and Health.

Vatan, M. (2017). Evolution of construction systems: Cultural effects on traditional structures and their reflection on modern building construction. In G. Koc, M.-T. Claes, & B. Christiansen (Eds.), *Cultural influences on architecture* (pp. 35–57). Hershey, PA: IGI Global. doi:10.4018/978-1-5225-1744-3.ch002

Wambua, M. P. (2015). *Environmental and socio-economic impacts of sand harvesting on the community in river kivou catchment, Mwingi sub county, Kitui County, Kenya. Thesis*. Nairobi, Kenya: Kenyatta University. Retrieved from http://ir-library.ku.ac.ke/handle/123456789/14023

Welland, M. (2009). *Sand: The never-ending story*. Berkeley, CA: University of California Press.

Wesley, D., Puffer, S. M., & Wade, B. (2018). *Mining and corporate social responsibility: Scotbar Proprietary Limited*. London, ON, Canada: Ivey Case Publishing.

White, N. (2017). Sydney's ghost homes: How 200,000 homes sit vacant in Australia's most expensive city - because foreign investors buy them and leave them to gather dust. *Associated Newspapers*. Retrieved from http://www.dailymail.co.uk/news/article-4270282/Sydney-200-000homes-foreign-investors-blamed.html

Williams, A., & Micallef, A. (2011). *Beach management: Principles and practice*. Earthscan. Retrieved from https://books.google.com/books?id=z55JHEwrlLMC

Wooldridge, T., Henter, H. J., & Kohn, J. R. (2016). Effects of beach replenishment on intertidal invertebrates: A 15-month, eight beach study. *Estuarine, Coastal and Shelf Science*, *175*, 24–33. doi:10.1016/j.ecss.2016.03.018

World Atlas. (2017). Retrieved from https://www.worldatlas.com/articles/top-20-sand-exporting-countries.html

Zhang, G., Song, J., Yang, J., & Liu, X. (2006). Performance of mortar and concrete made with a fine aggregate of desert sand. *Building and Environment*, *41*(11), 1478–1481. doi:10.1016/j.buildenv.2005.05.033

ENDNOTE

1 Silica, or silicon dioxide, is commonly used as an anti-caking agent in processed meats.

Chapter 2
Determining Architecture's Footprint:
Preliminary Methods for Measuring the True Environmental Impact of Buildings

Blaine Erickson Brownell
University of Minnesota, USA

ABSTRACT

Current approaches to designing sustainable buildings are inadequate for meeting environmental goals. Buildings continue to consume nearly half of all resources, and architects, engineers, and contractors remain complicit in their deficient environmental performance—as well as the consequential global overshoot of resource consumption. It is imperative that the AEC industry pursue an alternative approach to green rating systems with the intent to determine measurable, absolute outcomes. The most appropriate existing model is the ecological footprint (EF) method devised by Mathis Wackernagel and William Rees at the University of British Columbia in the early 1990s. EF quantifies the human demand on the environment in terms of both resources and waste, translating these impacts into land area equivalents. This chapter aims to evaluate EF methodology for buildings by analyzing existing models and proposing new approaches while identifying their respective opportunities and limitations.

DOI: 10.4018/978-1-5225-6995-4.ch002

INTRODUCTION

Current approaches to designing sustainable buildings are inadequate for meeting environmental goals. Despite the progress made by green certification programs such as BREEAM, LEED, and Green Globes, climate change continues to accelerate and human societies remain incapable of meeting shared carbon reduction targets. Buildings continue to consume nearly half of all resources, and architects, engineers, and contractors remain complicit in their deficient environmental performance—as well as the consequential global overshoot of resource consumption.

The problem is not the lack of intent, but the absence of a methodology based on definitive measures. The green building programs and checklists used by AEC professionals today are problematic in that they deliver a false sense of accomplishment. They are relative models that estimate environmental improvements over standard building practices. Such changes are measurable at a local, incremental level—such as improved insulating capacity or reduced use of VOCs—and the architects who attain green certification for their buildings gain a sense of accomplishment. Yet such accreditation is meaningless in the absence of absolute measures of building performance based on global environmental objectives.

For example, a building might be awarded "LEED Silver" certification yet still contribute significantly to carbon emissions, thus exacerbating the challenge of meeting greenhouse gas reduction targets. Indeed, is there a calculable link between "LEED Silver," "Three Green Globes," "Living Building Challenge Certification," or any other environmental building labels and allowable CO_2 emission levels worldwide? Using these platforms, is it possible to say with any confidence that a building is utilizing no more than its fair share of the Earth's resources? The answer is no. In fact, the possibility exists that even if all new buildings were designed and constructed to "LEED Platinum" standards (or their equivalent), they might exacerbate current climate goals.

The green rating systems in frequent use today are relative models. They are motley collections of best practices that are primarily atypical, aggregated into point-based checklists. For example, reducing nighttime light pollution, minimizing stormwater run-off, or providing bike storage facilities are not standard building design strategies. Indeed, one can argue that each of these approaches enhances the environmental responsibility of a project. Yet how is this responsibility measured? Although green rating systems offer individual points that culminate in a total score, the points are arbitrary. Can one argue, based on scientific evidence, that the provision of bike racks is equal to the minimization of stormwater run-off in terms of environmental performance? Absolutely not. In all likelihood, the point systems are so inherently skewed and internally inconsistent as to be unreliable, particularly given the complex nature of assessing ecological impact.

It is imperative that the AEC industry pursue an alternative approach based on measurable outcomes. Such a strategy would connect building design decisions with global implications, enabling architects to calculate environmental impacts authentically. The most appropriate existing model is the Ecological Footprint (EF) method devised by Mathis Wackernagel and William Rees at the University of British Columbia in the early 1990s (Wackernagel & Rees, 1996). EF quantifies the human demand on the environment in terms of both resources and waste, translating these impacts into land area equivalents. EF is typically calculated in terms of individual or national resource consumption. The outcome is simple and readily comprehensible: the number of Earths required to sustain present demand. If a country is using its fair share of one planet's worth of resources or less, environmental objectives are met. If not, the use pattern is by definition unsustainable. No "silver" or "gold" labels are required: one is either sustaining the Earth's healthy functioning or not.

Wackernagel and Rees have written little about EF related to individual buildings, and the AEC industry does not utilize the EF approach. Nevertheless, it is possible to make credible links between EF and life cycle assessment (LCA), the comprehensive method of environmental impact accounting that is being adopted within the construction industry. The key is to determine the land area-equivalency of LCA midpoint impacts, which can be accomplished at various scales including a building, an assembly, or a material unit.

A few scholars have begun to explore EF methodology for building design and construction with promising results. Researchers at the University of Siena and the University of Seville have published preliminary EF assessments of buildings based on specific case studies in Italy and Spain, respectively. These findings advance the scholarship on multi-family residential construction in particular, with some intriguing insights that have far-reaching implications. As compelling as these studies are, however, they represent only the early stages of EF for architecture. Much more work must be done to address regional and typological differences of building projects for broad applicability purposes.

This chapter aims to evaluate current EF methodology for buildings with such a goal in mind, analyzing existing models and proposing new approaches while identifying their respective opportunities and limitations. It highlights practices that may be readily applied regardless of location, based on universally accepted standards, and recognizes the additional steps that must be taken to establish a more widely applicable approach. The chapter also addresses the need for more creative means of visualizing footprint data in order to communicate this concept more effectively than numbers.

Ultimately, it is the author's hope that ecological footprint accounting can be developed into a reliable methodology for measuring the environmental performance of architecture. It will not be an easy goal to achieve given the intricacies, lack of knowledge, and value differences inherent in the project of ecological quantification. The broader adoption of EF will also invariably lead to disagreements concerning the concept of fair share. Nevertheless, the rapid pace at which the wealthiest societies accelerate their demand for planetary resources in blissful ignorance suggests that it is time for greater awareness and utilization of this concept.

BACKGROUND

The Ecological Footprint (EF) for a population is "estimated by calculating how much land and water area is required on a continuous basis to produce all the goods and services consumed, and to assimilate all the wastes generated, by that population" (Wackernagel & Rees, 1996, p. 61). The EF method is one of several widely adopted sustainability indicators including the Human Development Index (HDI), Index of Sustainable Economic Welfare (ISEW), and Sustainable Net Domestic Product (SDP), a few of which address quality of life or technology-related concerns. EF does neither; rather, it focuses exclusively on environmental issues—although these are indirectly related to the social and technological dimensions of sustainable development. The EF approach is simple and straightforward: it converts human consumption and disposal into land (and sea) area. According to Wackernagel and Rees,

The production and use of any good and service depends on the various type of ecological productivity. These ecological productive can be converted to land-area equivalents. Summing the land requirements for all significant categories of consumption and waste estimates the EF for the reference population. (Wackernagel & Rees, 1996, p. 67)

The original EF method proposed the following land categories based on use: 1) energy land; 2) consumed land; 3) currently used land, and; 4) land of limited availability (Wackernagel & Rees, 1996, p. 68). Energy land is the area of forest required to absorb the CO_2 produced by fossil fuels (another definition is the land area needed to grow biomass to replace fossil fuels). Consumed land is that occupied by the built environment. Currently used land is productive area that includes gardens, cropland, and pasture or managed forest (not to be double-counted with energy land above). Land of limited availability comprises untouched wilderness critical for sustaining biological diversity as well as non-productive areas such as deserts and icecaps. Since 1996, EF has evolved to include productive sea area to include seafood and other forms of marine consumption.

Although the Ecological Footprint concept was first published over two decades ago, there is much confusion today regarding "footprint" terminology. For example, a common label is "environmental footprint," which applies generally to consumption of natural resources. This term may be found in publications such as the article "Environmental footprint assessment of building structures: A comparative study" (Sinha, Lennartsson & Frostell, 2016). Despite the similarity of the labels, this study concerns life cycle assessment (LCA) rather than EF methodology. Other typical terms include "carbon footprint," which addresses CO_2 emissions related to energy consumption, and "water footprint," a measure of the volume of freshwater use. Although Galli et al. (2012) have attempted to integrate the various indices into a "Footprint Family" of indicators, such an approach is largely unknown within the AEC industry. It is therefore critical that footprint terminology is used with precision.

The Ecological Footprint calculation procedure involves the transformation of consumption into productive land area using the following equation:

$$aa_i = c_i / p_i$$

where aa is the productive land area for each category (ha), c is total consumption (kg), and p is productivity (kg/ha). The Ecological Footprint per capita (ef) is then calculated:

$$ef = aa_i / N$$

where N is the population included in the analysis. The total EF may also be obtained as follows:

$$EF = N(ef)$$

where N is the population size and ef is the average per capita footprint.

Buildings are referenced in EF methodology in two ways. The first is the territory labeled "consumed land," which refers to the literal footprint of the built environment. In addition to building structures, this area includes various transportation and service infrastructure. The second is one of the five major categories of consumption, which is housing—the others are food, transportation, consumer goods, and services (Wackernagel & Rees, 1996, p. 67). The fact that housing is a standalone category enables EF tracking of single and multifamily residential construction. Unfortunately, the many other building programs such as commercial, institutional, and educational are not addressed specifically as consumption categories and must be inferred from general data.

Conceptually, EF represents a natural fit with buildings and related forms of development. Unlike consumer products or vehicles, which are inherently mobile, buildings and the sites they occupy are for our purposes static and fixed to the land—allowing for contextually specific consumption and waste measurements. Buildings thus have two types of footprint: a literal, physical footprint (directly consumed land area) and a theoretical Ecological Footprint (a summary of all resource flows and waste sinks translated into land area). The EF is inherently comprehensive as it includes the consumed land area in its calculations.

For the purposes of informed architectural design and analysis, the EF may be applied at a variety of scales. Just as EF conventionally considers the widely divergent scales of nations and individuals, it may also be used to assess different orders of magnitude appropriate to the constructed environment—such as material units, material assemblies, buildings, campuses, neighborhoods, cities, and regions. Depending on the scale of concern, either a bottom-up or top-down approach—or some hybrid of the two—will be more appropriate. For example, a material unit EF might be calculated based on known ingredients and embodied CO_2 measurements—whereas a city EF might be estimated based on the "mirrored density" approach, which involves the extrapolation of baseline figures across an entire urban area (Wackernagel & Rees, 1996, p. 104).

Estimating the EF of buildings involves an assessment of two primary resources: materials and energy. Materials specified for a building design may in some cases be connected directly to particular land-use categories. Wood, for example, may be linked to managed forest land. Energy pertains to the energy consumed throughout a material's life cycle and is calculated as energy land. This area is also called "CO_2 land" since carbon emissions are a primary concern of energy consumption (Wackernagel & Rees, 1996, p. 68). Although EF conceptually includes areas required to absorb other forms of waste, EF methodology is not as developed in this regard. Building-related waste and emissions (aside from CO_2) will, therefore, remain beyond the scope of this chapter.

Building EF calculations can be made for different scales and life cycle stages. Therefore, the following four models are proposed: 1) embodied footprint; 2) operational footprint; 3) occupant footprint, and; 4) influence footprint. The embodied method concerns the material and energy EF associated with specified materials from harvesting through construction. The operational approach regards the EF resulting from use and maintenance throughout a building's lifespan. The occupant method focuses on per capita EF implications for regular building occupants and users. The influence method considers the broader audience affected by a building project, such as construction labor or indirect commercial activity. Although this list of models is not exhaustive, it is intended to encompass typical assessment scenarios in the AEC industry, such as product specification (embodied footprint) or programming (user footprint). The following sections outline the procedures for each model as well as their limits and implications.

Embodied Footprint

The embodied footprint of architecture is the easiest to conceptualize, given the visual evidence of the process of construction as well as the final "product" of the building itself. The research of Jaime Solis-Guzman and Madelyn Marrero (2015) at the University of Seville's Department of Building Construction focuses on this early phase of a building's life to inform better architectural design and construction practices. In *Ecological Footprint Assessment of Building Construction*, the authors outline an EF evaluation procedure that characterizes resources according to three primary categories: materials, manpower, and machinery (Solis-Guzman & Marrero, 2015). For clarity purposes, it is important to distinguish between the two life cycle stages considered in this work: 1) the product stage, including the harvesting of raw materials and manufacture of building products, and; 2) the construction stage, including prefabrication of building assemblies as well as the transportation of products to the construction site.

Embodied footprint calculations follow a bottom-up approach based on resource and emissions measurements that are tied to specific quantities of materials. Beginning with the product stage, we can quantify various resource inputs based on established industry data. Visualization of this information enables the representation of resources in physical terms, which is the goal of the EF approach. Due to geographic, technological, and other variations—in addition to the fact that the eco-properties of materials are not well characterized—it is wise to use quantitative ranges when possible to account for imprecision.

The first step is to consider the raw material used to make the product, such as timber for wood. Because the Ecological Footprint approach includes a distinct category for managed forest land, we can quickly estimate this portion of a wood product's area requirements. According to Wackernagel and Rees (1996), average managed forest productivity may be calculated at 2.3 m^3 of usable wood fiber per hectare annually. Because this figure is derived from temperate Canadian forests, we might also include the more conservative number of 2 m^3 per hectare annually from a Dutch Friends of the Earth report that averages global conditions (Buitenkamp, Venner, & Wams, 1993).

Another physical resource that we may estimate is water. A softwood like pine requires the use of between 500 and 750 Liters of water per kilogram of material, from cradle-to-gate (Ashby, 2012). Given a density range for softwood between 440 and 600 kg/m^3, one cubic meter of softwood utilizes 220,000–450,000 L of water. As stated above, the EF method does not yet have a precise method of accounting for water as a productive "land-area" (marine), and the water used for timber would come from freshwater sources that are part of other land designations—such as the managed forest itself. Nevertheless, it is helpful to visualize water consumption in some way, if only for comparison purposes. One approach is to translate the required volume of water from L into m^3, assuming an arbitrary depth of 1 m for the sake of simplicity. The resulting "water footprint" is between 220 and 450 m^2 in plan view (Figure 1).

After calculating physical resource inputs, we may estimate the physical impact of emissions. Although there are many material emissions worthy of monitoring, as when conducting a life cycle assessment, the EF approach focuses on CO_2 given its significance in climate change and its direct correlation with energy consumption. Until recently, energy-related sustainable design considerations have focused almost exclusively on operational energy or the everyday use of electricity for building functions. However, building scientists have begun to emphasize the embodied energy in buildings as they realize the significance of its impact. According to Giordano et al. (2015), for the materials specified for Nearly Zero Energy Buildings (NZEB), "Embodied Energy (EE) can be assumed significant from a life-cycle perspective

Figure 1. EF signature for 1 m³ of softwood: It is possible to combine and depict multiple measures of consumption and emissions for a single unit of material, at scale, in one diagram. The cubic volume of wood appears as a 1 m x 1 m white square in the center.

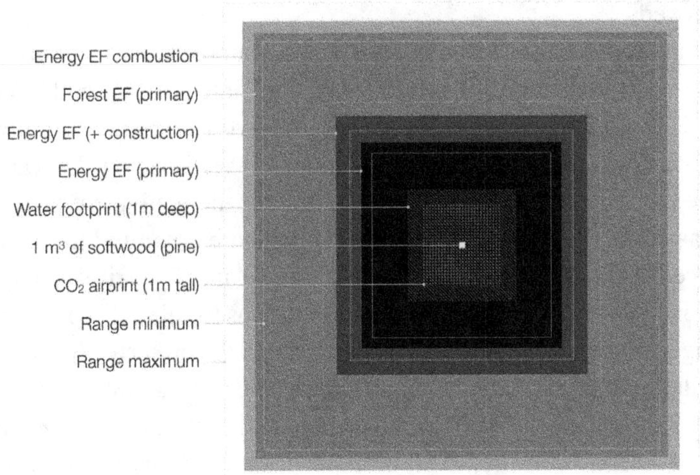

and in a NZEB approach." The authors reveal that NZEB building designs often increase the embodied energy of building products and systems as architects specify advanced insulation or energy monitoring technologies to reduce operational power. In this case, they may not be minimizing the building's footprint but rather deemphasizing its presence in the use phase of the building lifecycle. Structural engineer Bruce King advocates a greater awareness of architecture's embodied footprint, estimating that embodied carbon—or the CO_2 released as a result of the harvesting, processing, and transportation of materials—amounts to roughly ten percent of planetary emissions (King, 2017).

One simple way to approximate carbon emissions in physical terms is by representing the volume of CO_2 produced at each stage of a material's lifecycle. For example, Ashby (2012) estimates that softwood (pine) contributes the following CO_2 emissions in three stages: 1) 0.36–0.40 kg/kg in primary production, 2) 0.022–0.027 kg/kg in construction, and 3) 1.76–1.85 kg/kg in combustion (end of life). At room temperature and standard pressure, 1 kg of CO_2 occupies a volume of 559 L or 0.559 m^3. As mentioned above, 1 m^3 of softwood equals 440–600 kg. So, we may calculate resulting quantities of CO_2 produced in the three stages: 1) 88.55–134.16 m^3 in production; 2) 2.33–3.01 m^3 in construction, and; 3) 20.81-24.91 m^3 in combustion (end of life). Similar to the representation method used for water above, we can choose an arbitrary height of 1 m (the same height as a 1 m^3 cube of softwood) to calculate a physical area. This strategy gives us the resulting carbon "airprint" for each stage, which may be visualized individually or comprehensively. Note that the CO_2 emitted for combustion is equal to that stored within the material; thus this airprint may be represented as a positive carbon store so long as the material is not destroyed.

While this approach reveals the actual volume occupied by emitted CO_2, the Ecological Footprint method also translates this CO_2 into "energy land." Wackernagel and Rees (1996) outline three methods for converting fuel consumption (and its resulting emissions) into corresponding land area: 1) estimate the land area required to grow the quantity of biomass needed to deliver the equivalent energy as fossil fuel, such as corn for ethanol; 2) estimate the land area required to sequester fossil fuel-produced CO_2, as in forests and peat bogs; and 3) estimate the land area "required to rebuild natural capital at the same rate as fossil fuel is being consumed" (Wackernagel & Rees, 1996, p. 73). The authors call the preferred third strategy "the CO_2 assimilation method" and calculate the land-for-energy ratio of fossil fuel at 1 hectare per 1.8 metric tons of CO_2 emitted annually (Wackernagel & Rees, 1996, pp. 74-75). This ratio converts to 5.56 m^2 per 1 kg of CO_2 issued each year. Using the same three stages of softwood production results in the following: 1) 881–1,334 m^2 in primary production; 2) 54–90 m^2 in construction, and; 3) 4,306–5,000 m^2 in combustion (end of life).

Note that the CO_2 assimilation ratio can vary widely depending on fuel type and delivery. For example, electricity generated from fossil fuels with a 30% efficiency results in an area three times larger, whereas energy derived from renewable sources requires as little as ten percent of the land area (Wackernagel & Rees, 1996, 74). Nevertheless, the working ratio 5.56 m^2 per 1 kg of CO_2 is useful in the way it accurately connects energy land to timberland via treatment of wood as a carbon bank. Note the similarity between the energy land area resulting from the combustion of wood at the end of its life, or 4,306–6,172 m^2, and the forest area required to harvest the same volume of timber per year, or 4,348–5,000 m^2. Hence the rationale for taking a natural capital-regeneration approach, which directly links the quantity and rate of fossil fuel consumption to the area and replenishment rate of forests. Tracking the stored carbon in wood and other biomass-based products is a critical component of the material selection process. By revealing the extent to which the productive capacity of land may be "borrowed" in the form of embodied carbon, these materials reveal their environmental advantages immediately when compared with non-renewable products. (To use a financial analogy, one could say that the carbon has simply been diverted to another account—although this account no longer earns "interest" via growth as the previous one does.)

As with stored carbon, embodied resources and emissions data are particularly useful when making comparisons. Material components may be evaluated as declared or functional units, in keeping with life cycle assessment terminology. When the particular function of a material is unknown, or when the array of possible applications is too broad, one must use declared units. At the material scale, it is appropriate to make elemental calculations by mass, volume, density, and so on—as opposed to the building scale, where one might wish to estimate a quantity of material per unit area. A standard material comparison in architecture considers the three substances most frequently used in structural frames: steel, concrete, and wood (Figure 2). If these materials are compared by equivalent mass, their volumetric differences become apparent. Although 1 kg of virgin steel (low carbon, rolled) occupies a small volume of 127 cm^3, its embodied CO_2 for primary production and processing is 1.2 m^3. 1 kg of softwood, by comparison, has a much larger volume of 1,023 cm^3 yet has an embodied CO_2 that is five times smaller than that of steel. Meanwhile, 1 kg of concrete is 408 cm^3, yet its CO_2 "airprint" is about a quarter of softwood's. Using the CO_2 assimilation ratio, the embodied CO_2 of the materials translates into similar proportions of energy land footprints, with 1 kg of steel occupying nearly 12 m_2 of area.

Given the questionable environmental track record of concrete, readers may be surprised that its Ecological Footprint appears to be superior to that of softwood. However, a volumetric equivalency reveals concrete to be inferior. Considering 10

Figure 2. Three methods of comparing units of steel, concrete, and wood. Each process results in different environmental impacts, and functional equivalency is preferred when possible. (Not to scale)

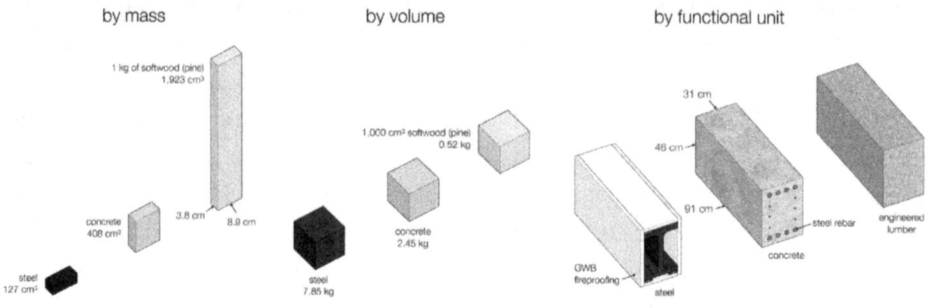

x 10 x 10 cm cubes of material, concrete would emit 0.15 m^3 of CO_2 versus 0.12 m^3 for softwood—and steel would release a CO_2 volume 78 times greater! The energy land footprints for these samples are similarly disparate, with the steel cube occupying over 93 m^2 of area.

AEC industry professionals would rightly argue that this is still not a fair comparison for selection purposes, given that the three materials have varying properties. Steel's high strength in compression and tension allow for it to be used in relatively small quantities, for example. Additionally, concrete is still not being accurately represented if it is not reinforced—a requirement for nearly all concrete applications in buildings. The inclusion of reinforcing steel predictably increases concrete's embodied carbon and energy land footprint. Thus, for architects to choose materials more equitably, they must be compared as functional units.

According to Kathrina Simonen (2014), a functional unit defines "a unit of analysis that includes quantity, quality and duration of the product or service provided." An example would be the consideration of equal lengths of structural beam. In this case, the steel would include a fireproof enclosure, and the concrete would require steel reinforcement. Given the heft of the first two examples, engineered lumber is selected for the softwood beam. A comparison of embodied CO_2 reveals a ratio of 1 to 1.6 to 8 for wood, concrete, and steel, respectively. The engineered lumber beam's energy land footprint is 256 m^2, and the one for steel is 2,081 m^2—still significantly large. Steel's high energy requirement reveals its underestimated importance in the embodied energy calculations for concrete. Building professionals typically find fault with the Portland cement ingredient of concrete, given its high energy requirements in processing; yet the reinforcing steel is also a critical factor in environmental performance.

Although this functional comparison is more objective, it is still open to criticism. What are the beams supporting? In reality, would they not assume different profiles, volumes, and mass when carrying a particular dead load distributed across the same tributary area? For that matter, would they not support different flooring systems in service of bearing this load? The crucial nuances of functional unit comparison expose the complexities of analyzing different building materials based on environmental performance. A more accurate analysis might, therefore, consider a simple structural bay composed primarily of each material, supporting the same tributary area. A comparison of a single-story frame with four columns on 20' centers supporting a 900 SF total floor area is revealing (Figure 3). Using Athena Impact Estimator for Buildings for LCA calculations, reinforced concrete is revealed to be a far worse performer than steel (with corrugated metal deck and topping slab) or wood (glulam structure with wood deck) (Athena Sustainable Materials Institute, 2017). With 5,050 kg of CO_2 emitted during primary production alone, a concrete structure has nearly three times the carbon footprint of steel and 6.5 times that of wood. The difference highlights the significance of functional equivalency based on structural performance: the same loading over the same tributary area requires significantly more volume of concrete than steel or wood. When one factors in the positive carbon stored in the wood, it vastly outperforms steel as well as concrete in terms of energy land footprint (Figure 4).

Operational Footprint

Many of the embodied footprint methods above are made without reference to time; yet time is required to estimate the rate of replenishment of productive land areas. The Ecological Footprint is typically calculated based on the quantity of resources used and waste produced annually. In the absence of specific scheduling information, one might approximate the EF of a particular building material or assembly from "cradle to occupation," or from harvesting through the end of construction, as if it occurred during one year's time. Given that many buildings are built in one year or less, this is not an unrealistic assumption; however, the actual duration of phases could be much longer. Nevertheless, when specifics are unknown during the design process, time is left out of embodied footprint calculations.

In contrast, operational footprint analysis accounts for life cycle realities, emphasizing the use phase of a building (hence the term operational), which typically has the most prolonged duration of the life phases. Like embodied footprint calculations, operations include both resource inputs and emissions outputs. Resources consist of materials and energy: building products used for maintenance as well as ongoing energy consumption. The embodied footprint reflects how architects think about design, in that it addresses the materials selected to construct a building.

Figure 3. Ecological Footprint (energy land) required for a structural bay of engineered lumber, steel, and reinforced concrete, from left to right, in order of impact. To communicate the physical extents of each footprint, the EF is calculated based on taking the length of a football field and expanding it to the required area, with each structure located on the centerline of that area. The viewpoint is located at the edge of the field.

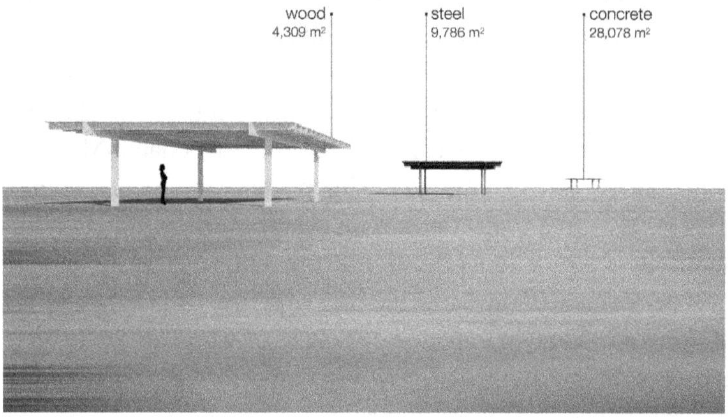

Figure 4. Material comparisons may be aggregated into a series of material EF signatures, based on one material representing 100 percent. This series reveals the significant differences between various methods of comparison, with the functional system being the most relevant.

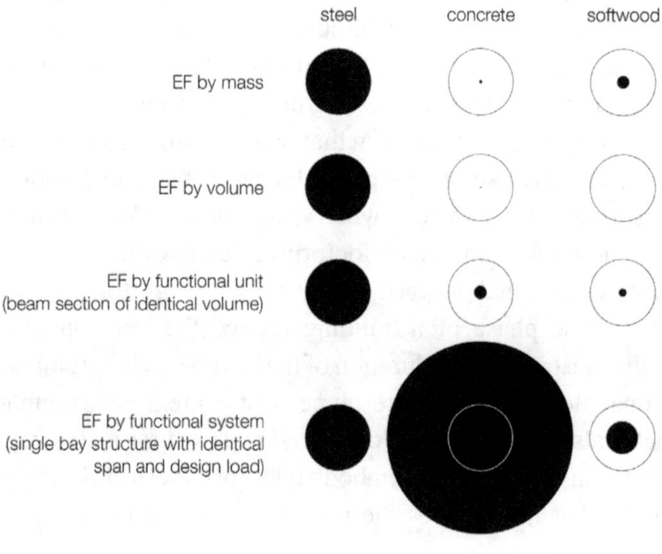

Meanwhile, the operational footprint reveals the potential significance of occupation-related environmental impacts.

Spending much of his career focused on the change rate of buildings, British architect Frank Duffy determined that the operational consumption of architecture can overwhelm its original consumption of resources (Brand, 1994). "Our basic argument is that there isn't such a thing as a building," Duffy claims. "A building properly conceived is several layers of longevity of built components" (Brand, 1994). In *The Changing City*, Duffy and co-author Alex Henney estimate that a typical office building requires three times the capital investment over a fifty-year period than the original structure (Duffy & Henney, 1989). The authors divide this investment in three categories based on duration: structure (50 years), services (15-20 years), and space plan (5-7 years). Although the building's structure may not require replacement during a 50 year period, the services and space plan (interior furnishings) will change many times (also called a recurring embodied footprint). Borrowing an expression from Bastioni et al (2006), buildings may be considered "open evolutionary systems" that require an ongoing influx of natural capital to function.

Program has a significant effect on the operational footprint. For example, a life cycle assessment of a typical, speculatively built home in Australia reveals a different story (Haynes, 2010). Within an equivalent lifespan of 50 years, the four-bedroom, brick-clad house is similarly anticipated to have ongoing resource demands. However, the most substantial quantity of carbon is embodied in the home's original construction materials—at 63 percent over the half-century lifespan (Haynes, 2010, p. 10). The process of construction accounts for 9 percent of the embodied carbon, leaving 28 percent for recurring embodied carbon. Note that the Haynes and Duffy examples are not equivalent: in addition to and geographic, temporal, and accounting differences (Duffy is focusing on capital), Haynes does not include the "space plan" of interior furnishings and various other material items that encounter the highest frequency of churn. Nevertheless, the comparison does raise the question of programmatic influences on material and energy flow rates.

Absent in these examples is the consideration of operational energy—the energy used for building electricity, HVAC, and lighting—which also varies by program. To estimate the efficacy of a building's energy consumption, a fundamental metric is energy use intensity (EUI), which is expressed in energy use per area per year. By normalizing all buildings by EUI, their performance can easily be compared regardless of size differences.

The U.S. Environmental Protection Agency (EPA) tracks the energy consumption of over a quarter of a million U.S. buildings via its Portfolio Manager program (U.S. EPA, 2012). Building activity is considered a "key contributing factor" in the EPA benchmark study, and the highest energy consumers include supermarkets, hospitals, office buildings, and senior care facilities. Supermarkets and convenience stores are top consumers due to high refrigeration loads, with an average EUI of 480–536 kBtu/ft^2 (U.S. EPA, 2016). In contrast, office buildings range between 117 and 183 kBtu/ft^2. Although the Portfolio Manager does not include single-family homes in its 2016 data, the EUI for multifamily housing is 128 kBtu/ft^2 (U.S. EPA, 2016) (Figure 5).

To estimate the operational footprint, the varied energy mix must also be considered. For example, a recent study of U.S. commercial buildings identified electricity as the most consumed form of energy (61%), followed by natural gas (32%), district heat (5%), and fuel oil (2%) (U.S. EIA, 2016). As mentioned previously, the energy use has a significant impact on a building's EF. Due to processing losses, Wackernagel and Rees' CO_2 assimilation ratio for electricity, a secondary energy, results in a footprint three times larger than that of natural gas, a primary fossil fuel.

Figure 5. Energy Use Intensity (EUI) of 100 building programs, organized according to U.S. Energy Star categories for property types. Long lines indicate source (power plant) EUI and short lines indicate onsite generation. The most significant consumer (upper right) is fast food restaurants, which average a massive 1,015 kBtu/ft^2 annually.

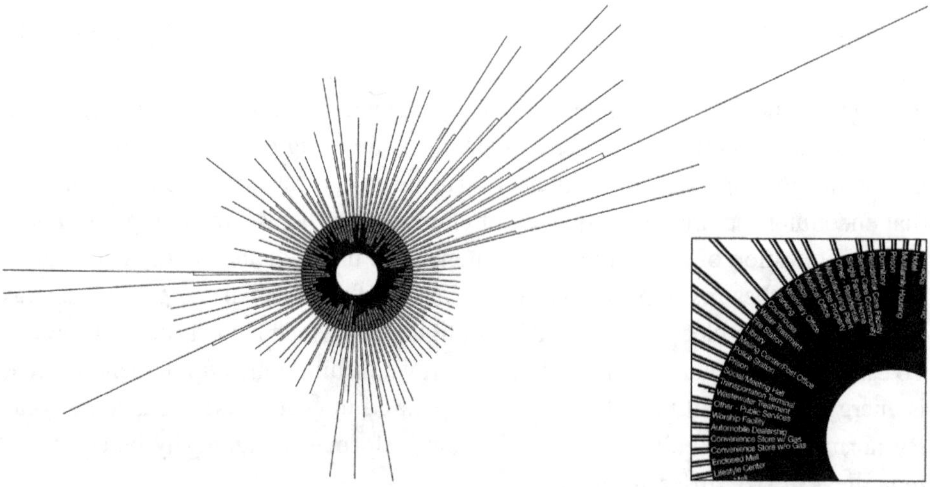

Occupant Footprint

Embodied and operational footprints account for the materials, energy, and related emissions throughout a building's lifecycle. However, the above examples do not account for the people who occupy buildings. Sustainability is, after all, a human-focused concept insofar as it considers the ability for current and future populations to meet their fundamental needs. Per capita metrics are also a cornerstone of Ecological Footprint accounting, which seeks to link the Earth's biocapacity in physical terms to the global population. These metrics are similarly important in the EF analysis of buildings.

Wackernagel and Rees remind us that sustainability depends upon an assessment of user consumption per time interval (Wackernagel & Rees, 1996, p. 118). "For example, it might be sustainable to operate a gas-guzzling Rolls Royce if it were shared among 20 friends, and maintained for a long time," they write. "On the other hand, it might be unsustainable for everybody to own an electric car" (Wackernagel & Rees, 1996, p. 118). According to this perspective, one cannot accurately assess the environmental performance of a product or building without knowing the variables of user consumption and time. An example building analogy might be a sizable family occupying an energy-inefficient McMansion over generations as being more sustainable than many individuals owning their own Passive Houses for a short period. This interdependency between buildings, occupants, and time points to another weakness in green rating systems, which typically do not account for per capita use and projected lifespan, and therefore can support poor environmental choices unwittingly.

Wackernagel and Rees' antidote is the approximation of fair Earthshare, which is "the area of ecologically productive land 'available' per capita on Earth" (Wackernagel & Rees, 1996, p. 119). Based on an estimation of 1.5 hectares per person, per capita consumption and emissions may be translated into time-based percentages of fair Earthshare. According to the U.S. Department of Energy's Office of Energy Efficiency and Renewable Energy, U.S. houses used an average of 101,800 Btu/ft^2 in 2012 (DeSilver, 2015). The average U.S. house size that year was 1,864 ft^2, or 173.2 m^2—translating to 189,755 kBtu or 200,357 MJ. Wackernagel and Rees use an annual productivity of 100 GJ/hr/year or 150 GJ/year per Earthshare. So if two people share the house, we calculate the hours of Earthshare as follows: (8,760 hr/yr x 200,357 MJ) ÷ (150,000 MJ/yr x 2 people) = 5,850 hr/yr, or 244 days each. Thus, regarding operational energy alone, each resident consumes 2/3 of their annual Earthshare just by living in the house. This figure does not include the embodied energy or materials for the building; nor does it include any other goods or services the residents will consume throughout the year. Now consider that the average U.S. house size in 1970 was 28% smaller, or 1,456 ft^2. This area would translate into the following, using 2012 energy figures: (8,760 hr/yr x 156,502 MJ) ÷ (150,000 MJ/yr x 2 people) = 4,570 hr/yr, or 190 days each—a gain of almost eight weeks of Earthshare per capita. Adding more residents can improve a score more than an area reduction. For example, a family of six people living in the 1,864 ft^2 house would use only 81 days' worth of Earthshare per person—half the amount per person of the couple living in the smaller house.

This formula, which we could call the per capita method, serves as a simple means of calculating the individual EF for any building. However, most of us spend our lives in more than one building. For example, the residents in the house above might spend an average of 40 hours at an office and 5 hours in other buildings, such as shops or restaurants, each week. To avoid double-counting, one could argue that their Earthshares should be allocated between the various buildings they occupy. Thus, we can employ a time allocation method to divide individuals' hours between the various edifices they inhabit. Thus, a couple who occupies the average-sized 2012 U.S. house 100 hours each week, or 60% of the time, would use about 145 days of operational Earthshare. Let us presume that they both work at a 45,000 ft^2 office building with a typical EUI of 148 kBTU/ft^2 and a population of 300 employees (using a conservative estimate of 150 ft^2 per employee). Each individual's per capita share would be: (8,760 hr/yr x 7,032,094 MJ) ÷ (150,000 MJ/yr x 300 people) = 1,369 hr/yr, or 57 days. At 40 hours per week, this amount translates to only about two weeks of Earthshare. The total between home and office is 145 + 14 = 159 days, or 85 days less than the original figure of 244 days. Of course, the equivalent Earthshare for time spent in other buildings—as well as commuting—must also be included to make a fair comparison.

However, this approach is also flawed insofar as buildings do not stop consuming and emitting resources when unoccupied. The home and office above are likely conditioned when uninhabited, and there are probably plug loads and other electrical loads running at all hours. So additional Earthshare time should be added to each individual's ledger for these unoccupied loads. Yet where should the system boundary be drawn? Is it fair to add a vacant building "Earthshare tax" for every building someone occupies—even for just a short period? If conducted on a time allocation basis, one could justify such an approach, which would expand the environmental consequences of people's personal choices (e.g., not just the purchases consumers make, but the very buildings they inhabit).

Although it may appear reasonable, the time allotment model points to a curious phenomenon we might call the "occupation paradox." The per capita Earthshare approach demonstrates that a building benefits from an increase in the number of occupants and the average time of occupation. Thus, when a building has fewer occupants, the per capita Ecological Footprint increases considerably; but if there are zero occupants, the per capita EF cannot be estimated using this approach (since one cannot divide by zero). Thus, we might choose to include some "ghost" occupants in a given building to calculate consumption and emissions when people are absent. Since these ghosts are likely the same individuals who are otherwise present, we return to our double-counting method above. This conclusion suggests two potentially problematic and contradictory incentives from an EF reduction standpoint: buildings "benefit" from a maximum occupancy all the time, and individuals benefit by occupying the fewest number of buildings with the smallest EFs.

An illustrative building typology is the hotel. Given the nature of transitory inhabitation and varying occupancy rates, the hotel has been the focus of early studies in occupant-related carbon footprint analysis. A compelling summary appears in the article "Reviewing the carbon footprint analysis of hotels: Life Cycle Energy Analysis (LCEA) as a holistic method for carbon impact appraisal of tourist accommodation" (Filimonau, Dickinson, Robbins, & Huijbregts, 2011). A collection of studies conducted between 1993 and 2009 in various geographic locations reveals an expansive EUI range between 280 and 2,570 MJ/m^2/year (Filimonau, Dickinson, Robbins, & Huijbregts, 2011, p. 1925). The authors translate this data into total annual GHG emissions per "1 guest night" in kg CO_2-eq. An in-depth comparison of two hotels located in Poole, Dorset (UK) indicates a carbon footprint per "1 guest night" stay of 11.65 kg CO_2-eq and 8.25 kg CO_2-eq for hotel 1 and 2, respectively (Filimonau, Dickinson, Robbins, & Huijbregts, 2011, p. 1927). Using the EF CO_2 assimilation conversion of 5.56 m^2 per kg CO_2 results in an EF per guest night of 64.77 m^2 and 45.87 m^2, respectively. Considering that the gross floor area (GFA) per guest room for each hotel is 38.82 m^2 (3,300 m^2 ÷ 85 rooms) and 24.09 m^2 (2,000 m^2 ÷ 83 rooms), the EF per guest night is 1.67 and 1.9 times greater than the actual area per room.

This calculation raises a notable point: despite the lower EF and GFA of hotel 2, its ratio of "1 guest night" EF to GFA per room is higher than that of hotel 1, revealing an occupancy footprint that is proportionately larger per area provided to guests. Additionally, it is reasonable to imagine the design of a hotel with a sufficiently low carbon (operational) footprint such that the EF is equivalent to the GFA per guest room. In this case, the proprietors could claim that the hotel has a neutral occupant footprint relative to its physical size (the GFA per room calculation takes into account all occupiable space within the building, including common and service areas). Such an aim would make a compelling goal from an occupant's perspective.

However, building owners would need to factor in the overall occupancy rate of the establishment for their EF calculations. In 2015, the occupancy rate in UK regional hotels was 76 percent (Statista, 2018). Thus, the operating footprint of the 24 percent of vacant rooms would impose an additional load on the occupied ones (although this could be minimized if the vacant rooms are not conditioned and not drawing electricity while vacant). Furthermore, "1 guest night" measurement does not account for the number of inhabitants per room. A true occupant footprint study would not only include the total population of guests but also the number of hotel staff present within the facility. In the likelihood that the total number of hotel inhabitants is higher than the total number of guest rooms, this approach would deliver a lower EF.

The per capita footprint picture becomes even more complicated when one considers large EF buildings with relatively small occupant loads, such as data centers or power plants. Should an individual who utilizes a building's services without necessarily occupying it—such as cloud computing or public sanitation—contribute an equivalent Earthshare? The logical answer is yes: if we benefit from a service that requires resources and adds waste, we must include it in our total EF. This approach suggests that the more carbon intensive the aggregate services provided to a user, the greater the individual's EF. Such functions include traditional, centrally managed infrastructure such as energy, water, and sewage treatment, as well as elective services like internet and telecommunications. Provisions vary widely in their physical proximity: a municipal water source might be located within a few miles of a user, whereas a web hosting provider might reside many states (or countries) away.

Such considerations might suggest that the term "occupancy footprint" be limited to physical occupancy—although this metric could be expanded to include the occupation of one's neighborhood or city. As service provisions expand beyond a tangible scale, we could consider a "service footprint" to account for the carbon associated with individual use yet not directly connected to one's physical location. Furthermore, the inherent complexity of the additional system boundary calculations suggests we should employ a compound method, or top-down calculation of input-output flows for a given population, versus a component method as discussed thus far (Solis-Guzman & Marrerro, 2015).

Influence Footprint

Such considerations lead to the question of a building's broader influence. The examples given above—services that are provided to individuals remotely and therefore augment their per capita Earthshare—apply here. Another aspect of influence relates to different stages of a building's lifecycle. Building product manufacturing, onsite construction, maintenance, and demolition all contribute to a building's EF based the labor populations involved in addition to the embodied, operational, and occupant footprints. As mentioned previously, Solis-Guzman and Marrerro (2015) distinguish between direct consumption and indirect consumption in their estimations of building construction EF. Direct consumption considers the energy used directly onsite and is thus part of the embodied footprint. Indirect consumption includes secondary resources and emissions including materials (also embodied, production stage) and "manpower" that consists of the transportation- and food-related footprints of the labor force (Solis-Guzman & Marrerro, 2015). A study of a typical multifamily residential typology in Spain, for example, revealed that the total building construction footprint included 1 percent for machinery and 3 percent for food (Solis-Guzman & Marrerro, 2015).

Major capital projects can exert an influence that is much farther reaching than their particular sites. According to Wackernagel and Rees,

All large-scale developments such as power plants and transportation infrastructure projects and even changes in zoning can have long-ranging effects on material and energy consumption, which are usually ignored in traditional environmental impact assessment. (Wackernagel & Rees, 1996, p. 109)

For example, the construction of a bridge can lead to significant changes in development resulting from enhanced mobility and accessibility. A study of the widening of the Lions Gate Bridge in Vancouver from 3 to 5 lanes resulted in a 200 km^2 expansion of the serviced area based on modified transportation and settlement patterns (Wackernagel & Rees, 1996, p. 110).

A culturally significant building can have a similarly outsized impact on EF. The Guggenheim Museum in Bilbao, for example, inspired the phrase "The Bilbao Effect" due to the enormous increase in international travelers visiting the building after its opening. In 2015, a total of 1,103,211 people visited the museum, with directly related expenditures of 363 million Euros and the support of 6,875 jobs (Guggenheim Bilbao, 2015). Such significant economic and social impact undoubtedly has a sizable Ecological Footprint that may be directly attributed to the building's remarkable influence.

A cultural destination of this significance influences both permanent as well as transient populations. The increase in the local permanent population may be calculated based on a building's service occupancy in addition to the population change in neighboring developments—minus the occupancy of any structures demolished to make way for the new construction. The increase in the transient (tourist) population may be estimated based on the additional number of visits by non-local travelers. A growing body of research concerning the carbon footprint of tourism provides supporting information in this regard.

For example, the Breda University of Applied Sciences in The Netherlands, which administers the Carbon Management for Tour Operators (CARMATOP) project, estimates that global tourism contributes approximately five percent of all human-made emissions (Centre for Sustainable Tourism and Transport, 2015). A 2016 study of Dutch tourists reported that the daily carbon footprint for outbound (international) holidays ranged between 58 and 61 kg CO_2-eq based on the length of stay (shorter stays result in higher footprints based in large part on air travel) (Centre for Sustainable Tourism and Transport, 2016). These quantities result in a large daily EF of 339 and 322 m^2, respectively. The same study included an intriguing determination related to domestic holidays: "On average, CO_2 emissions per day are slightly lower for domestic holidays than for staying at home (24.5 vs. 27.0 kg/day)" (Centre for Sustainable Tourism and Transport, 2016). This finding indicates that domestic travel in countries like The Netherlands is not only less resource intensive than international travel but also has a lower impact than staying put. It also points to the need for reciprocal calculations. Tourist destinations may have outsized resource use, but the non-local visitors are not consuming energy and materials at home. Thus, in the case of transient (tourist) impacts, the influence footprint should make provisions for the concomitant reduction in resident (e.g., stay at home) consumption. Such accounting will cast a more favorable light on cultural destinations which, in spite of exhibiting a high occupant footprint, may have a relatively lower influence footprint based on reconciling the reduced EF of tourists' vacant homes. In other words, travel may not be as resource-intensive as preliminary considerations might suggest.

The procedures outlined above may appear straightforward; however, calculating architecture's influence is inherently challenging. How many of Bilbao's 1,103,211 visitors in 2015 planned their travels solely based on the existence of the Guggenheim Museum? For many, the project may have inspired the idea of a visit, but most tourists likely add other destinations to their itineraries. Still others make unplanned visits to such destinations. Thus, in the absence of detailed survey data, EF calculations based on influence will remain rough approximations based on aggregate reporting.

At a global scale, Ecological Footprint accounting reveals the considerable resource imbalances that result from economic inequities. The most economically prosperous nations typically serve as model examples for developing countries, yet they are the worst offenders from an environmental perspective—"running massive unaccounted ecological deficits with the rest of the planet" (Wackernagel & Rees, 1996, p. 98). Economic disparity and the resulting environmental cost is also seen at the building scale and should be included in EF approaches. For example, the wealthy owner of a large estate might employ a large staff of housekeepers and gardeners, some of whom live on-site. Yet these employees' individual Ecological Footprints will likely be much smaller than that of the estate's proprietor—even when he or she is physically absent from the property—due to the inequitable distribution of resources. The environmental impact of income inequality suggests that economic factors will play a role in future building EF methodology, attributing Earthshare in proportion to wealth used to harness natural capital.

Integration and Limitations

Each of the four methods outlined above serves a different purpose in design, construction, and post-occupancy evaluation processes. The methods may also be combined to generate a holistic picture of a building's anticipated Ecological Footprint, similar to conducting a whole building life cycle assessment. (Figure 6) LCA data may be utilized for this purpose to determine embodied CO_2 from global warming potential, energy input sources, and a bill of materials. EF requires additional work to estimate land-area allocations, operating energy consumption, and occupancy loads. The advantage of visualizing the resulting EF is that it conveys quantities that are directly associated with the scale of a building and its site, giving the data physical attributes that Architecture Engineering Construction (AEC) industry personnel can appreciate more than numbers alone.

Although the Ecological Footprint method is effective in quantifying specific territorial impacts in buildings related to carbon emissions (energy land) and wood product consumption (timberland), other materials are not currently measured concerning productive land area. For example, the impacts of widely consumed

Figure 6. The holistic Ecological Footprint of a typical single-story concrete office building in Toronto. Based on the biological model of tree rings, this diagram depicts the building life cycle as an outwardly growing set of circles. Embodied and operational CO_2 inputs and emissions are combined in an arbitrary 50-year span. (Data source: Athena Impact Estimator.)

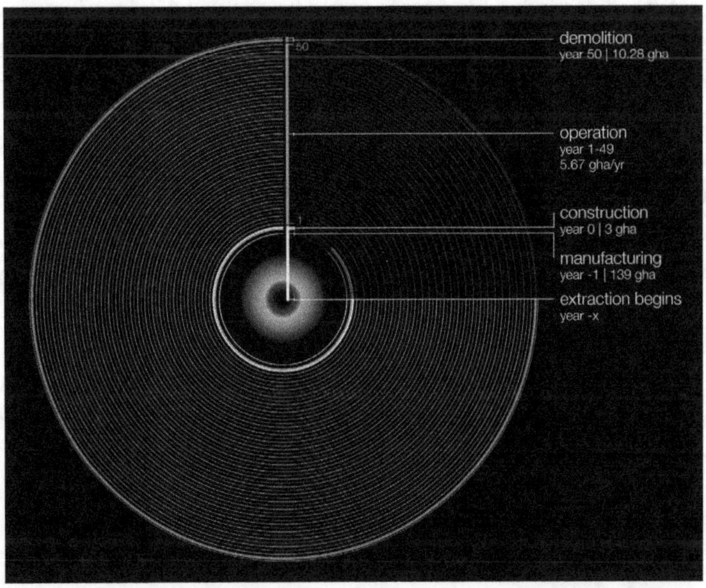

materials such as concrete and steel appear in the energy land designation of EF, but not in other territories. The reality is that activities like mining and excavation reduce bioproductive land areas by a measurable amount—but which areas? If the sites of operation are known, these can be categorized accordingly: a mine adjacent to a developed area could be considered degraded land, whereas sand excavation in a desert could be regarded as non-productive land (which is typically not counted, but should be in this case). Presumably, the inherent complexity in tracking material sourcing has made EF a blunt instrument in this regard. Nevertheless, the land impacts are significant, as the ecological rucksack—or the similar measure called Materials Intensity per Unit Service (MIPS)—of many nonrenewable materials is considerable (Kibert, 2013, p. 48). Virgin aluminum, for example, has an ecological rucksack of 1:85, meaning that 85 kg must be displaced for every kg of raw aluminum produced. This displacement must be accounted for in terms of lost bioproductive area in addition to the energy land required to absorb related CO_2 emissions.

EF land-use categories limit creative thinking about how the constructed environment can play a role in the production of natural capital. For example, the green roof portion of a building could theoretically function as "gardens" or "cropland," yet such a building would be wholly classified under "degraded land" in EF categorization. This label might be accurate with regard to the building's overall footprint, including all of its components located below its green roof. However, green roofs could be used to grow biomass for food, products, and building materials—particularly when implemented at scale—and should, therefore, factor into the "cultivated systems" calculations of the EF method. Roofs that act as water storage reservoirs could similarly benefit EF estimates for buildings. At the same time, underutilized degraded land areas could be occupied by productive operations such as urban agriculture and other low-impact resource harvesting activities.

Another limitation of the building EF approaches outlined above, which are primarily component or "bottom-up" driven, is that they fail to recognize the importance of interconnectedness between natural ecosystems. An entire neighborhood of buildings could be designed with a minimal footprint yet due to its placement impede the circulation of a vital wildlife migration corridor. EF is also conceptually hindered at the macro scale by its compartmentalization of bioproductive lands into distinct categories. This land-use model fits easily within our current land allocation paradigm, which is starkly defined by political and legal boundaries. Yet natural systems do not abide by property lines. Developing research on vagility, or the movement ecology of organisms, reveals the extent to which animal habitats have been adversely affected by human development. According to Tucker et al (2018),

Global loss of vagility alters a key ecological trait of animals that affects not only population persistence but also ecosystem processes such as predator-prey interactions, nutrient cycling, and disease transmission.

Despite the fact that urban ecosystems comprise only four percent of the Earth's total land area (Solis-Guzman & Marrerro, 2015), movement ecology research suggests that the ways in which developed land is deployed have an outsized negative influence on natural ecosystems. By compartmentalizing land uses into neat categories, the EF method artificially reduces the deleterious effects of human land management practices.

One of the most difficult challenges with EF calculations concerns the wide variation in emissions based on energy source. Wackernagel and Rees outlined the following levels of energy productivity in their original book (in Gigajoules per hectare, per year): fossil fuels (based on type), 80-100; hydro-electricity, 150-15,000 (based on altitude); solar hot-water, up to 40,000; photovoltaics, 1,000; and wind energy, 12,500 (Wackernagel & Rees, 1996, p. 69). Higher levels of productivity correspond to smaller EF values. As mentioned previously, energy land may be calculated in several different ways, and delivery efficiency is a fundamental factor in EF quantification (such as the reduced efficiency resulting from generating electricity from fossil fuels). The CO_2 assimilation method used throughout this text is therefore a crude approximation at best; in reality, every EF calculation would be made with an accurate and complete assessment of the energy mix used as well as the rate of efficiency of energy delivery. Such a calculation is easier to perform for an operational footprint, such as for an individual building with established energy sources, than for the embodied footprint of the many materials and assemblies specified for a typical building. Given this distinction, it is therefore no surprise that operational EF accounting is conducted on a more regular basis than embodied EF measurement.

Like whole building LCA, whole building EF methods are stymied by our inability to predict end-of-life. Typical approaches to estimating building longevity—such as the proven physical durability of materials or the timeframes of manufacturer warranties—expose the limits of prediction. A survey conducted by the Athena Sustainable Materials Institute revealed that most buildings are demolished prematurely due to unforeseen causes (O'Connor, 2004). Unanticipated changes in building program, land use, or desirability are a few reasons that buildings meet an untimely demise. "Building industry beliefs that some structural materials last longer than others are most likely confusing how long a building could last with how long it is actually kept in service," writes study author Jennifer O'Connor (2004, p.4). "In the worst case, designing for longevity can lead to design choices that are well-intentioned but, in fact, yield poor environmental results." The likelihood that a building may be demolished prior to reaching physical obsolescence suggests that architects should emphasize strategies for disassembly and material recovery over a maximum lifespan. Such approaches do not influence EF calculations, which quantify materials but not design adaptability.

EF is also limited in its accounting for second-life or repurposed materials in which recycling energy—but not primary energy and material—is quantified. The embodied footprint of materials is often reset to zero when they are repurposed, based on the logic that the substances' prior footprint was "paid for" by their first use and that they would otherwise go to a landfill. However, such thinking ignores the original lifespan of the materials relative to the rate of replenishment of natural capital. For example, a piece of lumber might be re-engineered after a brief initial use; meanwhile, the equivalent managed forest area required to produce the product would still be regrowing. It therefore seems logical to adopt a depreciation method for calculating material footprint rather than the cash-basis, to use accounting terminology, that is currently employed. Depreciation can be based on an estimate of a material's physical longevity, and the "unpaid balance" of the footprint should carry over to the subsequent life (or death) of a material. In this way, building EFs could be tracked via amortization-style ledgers, enabling current and future owners to calculate the "payoff" of embodied footprints and discouraging landfilling of the materials.

CONCLUSION

Despite these limitations, the Ecological Footprint method can be an effective tool for determining the environmental impact of buildings. Compared with green rating systems that assist in making relative adjustments to building designs based on departures from baseline practices, EF enables AEC professionals to focus on absolute impacts regardless of the baseline. Building EF is greatly facilitated by whole building LCA and operates best when using lifecycle data. However, it provides functionality not offered by LCA: it approximates the physical manifestation of consumption and emissions in terms of tangible area. This information may be represented in many ways for clear and accessible communication that is more effective than numbers alone.

Determining the EF of buildings will likely spark controversies that Ecological Footprinting has thus far avoided. Conventional EF methods approximate the footprints of nations and individuals. If a country or person is reported to use five Earths, for example, the outcome is negative yet remains mostly abstract. However, EF's adoption as a design tool for architecture may distress some building clients and occupants. Consider the hypothetical commission of a massive 20,000 ft^2 (1,858

m^2) mansion by a wealthy client pursuing environmental certification. According to most applicable green rating systems, such as LEED v4 for Homes Design and Construction, it is possible for such a house to receive the same rating as a 1,000 ft^2 (93 m^2) residence based on the adoption of similar strategies. Because green rating systems are unconcerned with the scale of consumption and emissions, and therefore disassociated from their physical consequences, they remain safely apolitical. As soon as architects link per capita footprint to design decisions, they expose inequities in the ways individuals of different socioeconomic strata intend to utilize natural capital. Suddenly, the mansion above might be shown to require 20 times the bioproductive land area of the small house, thus irritating the "well-intentioned" wealthy client who desires a green rating and angering the compact house client upon learning of the wealthy client's gross level of consumption. In this way, building EF may be called the "Tragedy of the Commons"-made-explicit—in that it emphasizes resource imbalances within a population at a palpable scale. In biologist Garrett Hardin's well-known evaluation of population growth and its consequences, he writes: "It is fair to say that most people who anguish over the population problem are trying to find a way to avoid the evils of overpopulation without relinquishing any of the privileges they now enjoy" (Hardin, 1968, p. 1243). Green rating systems perpetuate this situation for such individuals and therefore are beguiling tools—lulling their users into a false sense of accomplishment.

EF also exposes the fallacy of McDonough and Braungart's theory of abundance that is central to their environmental philosophy. In both *Cradle to Cradle* (2002) and *The Upcycle* (2013), the authors discourage resource conservation, arguing that well-designed products and processes will obviate the need to conserve. This reasoning is compelling, and we should certainly strive to create the kinds of toxin-free, renewable energy-fueled products and processes the authors advocate as much as possible. However, the promotion of abundance as a primary objective lends support to the current massive consumption rates of human society—and especially the wealthiest populations. According to McDonough and Braungart (2013),

Abundance—of us, of our products—is not the scourge: Society can accommodate and encourage even hundreds of thousands of products, from thousands of cultures, and even honor every one of 10 billion people predicted to be here later in this century.

Yet the Earth has physical limits, no matter how clean or reusable we make products. According to Ashby (2012), at the current rate of global utilization "we will use and—if we discard it—throw away as much stuff in the next 25 years as in the entire history of industrialization." Accelerating consumption means the continual harvesting of virgin resources; recycling cannot keep up, even at a 100 percent cradle-to-cradle rate, because consumption does not remain flat. Even renewable resources have their limits, based on replenishment time and (dwindling) land area required for cultivation. Thus, unchecked abundance is a hazardous proposition, and will ultimately lead to global demise. The EF concept reveals the flaws in McDonough and Braungart's thesis.

The optimal environmental approach combines aspirations of regenerative design with smart conservation strategies. The EF method accommodates this marriage in a measurable way. For example, a building supplied with an onsite gray water filtration and storage system that provides enough recirculated water for all occupant needs would have an ideal water footprint. It would also fulfill McDonough and Braungart's promise of abundance. However, if this building were a skyscraper built out of the world's entire supply of titanium—if such a feat were possible—it would have a colossal embodied footprint. No matter how clean, nontoxic, and recyclable this technical nutrient would be, such a form of abundance would be foolhardy in the extreme. (This building could also earn LEED certification for employing a sufficient number of point-worthy environmental strategies.)

Building EF uncovers insights that can drive informed design decisions. For example, González-Vallejo, Marrero, and Solís-Guzmán (2014) employed the EF method to analyze residential buildings in Spain (Figure 7). They determined that single-detached houses exhibit an EF that is about 1.3 to 1.6 times higher than multifamily residences, and a per capita EF that is about 2.6 times higher than multifamily housing based on lower occupation (González-Vallejo, Marrero, & Solís-Guzmán, 2014, p. 83). Furthermore, they determined that the optimal number of floors is four, based on a per capita EF calculated per floor above ground level (4.191 gha per person). A single story is inefficient given the embodied footprint of the materials (10.811 gha per person), whereas a tall building (10 stories or higher) indicates diminishing returns on efficiency (4.236+ gha per person) (González-Vallejo, Marrero, & Solís-Guzmán, 2014, p. 82). Could a widespread adoption of EF methodology influence future zoning laws to encourage four-story multifamily housing as the optimal dwelling format?

Much more work needs to be done before such adoption. A worthwhile endeavor would be the creation of a database of EF signatures for building materials and assemblies with regionally specific data. Another would be the development of design software that enables architects and engineers to determine optimal EF outcomes based on real-time, visually immersive feedback. Such a tool should quickly adapt to contextual circumstances such as varying sources of energy. At a sophisticated level, EF could have a reciprocal effect on material flows, shifting source feedstocks. For example, the desire for increased carbon storage in wood framing might influence a change in the species composition of managed forests. Yet another future task would be the alignment of all LCA midpoint impacts with EF methodology, attributing a land value to measures like acidification, smog, and so on (Figure 8).

Figure 7. A treemap diagram showing the proportionate EF land areas for the production and construction phases of a multifamily residential building in Spain. Note the relative size of the actual building site in the lower right-hand corner. (Data source: González-Vallejo, Marrero, & Solís-Guzmán, 2014.)

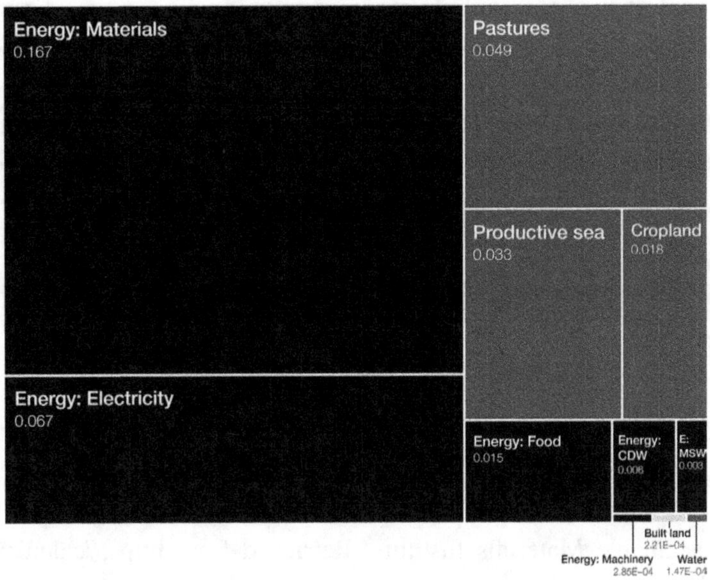

Figure 8. The range of potential connections between LCA impacts and EF land area categories. Unbroken lines indicate producers while broken lines indicate sinks.

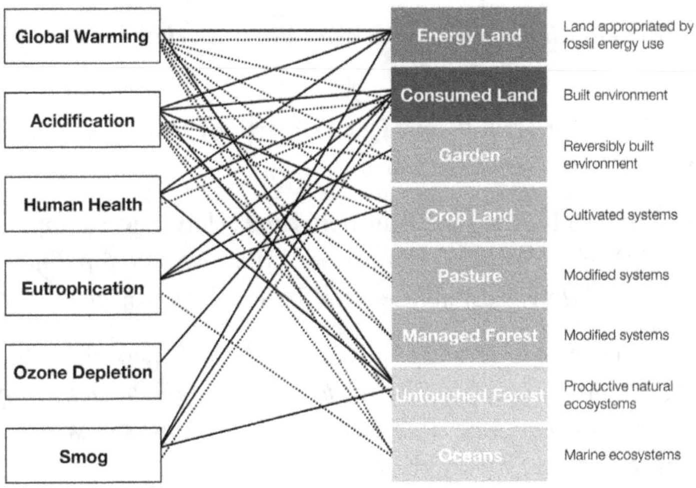

Architects and design teams are visual thinkers, and they imagine the constructed environment in the physical terms of structure, space, and territory. Although environmental building science has provided a wealth of information about how to design sustainably, the primary tools in use are either insufficient (green rating systems) or too abstractly numerical (LCA) to be tangibly meaningful. EF addresses these limitations and, with proper visualization, informs audiences about the real impact of buildings. Additionally, EF elucidates humanity's relationship with architecture insofar as it reveals more about the connection between indoor "dwell time"—including human occupation and duration—and buildings' ecological performance. If sufficiently developed, building EF has the potential to change status quo architectural design and specification practices in fundamental ways.

REFERENCES

Ashby, M. (2012). *Materials and the Environment: Eco-Informed Material Choice* (2nd ed.). Amsterdam: Elsevier Publishers.

Athena Impact Estimator for Buildings 5.2.01, Version 5.2 Build 0119. (2017). Athena Sustainable Materials Institute. Retrieved from https://calculatelca.com/software/impact-estimator/

Bastioni, S., Galli., A., Niccolucci, V., & Pulselli, R. M. (2006). The ecological footprint of building construction. *WIT Transactions on Ecology and the Environment, 93*.

Brand, S. (1994). *How Buildings Learn: What Happens After They're Built*. London: Penguin Books.

Buitenkamp, M., Venner, H., & Wams, T. (1993). *Action Plan Sustainable Netherlands*. Amsterdam: Dutch Friends of the Earth.

Centre for Sustainable Tourism and Transport, Breda University of Applied Sciences. (2015). *FACTSHEET CARMATOP: Carbon Management for Tour Operators*. Retrieved from https://www.cstt.nl/userdata/file/factsheetprojectcarmatop2014.pdf

Centre for Sustainable Tourism and Transport, Breda University of Applied Sciences. (2016). *Travelling large in 2016: The carbon footprint of Dutch holidaymakers in 2016 and the development since 2002*. Retrieved from https://www.cstt.nl/userdata/documents/travellinglarge2016-web.pdf

DeSilver, D. (2015). As American homes get bigger, energy efficiency gains are wiped out. *Pew Research Center.* Retrieved from http://www.pewresearch.org/fact-tank/2015/11/09/as-american-homes-get-bigger-energy-efficiency-gains-are-wiped-out/

Duffy, F., & Henney, A. (1989). *The Changing City.* London: Bulstrode Press.

Filimonau, V., Dickinson, J., Robbins, D., & Huijbregts, M. (2011). Reviewing the carbon footprint analysis of hotels: Life Cycle Energy Analysis (LCEA) as a holistic method for carbon impact appraisal of tourist accommodation. *Journal of Cleaner Production, 19*(17-18), 1917–1930. doi:10.1016/j.jclepro.2011.07.002

Galli, A., Wiedmann, T., Ercin, E., Knoblauch, D., Ewing, B., & Giljim, S. (2012). Integrating Ecological, Carbon and Water footprint into a "Footprint Family" of indicators: Definition and role in tracking human pressure on the planet. *Ecological Indicators, 16*, 100–112. doi:10.1016/j.ecolind.2011.06.017

Giordano, R., Serra, V., Totalla, E., Valentini, V., & Aghemo, C. (2015). Embodied Energy and Operational Energy assessment in the framework of Nearly Zero Energy Building and Building Energy Rating. *Energy Procedia, 78*, 3204–3209. doi:10.1016/j.egypro.2015.11.781

González-Vallejo, P., Marrero, M., & Solís-Guzmán, J. (2014). The ecological footprint of dwelling construction in Spain. *Ecological Indicators, 52*, 75–84. doi:10.1016/j.ecolind.2014.11.016

Guggenheim Bilbao. (2015). *Press release: 1,103,211 people visited the Guggenheim Museum Bilbao in 2015, the second-best figure in its history.* Retrieved from https://prensa.guggenheim-bilbao.eus/src/uploads/2016/01/NP_Balance-2015_EN.pdf

Hardin, G. (1968). The Tragedy of the Commons. *Science, 162*(3859), 1243–1248. doi:10.1126cience.162.3859.1243 PMID:5699198

Haynes, R. (2010). *Embodied Energy Calculations within Life Cycle Analysis of Residential Buildings.* Perth: ETool Global.

Kibert, C. (2013). *Sustainable Construction: Green Building Design and Delivery* (3rd ed.). Hoboken, NJ: John Wiley & Sons.

King, B. (2017). *New Carbon Architecture: Building to Cool the Planet.* Gabriola Island, BC: New Society Publishers.

McDonough, W., & Braungart, M. (2002). *Cradle to Cradle: Remaking the Way We Make Things.* New York: North Point Press.

McDonough, W., & Braungart, M. (2013). *The Upcycle: Beyond Sustainability—Designing for Abundance*. New York: North Point Press.

O'Connor, J. (2004). *Survey on actual service lives for North American buildings. Woodframe Housing Durability and Disaster Issues conference, Las Vegas, Nevada*. Vancouver, BC: Athena Sustainable Materials Institute / Forintek Canada Corporation.

Simonen, K. (2014). *Life Cycle Assessment*. Oxon, UK: Routledge Press.

Sinha, R., Lennartsson, M., & Frostell, B. (2016). Environmental footprint assessment of building structures: A comparative study. *Building and Environment*, *104*, 162–171. doi:10.1016/j.buildenv.2016.05.012

Solis-Guzman, J., & Marrerro, M. (2015). Ecological Footprint Assessment of Building Construction. Sharjah, UAE: Bentham Science Publishers. doi:10.2174/97816810809871150101

Statista. (n.d.). *Annual hotel occupancy rate in London (UK) from 2008 to 2018*. Retrieved from https://www.statista.com/statistics/323719/annual-hotel-occupancy-rate-in-london-uk/

Tucker, M., Böhning-Gaese, K., Fagan, W., Fryxell, J., Van Moorter, B., Alberts, S., ... Mueller, T. (2018). Moving in the Anthropocene: Global reductions in terrestrial mammalian movements. *Science*, *359*(6374), 466–469. doi:10.1126cience.aam9712 PMID:29371471

U.S. Energy Information Administration. (2016). *2012 Commercial Building Energy Consumption Survey: Energy Usage Summary*. Retrieved from https://www.eia.gov/consumption/commercial/reports/2012/energyusage/

U.S. Environmental Protection Agency. (2012). *Portfolio Manager Technical Reference: Energy Use Benchmarking*. Retrieved from https://www.energystar.gov/sites/default/files/buildings/tools/DataTrends_Energy_20121002.pdf

U.S. Environmental Protection Agency. (2016). *Portfolio Manager Technical Reference: U.S. Energy Use Intensity by Property Type*. Retrieved from https://portfoliomanager.energystar.gov/pdf/reference/US%20National%20Median%20Table.pdf

Wackernagel, M., & Rees, W. (1996). *Our Ecological Footprint: Reducing Human Impact on the Earth*. Gabriola Island, BC: New Society Publishers.

KEY TERMS AND DEFINITIONS

CO_2 Assimilation Method: A working ratio enabling the conversion of energy consumption into land area. In the case of fossil fuels, a typical ratio is 1 hectare per 1.8 metric tons of CO_2 emitted annually.

Ecological Footprint: The measure of human impact on the environment, expressed as the quantity of land necessary to meet the demand for natural resources.

Embodied Footprint: The area of land required to fulfill the resource needs of materials from harvesting through construction.

Energy Land: The land area theoretically required for the purposes of fuel consumption, such as the area of forest required to absorb a given amount of CO_2 annually.

Fair EarthShare: The area of available and ecologically productive land on earth, measured on a per capita basis.

Influence Footprint: The area of land required to provide resources for the broader audience affected by a building project, such as construction labor or indirect commercial activity.

Occupant Footprint: The area of land required to provide resources for the use and maintenance of a building, calculated on a per capita basis for the building's occupants and users.

Operational Footprint: The area of land required to provide resources for the use and maintenance of a building throughout its lifespan.

Chapter 3
Recycling and Reuse of Building Materials From Construction and Demolition:
An Environmental Evaluation for Sustainable Growth

Nadeem Faisal
Birla Institute of Technology, India

Kaushik Kumar
Birla Institute of Technology, India

ABSTRACT

Urbanization is creating enormous pressure for the effective utilization of the existing land with demolition of old structures for new and modern structures. The debris produced in demolition of these structures are in large amount and disposal of this waste in sustainable manner is the biggest challenge being faced today and should be considered as a resource. With the increasing waste production and public concerns regarding the environment, it is desirable to recycle these materials. If suitably processed in appropriate industrial plants, these materials can be profitably used in concrete. This chapter highlights the composition of construction and demolition waste, the necessity for its recycling, and possibilities that can be implemented for its resourceful use, further focusing on current trends in this field by elaborating various ways to use these waste from laboratory research scale to commercially available technologies around the globe. The chapter concludes with future research directions and guidelines for sustainable use of these wastes.

DOI: 10.4018/978-1-5225-6995-4.ch003

INTRODUCTION

Construction and Demolition (C & D) waste comprises of a noteworthy part of aggregate solid waste production on the planet. C&D waste is produced at a point when any construction/demolition action happens, for example, building, streets, flyover, metro, rebuilding and so forth these wastes are substantial, possessing high density, and possess extensive and humongous storage space either on communal waste bins or on the roads. It isn't surprising to see tremendous heaps of such waste, which is substantial also, stacked on streets particularly in vast projects, bringing about traffic blockage and interruption. It compromises of 10-20% of the metropolitan solid waste (barring vast construction ventures). Consequently, proper management of such wastes is required.

With a tremendous increment in the number of disposable materials on one hand and a with lack of dumping sites on the other, the waste disposal issues are quite serious and at certain times, at the quite alarming rate.

Protection of the earth and preservation of the quickly decreasing natural resources ought to be the core of sustainable development. Persistent industrial improvement postures difficult issues of construction and demolition waste disposal (Timothy, 1998).

Though then again, there is basic lack of natural aggregate for generation of new concrete, on the another, the huge volumes of demolished concrete produced from crumbled and outdated structures make serious natural and ecological problems (Schachermayer, 2000). Reusing of total materials from construction and demolition waste may lessen the request – supply gap in both these segment.

Cement and brick waste can be reused by sorting, crushing and sieving into the recycled total. This recycle total can be utilized to make concrete for street construction and building material.

As per an examination commissioned by Technology Information Forecasting and Assessment Council (TIFAC), New Delhi, 70% of the construction business doesn't know about reusing and recycling strategies. The examination prescribes establishments and creation of quality benchmarks of recycled total materials and recycled total concrete. This would help in setting up a target product quality for the maker and guarantee the client of a minimum quality necessity, consequently reassuring him to utilize it.

The prospering population is making pressure for better use of land in existing urban communities. To provide the need for lodging and commercial prerequisite new construction is being finished by annihilating the old structures and construction on the empty land. The change in economic conditions in the developing nations has caused the large-scale development in the construction industry. This construction activity is creating an enormous volume of Construction and Demolition (C&D)

Waste. Consistently around 3000 MMT (Million Metric tons) of waste is being created in European Union, out of which 30% of aggregate waste i.e. around 900 MMT is created by construction industry alone as C&D waste (Bravo, De Brito, Pontes, & Evangelista, 2015; Torgal, 2013). In the United States evaluate the formation of C&D waste is accounted for around 140 MMT every year (Torgal, 2013). In developing nations like India and China, there is a significant ascent in this C&D waste as detailed around 14 MMT is produced in Shanghai, China alone in 2012 (Ding & Xiao, 2014) Out of the aggregate waste around 80% comprises of bricks, blocks and concretes.

The illegal dumping and unarranged disposal of this C&D waste are causing the extreme biological and environmental concerns. Henceforth, the reuse and reusing of C&D waste is quite essential. Its planned utilization will not just decrease the misuse of virgin raw material yet additionally takes care of the issue of waste disposal. It additionally leads to greater accessibility of land by preventing dumping sites. Environmental effects, for example, deforestation, illicit mining of river beds for air and water contamination, utilization of fossil fuels and petroleum for transportation, topsoil loss and so forth is likewise diminished.

Presentation of green building concept has some way or another aided in using the waste produces amid construction yet a huge amount of awareness is required over the world in recycling and reuse of Construction & Demolition waste. Particularly in developing nations where construction is occurring on an enormous scale.

According to EU waste structure order 2008/98/EC least recycling level of C&D waste ought to be least 70% by 2020 (Ledesma, Jimenez, Ayuso, Fernandez, & De-Brito, 2015) which is an incredible step forward in focusing on the issue of Construction & Demolition waste. According to 2013 estimation aggregate reusing of this C&D waste in EU-27 is just 47%. (Torgal, 2013) This demonstrates expanding this rate to 70% is a huge undertaking which requires definite research at lab scale as well as advances in technology which are savvy to accomplish this objective.

The Construction & Demolition waste primarily comprises of concrete, ceramic, mortar, and brick together constitutes about 80%, though metal and wood about 10%. Metal and wood are by and large reused effortlessly, however, mortar, brick, concrete and ceramics are for the most part accessible in mixed form and need handling before being put to utilize. Despite the fact that utilization of Recycled Aggregates (RA) is being considered since most recent 50 years yet not being utilized as a part of new structures because of the absence of consistency in their properties and non-accessibility of an administrative system for their utilization. Different experimental outcomes have demonstrated that the utilization of these recycled aggregate prompts to poor concrete performance for the most part in durability and strength part. Reason revealed are higher porosity in these aggregates and high water absorption prompt poor water to cement (w/c) proportion. Likewise, amid demolition process, many micro-cracks generate in these aggregate prompting higher permeability and high penetration. (Soares, De Brito, Ferreira, & Pacheco, 2014). Many of processed reused aggregates are yet utilized as a part of nonstructural segments like paver blocks, Concrete pipes, Asphalt mix, Sub-base below street construction, Plain Cement Concrete, and fire protection blocks (Ledesma, Jimenez, Ayuso, Fernandez, & De-Brito, 2015; Meijide & Perez, 2014; Ozalþ, Yılmaz, Kara, Kaya, & S"ahin, 2016; Milicevic, Bjegovic, & Stirmer, 2015; Medina, Zhu, Howind, Frías, & Sánchez de Rojas, 2015; Leiva, Guzmán, Marrero, & Arenas, 2013; Ossa, García, & Botero, 2016; Rodríguez, Parra, Casado, Minano, Albaladejo, Benito, & Sanchez, 2016).

COMPOSITION OF C&D WASTE

In India according to (TIFAC, 2000) add up to the quantum of C&D waste created is about 15MMT anyway it isn't an exceptionally exact estimate since the vast majority of the C&D waste is dumped wrongfully which isn't represented in the reports. The masonry, cement, and mortar together constitute more than 65% of this C&D waste. (Thomas & Wilson, 2013). The ordinary structure of C&D waste created in India is shown in Figure 1.

CHARACTERISTICS

This class of waste is unpredictable because of the diverse kinds of building materials being utilized, however as a rule may include the accompanying materials (Table 1):

The EPA report grouped C&D Waste into 6 categories. These are shown in Figure 2 below.

ENVIRONMENTAL IMPACT

Improper disposal and unlawful dumping of these Construction & Demolition waste are creating an environmental degradation. This environmental effect is turning into a significant issue in urban areas and over the different region in dealing with their solid waste management. An immense amount of C&D waste causes ascends in flood levels of the waterways, scouring of the banks, exhaustion of resources, draining of hazardous materials in the water causing an effect on marine life. Illicit dumping around the streets causes traffic blockage, chocking the surface deplete causing flooding on the pavement and so forth. Construction & Demolition waste from small house demolition, for the most part, discovered its way into municipal

Figure 1. C&D waste arrangement in India (Bhattacharyya, Minocha, Garg, Singh, Jain, Maiti, & Singh, 2013)

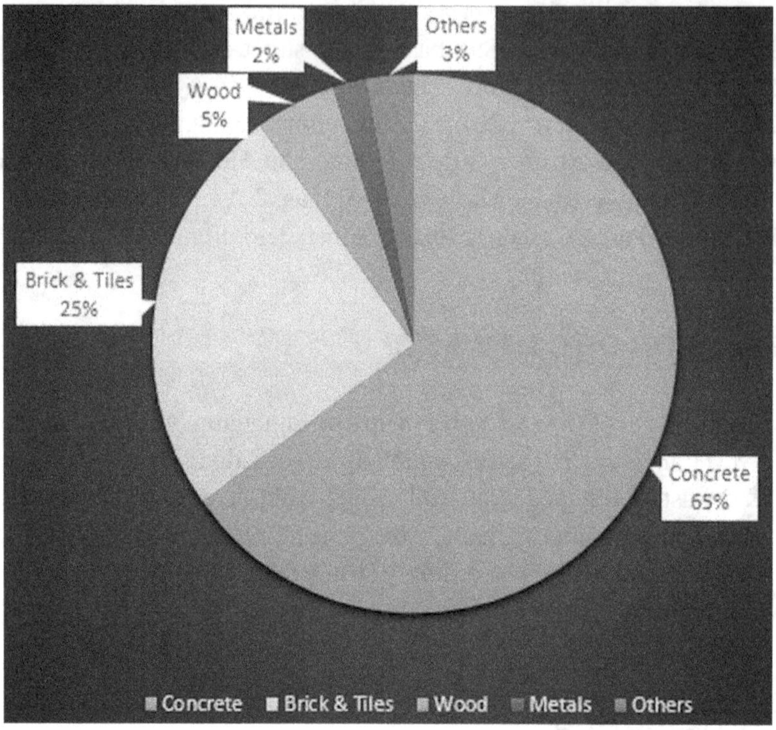

**For a more accurate representation see the electronic version.*

Table 1. Complex waste due to a different type of building materials

Major Components	Minor Components
Bricks	Panels (wooden, laminated)
Cement Concrete	Electrical fixtures (Copper/ aluminium wiring, wooden baton, bakelites/ plastic switches, wire insulation)
Stone (marble, granite, sandstone)	
Cement plaster	
Steel (from R.C.C, door / window frames, roofing supports, railings of staircase etc)	Pipes (GI, Iron, plastics
	Others (glazed tiles, glass panes, and paints)
Wood /Timber (especially demolition of old buildings)	
Rubble	

Figure 2. Characterization of Building-related C and D waste in USEPA 530-R-98-010, June 1998

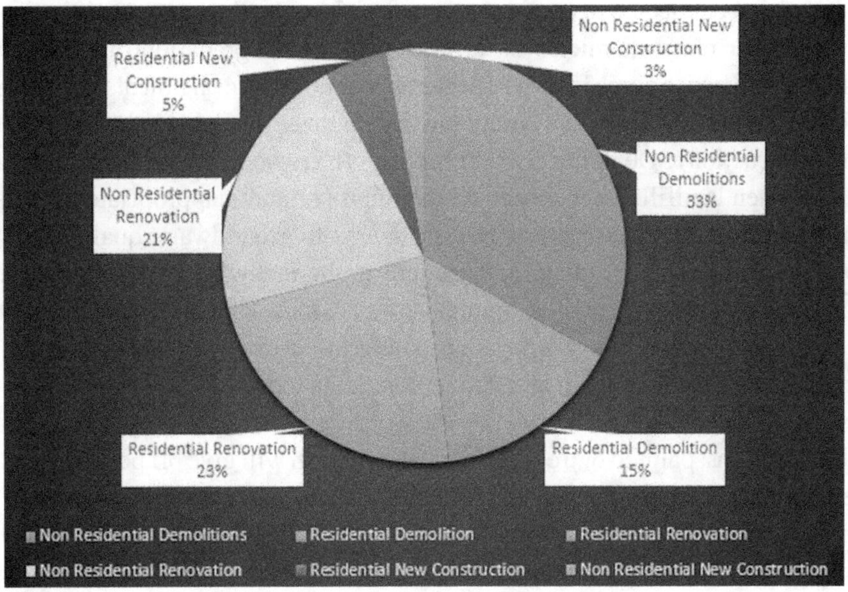

**For a more accurate representation see the electronic version.*

bin causing the issue in the treatment of solid municipal waste (Thomas & Wilson, 2013). These waste are once in a while covered in the site itself causing the formation of the impenetrable layer which doesn't permit the development of vegetation and stops the invasion of rainwater inside the ground.

Construction of any infrastructure has a significant ecological effect through extraction of raw materials, the utilization of energy in production procedures and transport, production of masses by byproduct waste, and the harm to environment and health in all periods of the life cycle of hazardous components.

The disposal of Construction & Demolition wastes has turned into a noteworthy problem in a recent year. Some building proprietors, waste haulers, and demolition contractual workers are discarding this waste improperly and wrongfully keeping in mind the end goal to maintain a strategic distance from transportation costs and tipping charges at waste disposal facilities. Unlawful disposal sites have been found in gravel pits and groundwater recharge regions, on cultivating land, and prime residential properties, and in borrow pits and low lying regions.

The land disposal of Construction & Demolition waste displays a risk of groundwater contamination as a result of trace amounts of dangerous constituents and hazardous materials, which are in some cases experienced. The potential for groundwater pollution comes about because of little amounts of hazardous materials, such as natural compounds or heavy metals that might be available in substances that have been connected to construction materials, or by the inappropriate disposal of buildups or mass chemicals in the waste stream. Degradation of groundwater quality may likewise come from a large number of non-toxic chemicals, for example, Chloride, Sodium, Sulfate, and Ammonia that might be available in leachate created from Construction & Demolition waste materials (Ex: wood, concrete, metal, drywall, asphalt) when landfilled. Accordingly, the improper and inappropriate disposal of Construction & Demolition waste poses a danger to groundwater quality.

An illicit disposal site may likewise pull in the unlawful disposal of different sorts of waste, including regular municipal waste, industrial and mechanical waste and hazardous waste this would additionally affect the site and increment the future cost for cleaning up an affected or contaminated site. The open burning and consuming of demolition material is a noteworthy concern and is prohibited. Plastic materials, foams, painted or treated wood, and so forth will give off poisonous fumes when burned. Demolition material from warehousing structures, agricultural and manufacturing facilities might be polluted with chemicals some spillage or from their ordinary operations and tasks. Leachate from the slag may affect the groundwater. Health Canada urges that "consuming any type of treated wood speaks to a health hazard and needs to be avoided".

RECYCLING OF C&D WASTE

Once the reusable material is taken out whatever is left of the material is for the most part comprises of Construction & Demolition Waste. According to BS 8500 (2002), this Construction & Demolition waste aggregate are ordered into two classifications one recycled concrete aggregate (RCA) and other crushed masonry based aggregate known as Recycled aggregate (RA) (Brito & Saikia, 2013). However the little measure of contaminants like wood particles, gypsum, paper, cardboard, glass, and plastics must be evacuated to get usable totals. Different systems, for example, eddy current magnetic detachment, Air shifting, dry density separation and spirals are few of the strategies pertinent today to naturally isolate the contaminants as said above

Creation of Construction & Demolition waste aggregate relies upon the original construction and demolition waste, on the demolition strategy utilized and the accumulation method from the site. Normally RCA contains around 70% of coarse and fine aggregate and 30% of concrete paste (Brito & Saikia 2013).

The nature of Construction & Demolition aggregate relies upon the quantity of mortar on that aggregate. Additionally, more the waste is dealt with, better the nature of aggregate is delivered. However higher processing expands the cost of aggregate and in this way making it financially unviable in places where natural aggregate is inexpensively acquired.

Construction & Demolition waste management might be characterized as the discipline related with the correct storage, accumulation and transportation, recuperation, recycling, processing, reusing and disposal of Construction & Demolition wastes in a way that is as per the best standards of human well-being, monetary, aesthetics and other ecological considerations (Rao, Jha, & Misra, 2006). The management approaches are unique in relation to one nation to another, similar to the levels of environmental protection. The greater part of the Construction & Demolition management systems looked into on the accompanying premise:

C & D waste management incorporates following advances.

1. Storage and isolation.
2. Collection and transportation.
3. Recycling and reuse.
4. Disposal.

STORAGE AND ISOLATION

Construction & Demolition wastes are best stored away at source i.e. at the point of generation. On the off chance that they are scattered around or tossed out on road, they make problems to traffic activity as well as add to the workload of the local body. A reasonable screen ought to be given with the goal that the waste does not get scattered and does not turn into an eyesore or a blemish.

Segregation can be done at source amid Construction & Demolition exercises or can be accomplished by preparing the mixed material to evacuate the outside materials. Segregation at source is most effective as far as energy use, financial aspects and time are considered. Net segregation of Construction & Demolition wastes into road work materials, basic building materials, rescued building parts and site leeway waste is important. Extra segregation is required to encourage reuse/recycling of materials like wood, glass, cabling, plastic, mortarboard etc. before demolition so as to create a recycled aggregate that will meet the specification. Figure 3 below shows a basic waste management model of C&D waste.

Figure 3. Construction & Demolition waste management model

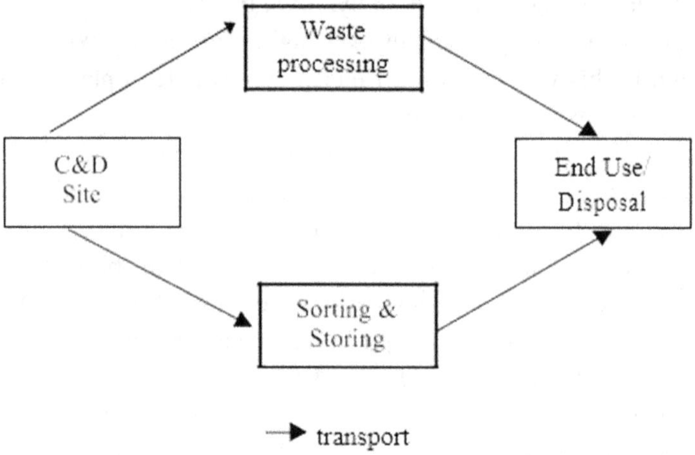

COLLECTION AND TRANSPORTATION

If Construction & Demolition debris put away in skips, at that point skip lifters fitted with hydraulic hoist system ought to be utilized for productive and incite evacuation. In the event that, trailers are utilized, at that point tractors may evacuate these. For dealing with extensive volumes, front-end loaders in combination with durable tipper trucks might be utilized so the time taken for stacking and emptying is kept to a minimum.

RECYCLING AND REUSE

Construction & Demolition waste is heavy and bulky and is, for the most part, unacceptable for the disposal by composting/incineration. The increasing populace and necessity of land for different utilizations has diminished the accessibility of land for waste disposal. Reutilization or reusing is an imperative system for management of such waste. Aside from mounting issues of waste management, different reasons which bolster appropriation of reuse/recycling technique are decreased extraction of raw materials, diminished transportation cost, improved benefits and lessened ecological effect. Most importantly, the quick exhausting reserves of natural aggregate has required the utilization of recycling/reuse innovation, keeping in mind the end goal to have the capacity to moderate the customary natural aggregate for other vital works.

In the present setting of expanding waste generation and developing open consciousness of ecological issues, recycled and reused materials from demolished concrete or masonry can be productively utilized as a part of various ways inside the building industries. The investigation overview shows the significant segments of the Construction & Demolition waste stream are excavation material, concrete, blocks and tiles, wood and metals.

Concrete

Concrete shows up in two structures in the waste. Basic components of the building have reinforced concrete, while establishments have mass non-reinforced concrete. Excavations deliver topsoil, mud, sand, and rock.

Excavation materials might be either reused as filler at a similar site after the culmination of work, in street construction or in stone, rock and sand mines, landfill construction, basic fill in low lying zones to aid future advancement, in the garden and land scraping.

Cement and workmanship constitute over half of waste created. It can be reused in slab/block form. Recycling of this waste by changing over it to aggregate offer double advantage of sparing landfill space and diminishment in the extraction of normal raw material for the new construction industry. A fundamental technique for reusing of cement and masonry waste is to crush the debris to deliver a granular product of given particle size. Plants for the handling of demolition waste are separated in light of mobility, kind of crusher and procedure of separation and partition.

There are three sorts of reusing plants viz. Portable, Semi-mobile and Stationary plant. In the mobile or portable, the material is squashed and screened and ferrous impurities are isolated by magnetic separation and partition. The plant is transported to the demolition site itself and is suitable to process just non-contaminated cement or masonry waste.

In the semi-portable plant, evacuation of contaminants is done by hand and the finished result is likewise screened. Magnetic separation and partition for the expulsion of ferrous material are done. Finished result quality is superior to that of a mobile plant.

Above plants are not competent to process a mix demolishing waste containing foreign materials like wood, metal, plastic, and so on. Stationary plants are prepared for doing pounding, screening and in addition, purification to isolate the contaminants. Issues important to be considered for the erection offer stationary plant will be Plant area, street infrastructure, accessibility of land space, the arrangement of Weigh Bridge, arrangement for storage area and so on.

Distinctive sorts of crushers are utilized as a part of recycling plant to be specific jaw-crusher, impeller-impact crusher, impact crusher, etc.

Applications

Recycled aggregates can be utilized as general mass fill, sub-base material in street construction, canal lining, play area, fills in drainage ventures and for making new concrete to a less degree. While utilizing recycled aggregate for filler application, care must be taken that it is free of contaminants to prevent danger of groundwater contamination utilization of recycled aggregate are sub-base for street construction is generally acknowledged in the greater part of nations.

Bricks and stonework emerge as waste amid demolition. These are by and large blended with cement, lime or in general mortar. It is utilized as a part of the construction of the street base and drayage layer, and mechanical soil stabilizers because of its inactivity after separation and crushing.

Tile materials reusing are relatively indistinguishable to bricks. Tile is regularly blended with a brick in the final recycled product.

Metal waste is produced amid demolition as pipes, light sheet material utilized as a part of ventilation framework, wires, and sanitary fittings and as reinforcement in the concretes. Metals are recouped and reuse by re-liquefying. Aluminum can be recouped without contamination; the material can be straightforwardly sold to a recycler.

Wood recuperated in great condition from window frames, bars, doors, partitions different fittings is reused. Though, wood utilized as a part of the construction is frequently treated with chemicals to anticipate termite pervasion and needs extraordinary care amid disposal since different issues related to wood waste are the incorporation of nails, jointing, fixings, and screws

Moreover, wood individuals have a high market an incentive for unique reuses (furniture, cupboards, and floor materials). Lower quality waste wood can be reused/consumed for energy recuperation. Scrap wood is destroyed in-site/in a unified plant. Shredded wood is magnetically arranged for scrap metal. Wood chips are put away in order to stay dry and can be utilized as fuel. Likewise, it is utilized as a part of the generation of different press boards and fiber boards and utilized for animal bedding and shelters.

Bituminous material emerges from street construction, breaking and burrowing of streets for administrations and utilities. Reusing of Bituminous material can be done by hot or cold mixing strategies either at the location or at an asphalt mixing plant it offers benefits like sparing being used of asphalt, saving of energy, lessening in total prerequisite and so forth.

Different varieties materials that emerge as waste incorporate glass, paper, plastic, and so on can be recouped and reused.

DISPOSAL

Being dominatingly idle in nature, Construction & Demolition waste does not make chemical or Bio-concoction contamination and pollution. Henceforth, greatest exertion one should make is to reuse and recycle them as clarified previously. The material can be utilized for filling/leveling of low-lying territories. In the industrialized nations, special landfills are at times made for latent waste, which are regularly situated in abandoned mines and quarries.

Different investigations and tests are done on these Construction & Demolition Waste aggregate to decide its properties and potential employment. A large portion of the work indicates great execution in mechanical property, however, the durability properties are real concern that should be addressed.

MECHANICAL PROPERTIES

All of the experimental works had presumed that the compressive strength estimation of cement prepared with Construction & Demolition Waste aggregate declined by expanding the replacement ratio with natural aggregate. The primary representing factor is poor Interfacial Transition Zone (ITZ) between the aggregate part and the paste for the recycled aggregates (Medina, Zhu, Howind, Frías, & Sánchez de Rojas, 2015; Milicevic, Bjegovic, & Stirmer, 2015; Ozalp̆, Yılmaz, Kara, Kaya, & S"ahin, 2016; Ledesma, Jimenez, Ayuso, Fernandez, & De-Brito, 2015).

In an investigation by Milicevic et al., (2015) around 62 tests of cement were created utilizing pulverized brick and rooftop tiles as aggregate with various substitution level and it was discovered that with reference to concrete mixture with natural aggregate, density and modulus of elasticity of Construction & Demolition waste concrete were 30% lower. It was additionally found that practically identical compressive strength was accomplished at low substitution levels.

The water absorption limit of RCA and RA are high when contrasted with regular aggregate. As revealed in RA adhere mortar is about 16-17% and for RCA it is about1-13% with an average of 5.6% (Brito & Saikia, 2013). This high water absorption prompt low workability for same water/cement (w/c) proportion when contrasted with ordinary concrete. To alleviate this issue high measurements of added substances are expected to compensate for the loss of the workability while utilizing RCA in concrete. This prompt poor mechanical execution because of development of weaker ITZ.

In other analysis by Ledesma et al. (2015) natural fine sand were supplanted by reused sand from C&D waste and utilized for M10 concrete creation with various substitution level from 0 to 100% and it was inferred that compressive quality was found over 10 Mpa for substitution up to half however drying shrinkage and water absorption expanded significantly. The author prescribed utilization of recycled sand for an indoor condition for nonstructural concrete for substitution up to 50%.

DURABILITY PROPERTIES

It has been noticed that the concrete made with Construction & Demolition waste aggregate indicate exceptionally poor durability execution when contrasted with concrete with natural aggregate.

In experimentation by (Bravo, De Brito, Pontes, & Evangelista, 2015) Construction & Demolition waste from five diverse recycling plant in Portugal were acquired. The impact of RA on strength execution and durability of concrete were tried both for the coarse and fine aggregate substitution.

It was presumed that carbonation profundity expanded up to 180% with reference to natural aggregate concrete. It was additionally discovered that chloride dispersion coefficient expanded up to 130% with respect to reference concrete. However, these outcomes were for 100% supplanting of natural fine aggregate with reused fine aggregate. Different researchers have likewise discovered the high chloride conductivity in the scope of 40% to 87% higher when recycled aggregates were utilized when contrasted with natural aggregate concrete (Brito & Saikia, 2013). Regarding shrinkage, it was accounted for in a review paper by Silva et al., (2015) that 100% coarse RCA in concrete can expand shrinkage by up to 80% when contrasted with normal aggregate concrete. Comparative outcomes were acquired by a number of researchers between 60-90% higher shrinkage when 100% substitution is done is improved the situation natural or regular aggregates with Recycled aggregates (Brito & Saikia, 2013). This high shrinkage characteristic prompts poor execution of concrete because of the more prominent degree of splitting and cracking, subsequently poor durability to aggressive conditions.

CASE STUDY: DELHI, INDIA

In a joint effort with Municipal Corporation of Delhi (MCD), Infrastructure Leasing and Financial Services Limited (IL&FS) set up India's first Construction and Demolition (C&D) waste recycling plant in Burari, New Delhi in 2009 with working limit of 500 tons for every day (TPD). The plant has been effectively preparing C&D waste into recycled aggregates which can be utilized for brick construction and

Figure 4. A comparison on durability for different materials on different parameters – Key 1. Poor 2. Below Average 3. Average 4. Above Average 5. Outstanding

	Ceramic Tile with Grout	Concrete (Cast-in-Place Sealed)	Granite (Sealed)	Laminate	Marble	Paper Composite	Solid Surface	Stainless Steel	Terrazzo	Quartz	Wood
Durability											
Scratch Resistant	5	5	5	3	3	3	3	2	5	5	2
Stain Resistant	1	5	4	3	1	3	5	5	5	4	1
Chip Resistant	1	3	3	5	3	5	4	5	2	4	5
Heat Resistant	5	5	5	3	4	3	3	5	5	4	1
Household Chemical Resistant	5	5	2	3	3	Depends	3	5	2	3	1
Easy to Maintain	2	3	4	4	2	4	4	4	3	5	2
Repair-ability	4	2	1	1	1	1	5	5	1	3	3
Non-Porous (Will not absorb moisture)	1	5	4	3	1	5	5	5	4	5	1
Renewable Surface	2	2	2	1	2	5	5	3	2	3	5

building streets. This pilot venture was set up on Public Private Partnership (PPP) premise to show the potential advantage of utilization of Construction & Demolition waste in urban city Delhi. Around 7 acres of land was given by MCD for a time of 10 years and the plant was commissioned in 2009.

This plant is a fixed sort recycling and reusing plant. Following procedures are engaged with reusing of Construction & Demolition Waste.

To start with isolation done for undesired things like plastic, FRP sheet, clothes, metal, and so on with mechanical and manual means at that point staying waste is isolated into three sections: 1) Whole blocks; 2) Big concrete pieces; 3) Mixed Construction & Demolition waste. Entire Bricks are sold independently, extensive solid blocks are broken into little pieces (200-400 mm size) utilizing rock breaker and mechanical sledge. These are then prepared and broken into smaller aggregate reasonable for making concrete. This concrete is utilized for influencing nonstructural materials like paving blocks, pieces, kerbstone, and tiles.

CASE STUDY: JAPAN AND KOREA

Different goal was set down in Korea's second Construction & Demolition Waste management plan (2012-2016) which predominantly stress on change in waste management data framework which emphasizes on online record of Construction & Demolition waste by both contractual worker and treatment organizations, considerable diminishment in the volume of mixed waste and usage of life – cycle inventory information on Construction & Demolition waste.

In Japan, high attention is given to ecological effect thus while deciding the cost or nature of item the environmental protection costs are included. Japan characterizes concrete class in view of properties of recycled aggregates. Class H concrete for quality up to 45 Mpa which uses great quality Construction & Demolition aggregate, Class M concrete which are not presented to serious environmental condition lastly class L concrete using low quality Construction & Demolition aggregate having high water absorption and utilized just for refill and leveling concrete (Bansal & Singh, 2014; Brito & Saikia, 2013)

FUTURE RESEARCH WORK

In an examination by Leiva et al. (2013), it has been discovered that there is a change in fire insulation attribute for blocks containing Construction & Demolition waste. Concrete blocks were readied utilizing recycled aggregate from 20 to 100% substitution of normal aggregate. These blocks were tried and found to have enhanced properties with reference to reference concrete for resistance to fire, heat protection, and acoustic protection. The explanation behind this enhanced property was a low density of blocks and more voids in these pieces along these lines making them appropriate for nonstructural utilization, for example, blocks and pre-assembled concrete panels.

In another examination in Mexico by Ossa et al (2016) the utilization of recycled aggregate up to substitution of 20% was prescribed in hot asphalt mix for paving urban streets. The comparative outcome was achieved by Meijide and Perez et al. (2014) for utilization of Construction & Demolition waste aggregate in the cold asphalt mix.

As we have seen the utilization of RA as Subbase has a tremendous potential for street construction. These reused material are not affected by weathering, scraped area, physical and concoction change henceforth particularly appropriate for sub-base layer in pavement construction (Jimenez, 2013).

More awareness about construction and demolition wastes can be created with the help and aid from the government. Research work emphasizing on C&D wastes should be given significant importance and proper funding's and allocation should be given by government and other agencies.

GUIDELINES

It has been discovered that there is no agreement on institutionalization process for reuse and recycling of Construction & Demolition waste. Every nation has its own particular standard and technology because of immense variety in the properties of Construction & Demolition waste over the areas. However, there is incredible consciousness around the world with respect to the appropriate use of this waste to save the earth. Following rules are suggested for sustainable utilization of these waste.

- One ought to follow a policy of Use-Reuse-Recycle-Landfill. To start with we ought to use the natural resources to its greatest potential by limiting the waste. At that point, the waste must be reused however as much as possible if not in a similar venture than in the other task. After reuse, the waste ought to be recycled utilizing environmental friendly procedures for its most maximum usage in different items. Finally, Landfill will be done in arranged and assigned areas as planned.
- In developing nations like India where natural aggregate is accessible at modest costs incentive must to be given by the administration/government for using Construction & Demolition waste aggregate alongside subsidy to organizations reusing and recycling these Construction & Demolition waste by giving them land free of cost, interest free loan on buy of plant and apparatus and making it obligatory for government venture to use items from C&D waste.
- More recycling plant ought to be proposed on PPP premise as the instance of Delhi, India in light of the detailed review about reasonability of these plants. Learning from the best procedures around the world can help in shaping rules for utilizing these waste.
- Strict direction and regulation to prevent illegal and illicit dumping by forcing high fines will forestall ecological degradation. Additionally, giving incentive for transportation of Construction & Demolition waste to the closest recycling facility in light of nature of waste will upgrade appropriate disposals.
- Selective demolition works on comprising of the efficient expulsion of different reusable parts like windows and doors, basic steel, metallic segments and so forth. From that point appropriate segregation and sorting at the site itself before sending to the recycling plant.

CONCLUSION

In the wake of distinctive innovations and technologies, different research work at the usage of Construction & Demolition waste it is inferred that the C&D waste reuse and recycling has an incredible business potential. Its prosperity relies upon the incentive by the government and development of measures, strict laws and regulation, controls and better state of an art technology for reusing of these wastes.

It has, for the most part, observed that the utilization of Construction & Demolition waste is limited for use in green structures which requires affirmation. In developing, nations absence of control measures and laws prompt illegal dumping causing ecological degradation. It is often seen that there is high diversification in properties of Construction & Demolition waste accessible over the area. Subsequently problems in standardization and classification. Many of mixed Construction & Demolition waste has a huge amount of clay-based material in it which makes is not appropriate for recycling as the cost of handling shoots up. Because of unsatisfactory quality, this sort of Construction & Demolition waste goes to landfill. With new and most recent research it is being presumed that the issue of Construction & Demolition waste use can be managed effortlessly given all segment of society comes forward to take promise for protection of the environment and healthy future for coming age.

ACKNOWLEDGMENT

The authors sincerely acknowledge the comments and suggestions of the reviewers that have been instrumental for improving and upgrading the chapter in its final form.

REFERENCES

Bansal, S., & Singh, S. K. (2014). A Sustainable Approach towards the Construction and Demolition Waste. *IJIRSET*, *3*(2), 9226–9235.

Bhattacharyya, S.K., Minocha, A.K., Garg, M., Singh, J., Jain, N., Maiti, S. & Singh, S.K. (n.d.). Demolition Wastes as Raw Materials for Sustainable Construction Products. *CSIR-CBRI News Letter*, *33*(2), 1-2.

Bravo, M., De Brito, J., Pontes, J., & Evangelista, L. (2015). Durability performance of concrete with recycled aggregates from construction and demolition waste plants. *Construction & Building Materials*, *77*, 357–369. doi:10.1016/j.conbuildmat.2014.12.103

Brito, D. J., & Saikia, N. (2013). *Recycled concrete in aggregate*. Springer. doi:10.1007/978-1-4471-4540-0

Coelho, A., & De-Brito, J. (2013). *Preparation of concrete aggregate from construction and demolition waste (CDW). In Handbook of recycled concrete and demolition waste*. Woodhead Publishing Limited.

Ding, T., & Xiao, J. (2014). Estimation of building related construction and demolition waste in Shanghai. *Waste Management (New York, N.Y.)*, *34*(11), 2327–2334. doi:10.1016/j.wasman.2014.07.029 PMID:25164857

Jimeniez, J. R. (2013). *Recycled aggregate for roads. In Handbook of recycled concrete and demolition waste*. Woodhead Publishing Limited.

Ledesma, E. F., Jimenez, J. R., Ayuso, J., Fernandez, J. M., & De-Brito, J. (2015). Maximum feasible use of recycled sand from construction and demolition waste for eco-mortar production e Part-I: Ceramic masonry waste. *Journal of Cleaner Production*, *87*, 692–706. doi:10.1016/j.jclepro.2014.10.084

Leiva, C., Guzmán, J. S., Marrero, M., & Arenas, C. G. (2013). Recycled blocks with improved sound and fire insulation containing construction and demolition waste. *Waste Management (New York, N.Y.)*, *33*(3), 663–671. doi:10.1016/j.wasman.2012.06.011 PMID:22784475

Majumdar, N.B. (2014). *Experience of the first commercial scale pilot project for C &D Waste Management in India*. IL&FS Environmental Infrastructure & Services Ltd.

Medina, C., Zhu, W., Howind, T., Frías, M., & Sánchez de Rojas, M. I. (2015). Effect of the constituents (asphalt, clay materials, floating particles and fines) of construction and demolition waste on the properties of recycled concretes. *Construction & Building Materials*, *79*, 22–33. doi:10.1016/j.conbuildmat.2014.12.070

Mehta, P. K., & Monteiro, P. J. M. (2006). *Concrete, Microstructure, Properties and materials* (3rd ed.). McGraw-Hill. doi:10.1036/0071462899

Meijide, B. G., & Pérez, I. (2014). Effects of the use of construction and demolition waste aggregates in cold asphalt mixtures. *Construction & Building Materials*, *51*, 267–277. doi:10.1016/j.conbuildmat.2013.10.096

Milicevic, I., Bjegovic, D., & Stirmer, N. (2015). Optimisation of concrete mixtures made with crushed clay bricks and roof tiles. *Magazine of Concrete Research*, *67*(3), 109–120. doi:10.1680/macr.14.00175

Ossa, A., García, J. L., & Botero, E. (2016). Use of recycled construction and demolition waste (CDW) aggregates: A sustainable alternative for the pavement construction industry. *Journal of Cleaner Production*, *135*, 379–386. doi:10.1016/j.jclepro.2016.06.088

Ozalp̈, F., Yılmaz, H. D., Kara, M., Kaya, O., & Şahin, A. (2016). Effects of recycled aggregates from construction and demolition wastes on mechanical and permeability properties of paving stone, kerb and concrete pipes. *Construction & Building Materials*, *110*, 17–23. doi:10.1016/j.conbuildmat.2016.01.030

Rao, A., Jha, N. K., & Misra, S. (2006). *Use of Aggregates from recycled Construction and Demolition waste in Concrete. In Resources Conservation & Recycling*. Elsevier.

Rodríguez, C., Parra, C., Casado, G., Minano, I., Albaladejo, F., Benito, F., & Sanchez, I. (2016). The incorporation of construction and demolition wastes as recycled mixed aggregates in non-structural concrete precast pieces. *Journal of Cleaner Production*, *127*, 152–161. doi:10.1016/j.jclepro.2016.03.137

Schachermayer, E., Lahner, T., & Brunner, P. H. (2000). Assessment of Two different separate techniques for Building wastes. *Waste Management & Research*, *18*(1), 16–24. doi:10.1177/0734242X0001800103

Silva, R. V., De Brito, J., & Dhir, R. K. (2015). Prediction of the shrinkage behaviour of recycled aggregate concrete: A review. *Construction & Building Materials*, *77*, 327–339. doi:10.1016/j.conbuildmat.2014.12.102

Soares, D., De Brito, J., Ferreira, J., & Pacheco, J. (2014). *Use of coarse recycled aggregates from precast concrete rejects: mechanical and durability*. Academic Press.

Thomas, J., & Wilson, P. M. (2013). Construction waste management in India. *American Journal of Engineering Research, 2*, 6-9.

Timothy, G. (1998). *The management and Environmental Impacts of C&D waste in Florida*. University of Florida.

Torgal, F. P., Tam, V. W. Y., Labrincha, J. A., Ding, Y., & De Brito, J. (2013). *Handbook of recycled concrete and demolition waste*. Woodhead Publishing Limited. doi:10.1533/9780857096906

Treating Construction and Demolition. (2004). *Waste Israel Environmental Bulletin*, 27.

Winkler, G. (2010). *Recycling construction and demolition waste- A LEED based toolkit*. McGraw Hill.

Chapter 4
Eco-Friendly Construction

Meghmala S. Waghmode
Annasaheb Magar Mahavidyalaya, India

Aparna B. Gunjal
Asian Agri Food Consultancy Services, India

Namdeo N. Bhujbal
Annasaheb Magar Mahavidyalaya, India

Neha N. Patil
Annasaheb Magar Mahavidyalaya, India

Neelu N. Nawani
Dr. D. Y. Patil Biotechnology and Bioinformatics Institute, India

ABSTRACT

Increase in urbanization leads to more construction of houses, dams, and streets. Reduction of the global warming effects can be carried out by recycling of construction material and searching for eco-friendly construction material. Greenhouse gas emissions can be reduced with the help of construction material which requires less energy for their production. The concept of eco-friendly construction is based on biomimetic (i.e., finding natural material with potential of endurance and self-cleaning properties). Construction materials like Portland cement and concrete can be replaced by eco-friendly biocement and bioconcrete. Production of biocement and bioconcrete can be done by using plants, algae, and bacteria. Use of less cement in concrete leads to less pollution. Concrete is the mixture of cement, sand, gravel, and water. By addition of pozzolan in concrete, the requirement of cement will be reduced. In the current review, major emphasis is given to eco-friendly construction material.

DOI: 10.4018/978-1-5225-6995-4.ch004

INTRODUCTION

Constructions are increasing due to urbanization and increase in civil standards. Construction materials are responsible for emissions of green house gases and thus leads to global warming. To avoid environmental issues, it is recommended to go for green building where construction material will be of biological origin. Considering the 5% CO_2 emission generated by cement industry, it is primary requisite to replace cement with biocement (Worrell, Price, Martin, Hendriks, & Meida, 2001). Blend of biosilica is biocement. Sand (SiO_2), clay (SiO_2,Al_2O_3, and Fe_2O_3), limestone or chalk ($CaCO_3$), iron ore (Fe_2O_3) are used for cement production. Plants or its material enriched with silica content will be the suitable candidate for biocement production. In the case of bacteria, ureolytic bacteria are mainly focused for precipitation of $CaCO_{3.}$ Green building index concept is started in some countries like Malaysia. Durability of structures exposed to harsh environmental structures could not be fulfilled by ordinary Portland cement (Achal, Mukherjee, Basu, & Reddy, 2009). Irreversible damage is caused by physical and chemical reactions of gases and / or liquids (Claisse, Elsayad, & Shaaban, 1997). To reduce environmental impact of construction industry, green building will be better choice (Chatterjee, 2009). Cement strength can be maintained by using materials which is having pozzolanic reaction. Selection of pozzolan depends on its reactivity, more cementitious strength is attributed with high reactivity. Any siliceous or aluminous material which can react with lime forming cementitious hydration products possessing cementitious properties. Both natural (calcined clay, calcined shale, and metakaolin) and artificial pozzolan (calcined clay, calcined shale, and metakaolin) can be used with Portland or blended cement which by hydraulic or pozzolanic activity, contribute to hardened cement. The main objectives of this review are viz., impact of construction on environment; replacement of construction materials; microorganisms in bioinspired construction and benefits of biotechconcrete.

Impact of Construction on Environment

In India, cement producing industries are important industries causing pollution. Cement production industries are responsible for air pollution due to emission of CO_2. Heavy metals like chromium, nickel, cobalt, lead and mercury are found in cement dust, affecting environment (Mujah, Shahin, & Cheng, 2016). Cement dust affect plants by reducing growth, decrease in chlorophyll, low starch content and lowering fruit setting (Pandey & Simba, 1990). Cement industry consumes high energy, causes depletion of natural resources and releases greenhouse gases (1 ton of CO_2 from 1 ton of cement) (Barker, Turner, Napier-Moore, Clark, & Davison, 2009). After water, concrete is the second most consumed substance. The dust particles arised from cement industry are present in the nearby trees (Farzadkia, Gholami, Abouee, Asadgol, Sadeghi, Arfaeinia, & Noradini, 2016). Aquatic ecosystem gets affected if concrete wastewater (with alkaline p^H) is discharged in waterbodies. Efficient recycling of construction and demolition debris can reduce waste management problems. According to Indian cement industry analysis, cement production in India accounts for 6.7% of world's cement output. India is aiming for 100 smart cities, which will lead to increased construction activity. Biocement is defined as composites of enzymes and microbial biomass with inorganic chemicals. Extensive research is focused on ureolytic bacterial mediated calcium carbonate deposites by the mechanism of microbial initiated carbonate precipitation (MICP). Main categories of organisms involved are viz., photosynthetic bacteria, sulfate reducing bacteria and nitrifying bacteria (Khanafari, Khams, & Sepahy, 2011). Precipitation of calcium carbonate depends on concentration of calcium, dissolved inorganic carbon, nucleation sites and pH. Calcium carbonate polymorphs are calcite, argonite and vaterite crystals, two hydrated crystalline phases (monohydrocalcite [$CaCO_3 \bullet H_2O$], ikaite [$CaCO_3 \bullet 6H_2O$] and amorphous calcium carbonate (ACC) (Rodriguez-Navarroa, Jroundib, Schiroa, Ruiz-Agudoc, & González-Munozb, 2012). Identification of crystals can be done physically and chemically based on solubility in HCl, Meningen reaction, infrared spectroscopy, petrographic microscopy and X-ray crystallography (Bouquet, Boronat, & Ramos-Cormenzana,1973). For MICP among three calcium sources viz., calcium acetate $Ca(C_2H_3O_2)_2$, calcium chloride ($CaCl_2$) and calcium nitrate $Ca(NO_3)_2$, calcium acetate showed maximum calcite formation from *Arthrobacter sulfurous, Bacillus muralis and B. atrophaeus strains* (Otlewska & Gutarowska, 2016). Ureolytic bacteria, acidogenic, halophilic, alkaliphilic, denitrifying, iron and sulfate reducing bacteria, cyanobacteria, algae, microscopic fungi are mainly concerned in the process of biomineralization. Construction by biological agent includes processes like biocementation, bioaggregation, biogrouting, biosaturation, bioclogging, bioencapsulation, biocoating and biorepair of concrete surface.

REPLACEMENT OF CONSTRUCTION MATERIALS

Biocementation

Biocementation is a process to produce binding material (biocement) based on MICP mechanism. The mechanical processes which lead to biocementation is shown in Table 1. The process can be applied in many fields viz., construction, petroleum, erosion control, and environment. These include wall and building coating method, soil and sand stabilizing. Primary role of microorganisms in carbonate precipitation is due to their ability to create an alkaline environment through various physiological activities. Microorganisms which induce the carbonate precipitation are those involved in nitrogen cycle; sulphate reducing bacteria; and photosynthetic microorganisms such as cyanobacteria and microalgae. Photosynthetic microorganisms produce urease or urea amidolyase enzyme to use urea, based on which biocementation is possible. During field application, the precipitation of calcium carbonate (biocement) is mixed with other supporting material such as sand. Photosynthetic and heterotrophic microorganisms have the potential to accelerate calcium carbonate precipitation. In biocementation, microorganisms used as media must meet the specific requirement, since the process creates a high pH in the environment and high concentration of calcium ion (Ariyanti, Handayani, & Hadiyanto, 2011). The various construction materials made from biocementation is represented in Table 2.

Microalgae in Biocementation

Microalgae due to its photosynthetic metabolism, have the potential to be used in biocementation. The species like *Spirulina, Arthrospira plantensis, Chlorella vulgaris, Dunaliella salina, Haematococcus pluvialis, Muriellopsis* sp., *Porphyridium cruentum* are autotrophic microorganisms which live through photosynthetic process. Several types of microalgae have the requirement of nitrogen which can be attained by urea hydrolysis mechanism. Microalgae are type of renewable resources which can be easily cultivated and grown easily (Ariyanti, Handayani, & Hadiyanto, 2011).

Calcium-Urea Based Biocementation

The most common type of biocementation is microbially-induced calcium carbonate precipitation (MICCP), which is the formation of calcium carbonate minerals such as calcite, vaterite, or aragonite on the surface of soil particles due to adhesion of cells of urease-producing bacteria on the surface of particle and a microgradient of concentration of carbonate and pH in the site of cell attachment due to hydrolysis of urea by urease from urea-producing bacteria (UPB). The biogeochemical reactions of this biocementation process are as follows:

$(NH_2)_2CO + 2H_2O$ urease $CO_2 + 2NH_4OH$

$CO_2 + H_2O$ carbonic anhydrase $H^+ + HCO_3^-\ 2\ H^+ + CO_3^{2-}$

$CaCl_2 + H_2CO_3\ CaCO_3 + 2HCl$

$2\ HCl + 2NH_4OH\ 2\ NH_4Cl + 2H_2O$

Calcium-Phosphate Based Biocementation

Calcium phosphate precipitation from calcium phytate solution using the phytase activity of microorganisms produces a mixture of the crystal forms viz., monetite, whitlockite and hydroxyapetite.

Monetatite precipitation

$Ca(H_2PO_4)_2 + CO(NH_2)_2 + H_2O$ + acid urease $CaHPO_4 + CO_2 + (NH_4)_2HPO_4$

Hydroxyapatatite precipitation

$5Ca(H_2PO_4)_2$+8 $CO(NH_2)_2$+8H_2O + acid urease $Ca_5(PO_4)_3(OH)$ + 2NH_4HCO_3 + 6CO_2 + 7$(NH_4)_2HPO_4$

Calcium-Bicarbonate Based Biocementation

The calcium-bicarbonate biocementation involves the precipitation of calcite with removal of CO_2 from solution of calcium bicarbonate (Ehrlich, 1999).

$Ca(HCO_3)_2$ $CaCO_3 + CO_2 + H_2O$

Iron-Based Biocementation

Iron-based biocementation is also performed with soluble chelates of ferrous or ferric iron precipitating to ferrous or ferric hydroxides. The reactions are as follows:

Fe^{2+} + 1.5 $(NH_2)_2CO$ + 0.25 O_2 + 5.5 H_2O + UPB Fe $(OH)_3$ + 1.5 $(NH_4)_2CO_3$ + 2H^+

The advantages of using iron hydroxide as the clogging compound is that the soil treated by iron minerals is more ductile and able to resist low pH conditions.

Calcium-Magnesium Based Biocementation

The biocement for calcium-magnesium based biocementation can be produced through dissolution of dolomite (a common raw material for the production of cement) in hydrochloric acid:

$Ca^{2+} + Mg^{2+} + CO(NH_2) + UPB\ CaMg(CO_3) + 2NH_4^+$

Biogrouting

Minerals like calcium carbonate, calcium phosphate, calcium oxalate, silicate, and iron oxide are precipitated in nature by microorganisms (Mann, 1993). These biominerals provide adequate strength and low environmental impact. In order to reduce environmental impact and cost in the future, grouting method utilizes calcium carbonate precipitated by microbial cells. This grout is commonly called "biogrout," which is a product of biomineralization. In biogrouting process, the depth of penetration relies on the bacterial size and optimal conditions needed for bacterial activity such as pH, salinity, oxidation-reduction potential, concentrations of nutrients, and content of water. Biogrout is cheaper than chemical grouting techniques. Raw materials used in chemical grouts range from \$2 to 72 per m^3 of soil, while biogrouts in the range of \$0.5 to 9.0 per m^3 of soil (Ivanov & Chu, 2008).

Bioclogging

Bioclogging is the process by which bacterial activity is involved in reduction in the hydraulic conductivity of soil and rock porosity. It can be used to reduce drain channel

Table 1. Mechanical processes which lead to biocementation Source: (Ivanov & Chu, 2008)

Group of Microorganisms	Biocementation Mechanism	Essential Conditions for Biocementation	Geotechnical Applications
Sulphate reducing bacteria	Production of undissolved sulphides of metals	Anaerobic conditions; presence of sulphate and carbon source in soil	Enhance stability for slopes and dams
Ammonifying bacteria	Formation of undissolved carbonates of metals in soil due to increase of pH and release of CO_2	Presence of urea and undissolved metals salt	Enhance stability for retaining walls, embankments and dams. Increase bearing capacity of foundations
Iron-reducing bacteria	Production of ferrous solution and precipitation of undissolved ferrous and ferric salts and hydroxides in soil	Anaerobic conditions changed for aerobic conditions; presence of ferric minerals	Reduce liquefaction potential of soil

Table 2. Various construction materials made from biocementation Source: (Ariyanti, Handayani, & Hadiyanto, 2011)

Microorganisms	Metabolism	Solution	Application	Reference
Bacillus cereus	Oxidative deamination of aminoacids	Growth media (peptone, yeast extract, KNO_3, NaCl)+ $CaCl_2.2H_2O$, Actical, Natamycine	Biological mortar	Muynck *et al.*, (2010)
Bacillus pasteurii	Hydrolysis of urea	Nutrient broth, urea, $CaCl_2.2H_2O$, NH_4Cl, $NaHCO_3$	Crack in concrete remediation	Santhosh *et al.*,(2001)
Bacillus sphaericus	Hydrolysis of urea	Yeast extract, urea, $CaCl_2.2H_2O$	Crack in concrete remediation	Belie (2010)
Bacillus pasteurii	Hydrolysis of urea	Nutrient broth, urea, $CaCl_2.2H_2O$, NH_4Cl, $NaHCO_3$	Bacterial concrete	Santhosh *et al.*, (2001)

erosion, form grout curtains to reduce the migration of heavy metals and organic pollutants, and prevent piping of earth dams and dikes. Gaps in construction pit, landfill, or dikean be sealed through bioclogging. *In-situ* production of water-insoluble polysaccharides is involved in bioclogging process. Stimulation of polysaccharide production can be done by the addition of carbon source and microorganisms to soil. The growth of microorganisms and accumulation of microbial *in-situ* is responsible for bioclogging. Oligotrophic bacteria from genus *Caulobacter* (Tsang, Li, Brun, Freund, & Tang, 2006), aerobic Gram-negative bacteria from genera *Acinetobacter, Agrobacterium, Alcaligenes, Arcobacter, Cytophaga, Flavobacterium, Pseudomonas, and Rhizobium* are reported to be involved in the bioclogging process through binding of soil with insoluble polysaccharides (Portilho, Matioli, Zanin, de Moraes, & Scamparini, 2006). Cellulose-degrading bacteria like *Cellulomonas flavigena* is also involved in polysaccharide mediated clogging process (Kenyon, Esch, & Buller, 2005). The mechanical processes which lead to bioclogging is shown in Table 3.

Bioencapsulation

Bioencapsulation is thin layer of composite bioprecipitated materials around solid particles used in many areas of civil. It is used for coating of concrete engineering objects to improve their asthetics.

Table 3. Mechanical processes which lead to bioclogging Source: (Ivanov & Chu, 2008)

Group of Microorganisms	Mechanism of Bioclogging	Essential Conditions for Bioclogging	Geotechnical Applications
Algae and cyanobacteria	Formation of impermeable layer of biomass	Light penetration and presence of nutrients	Reduce of water infiltration into slopes and control seepage
Aerobic and facultative anaerobic heterotrophic slime-producing bacteria	Production of slime in soil	Presence of oxygen and medium with ratio of C:N > 20	Control soil erosion and protects slope
Oligotrophic microaerophilic bacteria	Production of slime in soil	Low concentration oxygen and medium with low concentration of carbon source	Reduce drain channel erosion and control seepage
Nitrifying bacteria	Production of slime in soil	Presence of ammonium and oxygen in soil	Reduce drain channel erosion
Sulphate-reducing bacteria	Production of undissolved sulphides of metals	Anaerobic conditions; presence of sulphate and carbon source in soil	Form grout curtains to reduce the migration of heavy metals and organic pollutants
Ammonifying bacteria	Formation of undissolved carbonates of metals in soil due to increase of pH and release of CO_2	Presence of urea and dissolved metal salt	Prevent piping of earth dams and dikes

MICROORGANISMS IN BIOINSPIRED CONSTRUCTION

Anaerobic fermenting bacteria may be involved in cementation of soil particles under the presence of calcium, magnesium, or ferrous ions. This cementation can be due to the increase in pH caused due to release of ammonia and carbondioxide production in soil added with urea or waste protein (Kucharski, Winchester, Leeming, Cord-Ruwisch, Muir, Banjup… Mutlaq, 2005). The insoluble carbonates and hydroxides of metals will precipitate at alkaline pH and help in clogging soil. This can be useful in bioclogging and biocementation to precipitate silicates from colloidal silica suspension. Anaerobic respiring bacteria which can be used in bioclogging and biocementation are sulphate-reducing bacteria. Sulphate reducing bacteria make dihydrogen sulphide using organic acids, hydrogen, or alcohols as electron donors and sulphate as electron acceptor. Sulphide reacts with iron to form insoluble suphides of metals, which binds the soil particles. Facultative anaerobic bacteria are considered as the most suitable bioagents for soil bioclogging and biocementation because many species produce huge quantity of exopolysaccharides, which usually promote formation of cell. Aerobic bacteria can also be used for soil bioclogging, biocementation, and biobinding of soil particles because many species produce slime, form chains and filaments, increase pH, and oxidize different organic and inorganic substances. Actinomycetes, a group of Gram-positive bacteria, form particles-binding mycelium and produce particles binding slime in soil (Dworkin, Falkow, Rosenberg, Schleifer, & Stackebrandt, 2006). These bacteria are future candidate for the aerobic soil bioclogging, biocementation, and biobinding. The insoluble carbonates and hydroxides of metals precipitate at high pH and bind the soil particles, thus clogging the soil pores.

Benefits of Biotechconcrete

The benefits of biotechconcrete are viz., low production cost as compared to conventional cement; increased sustainability and low viscosity; eco-friendly due to less emission of green house gases, aids in reduction in global warming effect; regeneration of the sandstone formation within short time; *in-situ* application and soil strengthening, stabilizing and sand stabilizing in regions which are earthquake affected. The microbial urease enzyme catalyses the hydrolysis of urea into ammonium and carbonate. Ammonia released into the surroundings increases the pH which helps in accumulation of insoluble $CaCO_3$ in a calcium-rich environment. It has been reported that anaerobic hot-spring bacteria leach silica and help in the formation of new silicate phases that fill the micropores (Ghosh, Biswas, Chattopadhyay, & Mandal, 2009). The concentration of 10^5 cells ml^{-1} optimizes the microstructure of cementitious composites. It has been reported that since bacteria consumes oxygen, it may provide an additional benefit associated with the potential to inhibit reinforcement corrosion.

Performance Index and Application of Biocement

The microbial biocement will improve the physico-chemical properties of cement motor and also reduce the cracks. The biocement also has medical applications viz., to fill the gaps in the bones, treatment of problems due to skull defects. It also removes calcium from the waste and reduces the cost of waste water treatment.

FUTURE RESEARCH DIRECTIONS

The biocement should be used for the construction of eco-friendly buildings on a very large scale.

CONCLUSION

Biocement and bioconcrete are eco-friendly constructions methods. The green eco-friendly buildings will also reduce the problem of pollution which is very important. The biocement will have many applications and it can be used in many cases. Biocement will have advantages over the ordinary cement viz., for production, biocement will require less time; suitable for *in-situ* process and raw material are produced at low temperature. Biocement will contribute to sustainable construction as there will be no release of toxic emissions, no CO_2 emission and therefore, no pollution problem.

REFERENCES

Achal, V., Mukherjee, A., Basu, C., & Reddy, S. (2009). Lactose mother liquor as an alternative nutrient source for microbial concrete production by *Sporosarcina pasteurii. Journal of Industrial Microbiology & Biotechnology*, *36*(3), 433–438. doi:10.100710295-008-0514-7 PMID:19107535

Ariyanti, D., & Handayani, N. (2011). An overview of biocement production from microalgae. *International Journal of Science and Engineering*, *2*(2), 30–33.

Barker, J., Turner, A., Napier-Moore, A., Clark, M., & Davison, E. (2009). CO_2 capture in the cement industry. *Energy Procedia*, *1*(1), 87–94. doi:10.1016/j.egypro.2009.01.014

Belie, N. (2010). Microorganisms versus stony materials: A love-hate relationship. *Materials and Structures*, *43*(9), 1191–1202. doi:10.161711527-010-9654-0

Bouquet, E., Boronat, A., & Ramos-Cormenzana, A. (1973). Production of calcite (calcium carbonate) crystals by soil bacteria is a general phenomenon. *Nature*, *246*(5434), 527–529. doi:10.1038/246527a0

Chatterjee, K. (2009). Sustainable construction and green buildings on the foundation of building ecology. *Indian Concrete Journal*, *83*(5), 27–30.

Claisse, A., Elsayad, A., & Shaaban, G. (1997). Absoprtion and sorptivity of cover concrete. *Journal of Materials in Civil Engineering*, *9*(3), 105–110. doi:10.1061/(ASCE)0899-1561(1997)9:3(105)

Dworkin, M., Falkow, S., Rosenberg, E., Schleifer, K., & Stackebrandt, E. (2006). The prokaryotes: A handbook on the biology of bacteria. In M. Dworkin, S. Falkow, E. Rosenberg, K. Schleifer, & E. Stackebrandt (Eds.), *Archaea Bacteria: Firmicutes, Actinomycetes* (3rd ed.; Vol. 3). New York: Springer-Verlag.

Ehrlich, H. (1999). Microbes as geologic agents: Their role in mineral formation. *Geomicrobiology Journal*, *16*(2), 135–153. doi:10.1080/014904599270659

Farzadkia, M., Gholami, M., Abouee, E., Asadgol, Z., Sadeghi, S., Arfaeinia, H., & Noradini, M. (2016). The impact of excited pollutants of cement plant on the soil and leaves of tree species: A case study in Golestan Province. *Open Journal of Ecology*, *6*(7), 404–411. doi:10.4236/oje.2016.67038

Ghosh, P., Biswas, M., Chattopadhyay, B., & Mandal, S. (2009). Microbial Activity on the microstructure of bacteria modified mortar. *Cement and Concrete Composites*, *31*(2), 93–98. doi:10.1016/j.cemconcomp.2009.01.001

Ivanov, V., & Chu, J. (2008). Applications of microorganisms to geotechnical engineering for bioclogging and biocementation of soil in situ. *Reviews in Environmental Science and Biotechnology*, *7*(2), 139–153. doi:10.100711157-007-9126-3

Kenyon, W., Esch, S., & Buller, C. (2005). The curdlan-type exopolysaccharide produced by *Cellulomonas flavigena* KU forms part of an extracellular glycocalyx involved in cellulose degradation. *Antonie van Leeuwenhoek*, *87*(2), 143–148. doi:10.100710482-004-2346-4 PMID:15723175

Khanafari, A., Khams, N., & Sepahy, A. (2011). An investigation of biocement production from hard water. *Middle East Journal of Scientific Research*, *7*(6), 964–971.

Kucharski, E., Winchester, W., Leeming, W., Cord-Ruwisch, R., Muir, C., Banjup, W., . . . Mutlaq, J. (2005). *Microbial biocementation*. Patent Application WO/2006/066326; International Application No. PCT/ AU2005/001927.

Mann, S. (1993). Molecular tectonics in biomineralization and biomimetic materials chemistry. *Nature, 365*(6446), 499–505. doi:10.1038/365499a0

Mujah, D., Shahin, M., & Cheng, L. (2016). State-of-the-art review of biocementation by microbially induced calcite precipitation (MICP) for soil stabilization. *Geomicrobiology Journal, 34*(6), 1–14.

Muynck, W., Belie, N., & Verstraete, W. (2010). Microbial carbonate precipitation in construction materials: A Review. *Ecological Engineering, 36*(2), 118–136. doi:10.1016/j.ecoleng.2009.02.006

Otlewska, A., & Gutarowska, B. (2016). Environmental parameters conditioning microbially induced mineralization under the experimental model conditions. *Acta Biochimica Polonica, 63*(2), 343–351. doi:10.18388/abp.2015_1172 PMID:26894236

Pandey, D., & Simba, K. (1990). Effect of cement kiln dust on chlorophyll in wheat leaf. *Environment and Ecology, 8*(1B), 461–463.

Portilho, M., Matioli, G., Zanin, G., de Moraes, F., & Scamparini, A. (2006). Production of insoluble exopolysaccharide *Agrobacterium* sp. (ATCC 31749 and IFO 13140). *Applied Biochemistry and Biotechnology, 131*(1-3), 864–869. doi:10.1385/ABAB:131:1:864 PMID:18563660

Rodriguez-Navarroa, C., Jroundib, F., Schiroa, M., Ruiz-Agudoc, E., & González-Munozb, M. (2012). Influence of substrate mineralogy on bacterial mineralization of calcium carbonate: Implications for stone conservation. *Applied and Environmental Microbiology, 78*(11), 4017–4029. doi:10.1128/AEM.07044-11 PMID:22447589

Santhosh, K., & Ramachandran, V., & Ramakrishnan, S. (2001). Remediation of concrete using microorganisms. *ACI Materials Journal, 98*(1), 3–9.

Tsang, P., Li, G., Brun, Y., Freund, L., & Tang, J. (2006). Adhesion of single bacterial cells in the micronewton range. *Proceedings of the National Academy of Sciences of the United States of America, 103*(15), 5764–5768. doi:10.1073/pnas.0601705103 PMID:16585522

Worrell, E., Price, L., Martin, N., Hendriks, C., & Meida, O. (2001). Carbon dioxide emissions from the global cement industry. *Annual Review of Energy and the Environment, 26*(1), 303–329. doi:10.1146/annurev.energy.26.1.303

KEY TERMS AND DEFINITIONS

Eco-Friendly: Safe to the environment.

Global Warming: Rise in the temperature of the earth's surface.

Oligotrophic: Microorganism which is able to grow in environment which has low level of the nutrients.

Pollution: The presence of contaminants in the environment which has adverse effects.

Remediation: Cleanup or removing hazardous material (organic and inorganic) from the environment.

Sustainable: Steady state without exhausting natural resources and without causing any damage.

Chapter 5
Rethinking Waste Through Design

Caroline O'Donnell
Cornell University, USA

Dillon Pranger
Cornell University, USA

ABSTRACT

This chapter will study the proliferation of architectural follies that use recycled or recyclable materials in a move to promote better practices in waste and recycling. Given the slow uptake of this impetus in the architectural world proper, the text will investigate the obstacles in engaging in materially sustainable practices in the construction industry as well as case studies for rethinking currently problematic materials. However, while some improvements have been made in the construction industry's use of recycled materials, the industry often dismisses the afterlife of materials used throughout the process. What are the motivations of the industry and how can we incentivize circular thinking in an industry that produces hundreds of millions of tons of waste per year in the US?

CURRENT STATES

Global material crises are imminent. In the very near future, recycling will no longer be a choice made by those concerned about the environment, but a necessity for all. At the current rate of mining, for example, it is estimated that approximately 19 years of copper, 10 years of tin, and 10 years of zinc remain in the earth's crust (Frondel, 2007). Materials are finite and we will soon need to find alternative solutions to the acute problem of global consumption and disposal.

DOI: 10.4018/978-1-5225-6995-4.ch005

Meanwhile, global production of waste continues to climb. Each year, 1.2 billion metric tons of waste are generated globally, a figure expected to rise exponentially to 3.6 billion metric tons by 2100, as developing countries begin to urbanize, industrialize, and consume (Simmons, 2016). And while the consequences of waste on the environment have been well documented—toxins from plastics and other solid waste leach into our food and water-systems, massive amounts of energy are consumed through its global transport, marine and wildlife systems are disrupted by foreign materials—we choose, en-masse, to proceed with our disposable lifestyles.

As the world's most wasteful nation, the United States produces 3.2 kilograms of trash per capita per day, a total of 92 metric tons of garbage per lifetime (Humes, 2012). Of the 234 million metric tons of Municipal Solid Waste (MSW) generated in the U. S. in 2014, over 80 million metric tons were recycled and composted, equivalent to a 34.6% recycling rate (United States Environmental Protection Agency, 2016).

The apparent lack of concern for the problem of waste can be attributed in part to policies that remove waste from our everyday experience. Besides single stream waste systems, which reduce our interaction with our household waste to a minimum, many U.S. cities charge a flat fee or tax to pay for the voluntary disposal of trash and have little incentive or disincentive to motivate participation (Simmons, 2016). Far beyond the prying eyes of the cities, landfills fill up, resources are squandered, and environments are contaminated: but the consequences of these actions are not significant in our daily lives. This invisibilization of trash cannot but promote passive behaviors in the home, in industry, and in the specifications of designers.

U.S. policies around recycling differ significantly from their European counterparts and result in markedly different statistics. The fact that European countries recycle significantly more can be attributed to cultural differences, education, and any number of incalculable factors, but, in general, European countries have stricter policies that involve increased engagement with recyclable materials in the form of multi-stream separation. In Germany, for example, the consumer separates glass into colors as part of the recycling practice. In England, separate bins are provided by City Councils to collect food-scraps at a municipal level (The 7th Environment Action Programme (EAP), 2013).

But one need not look so far afield for a how-to guide. Some west coast cities have recycling rates approaching 80%. This high figure is partially attributable to California's state mandated waste diversion quotas: a 1989 law that required cuts in landfill loads by 50% before 2000; and another law in 2011, which increased that goal to 75% before 2020 (AB-341, 2011). In 2007, San Francisco completely banned disposable plastic bags (later a statewide mandate) and in 2009, the city enacted laws which made recycling and composting mandatory. Across the west coast, cities from San Jose to Portland have increased their recycling rates to above

70% through explicit public programs and policy change (Bureau of Planning and Sustainability, 2016).

Rethinking Recycling

In 1976 Swiss Architect, Walter R. Stahel proposed, in the report, "Potential for Substituting Manpower for Energy," the three key principles of recycling: *Reuse, Repair,* and *Remanufacture*. Stahel hoped to rethink the categories of waste management as processes that turn end-of-life goods into as-pure-as-new resources. William McDonough and Michael Braungart would later amend this list to include four additional principles, in their 2002 text, "Cradle to Cradle: Remaking the Way We Make Things:" *Reduce, Reuse, Recycle, Recover, Rethink, Renovate,* and *Regulate*.

Recently, however, after 40 years, and considering contemporary production, Stahel completely revised his original stance, and proposed an entirely new set of principles that directly target more contemporary issues of composite production. His new '7Ds' comprise imperatives to: *Depolymerize, Dealloy, Delaminate, Devulcanize, Decoat, Deregulate*, and *Deconstruct*, a clear shift from dealing with the consequences of waste towards an engagement with manufacturing and domestic practices. We can *reuse* a shoe for some time, for example, but if it has not been designed to be *deconstructed*, it will never be able to participate in a circular ecosystem (Stahel, 2016). This pivot in focus returns us to a position where the original design becomes as important as the subsequent actions of the consumer.

While these mantras have flourished in many industries, the building industry has been relatively slow on the uptake. Environmentally aware materials specified by architects are still by and large more expensive than the majority of products. Until there is an economic benefit, it is unlikely that these materials will become mainstream. Nevertheless, there have been breakthroughs in the sustainable construction product market, despite higher prices. Denim insulation, for example, while still almost twice the price per square foot than conventional fiberglass insulation, offers an advantage in acoustical performance and user-friendliness to offset its significant increase in pricing (HGTV, 2017). In addition, denim does not contain any chemical irritants or produce "fiberglass itch" upon contact with the skin during installation. These characteristics, as well as the feel-good factor associated with environmentally aware purchases, provide enough consumer appeal to justify the products sale in major home and appliance stores across the US.

Even if architects specify 100% recyclable elements, it is in fact the joining of materials that is often the most problematic aspect. A modern conventional window contains over 200 individual components that comprise its entire assembly, each requiring their own method of adhesive, mechanical, and/or friction connections (Andersen Windows and Doors, 2012). The joinery embedded in something as mundane as a window assembly grows exponentially in complexity when considering how this object is then joined to the adjacent material in the wall of a building and so on. Therefore—and perhaps this is the good news for designers—the problem becomes one of design. That is to say, we can no longer design buildings and details in the way that we have become accustomed to. We must completely rethink our design practices.

Such adjustments are taking place in many industries such as the Sports Shoe industry. In this industry, the sports shoe is one of the most difficult-to-recycle products due to its glued composition of different materials. Some of the world's major companies have launched products which are composed of recycled plastic for example, the *Adidas UltraBOOST Uncaged Parley Shoe*. For this particular sports shoe, the company claims that it has an upper sole composed of a high percentage of recycled plastic (Baker-Brown, 2017). More so, Nike has recycled sports shoes in which sports shoes are ground down after use to produce springy sports-field surfaces as well as new footwear (Nike, 2018). In the lighting and flooring industries, companies have developed innovative new leasing systems, whereby customers rent the product for a given time, after which the product can be rehabilitated and reused. One example in which this has been implemented is lighting company Philips' new pay-per-lux service leases lighting to Schiphol's new terminal 2 lounge in Amsterdam.

The Philips project led to a redesign of lighting, in order to increase efficiency and simplify maintenance. This example can be applied across industries, as designers of these products have had to change the way they think and design: priorities have had to shift from a focus on performance and quality to a focus on material providence and destiny (Philips, 2018). Designers have found a way to satisfy both of those goals and, perhaps most importantly, they have found a viable and growing market. Can designers of architecture make such adjustments?

In one area of architecture, the 7Rs and 7Ds *are* already thriving. The following case studies examine the role of experimental pavilions in the promotion of sustainable practices in future architecture.

Architectural Disruptors

In 2013, in response to Museum of Modern Art's annual call for a temporary installation, a monstrous structure rose in the courtyard of the museum's annex space at PS1, in Long Island City: "Party Wall," a large pavilion, using the byproducts from skateboard manufacturing and leftover steel. According to the jury, the success of this competition-winning entry was its unabashed use of borrowed materials from recycling loops, which sent a message that "A world in crisis needs disrupting responses to actual problems" (Gadano, 2016).

The project brief asked for a "design proposal for an environmentally friendly urban landscape" which would "focus on themes of sustainability, recycling, and reuse".[1] This sustainable language is a trend echoed in many briefs that proliferate in the U.S. architectural competition scene today: the 'City of Dreams' Competition for a pavilion on Governor's Island, New York, for example, requires designers to "think about designing for the future with the limited economic and natural resources at hand, and to create a sustainable design" (Warerkar, 2018); the 'Ragdale Ring' competition for a pavilion on the Ragdale Campus, just north of Chicago in Lake Forest, Illinois, asks that proposals address, "plans for the reuse, recycling, and dispersal of all material used in the project" (The Ragdale Foundation, 2018).

Certainly, part of this renewed interest must be attributed to a response to the criticism of the early 21st century's follies for being excessive and wasteful. Dubbed as "a thoroughly wasteful endeavor" (Chan, 2012), or "wasteful and untenable" (Coaker, 2013), and many other derogatory terms of extravagance, the use of raw materials for high-profile work of a temporary nature seemed to stick in several craws. Indeed, temporary pavilions, given their limited budgets and timeframes, have tended to focus on production, at the expense of the after-life of materials used. For example, as brought to light through Alejandro Aravena's installation at the 2016 Architecture Biennial, the previous year's Art Biennial consumed 14 kilometers of steel section, 10,000 square meters of plasterboard, and 100 tons of waste in total. Aside from the volume used and the energy necessary to mine, transport, and fabricate the raw materials used, all these materials were destined for the landfill after the event (Mairs, 2016).

The positive effect of such language in an increasing number of competition briefs, is that the innovative young architects participating in these competitions are pushing the disciplinary norms in response to these calls. In 2014, The Living's *Hy-Fi*, (the Young Architects' Program winner, succeeding *Party Wall*), built a tower of bricks grown from mycelium, a new 'cultivated' material, produced from the expansion of mushroom spores in a mold, that does not waste finite resources and can be decomposed after its useful lifetime (Baker-Brown, 2017).

A different response to the rising call for sustainability in pavilion design was conducted by T+E+A+M in their 2017 installation: *Clastic Order.* The designers erected a series of freestanding self-load bearing columns using plastiglomerate, consisting of a mixture of locally sourced construction debris, glass cullet, and post-industrial plastic waste. The columns used each component's properties to their advantage: plastic as a binding agent, and construction debris as mass and structure. By using materials that are typically destined for the landfill, the project drew awareness to the wastefulness of regional waste streams (T.E.A.M., 2017).

In addition to material innovation, several architects have begun to rethink the construction of the pavilion itself, rather than simply the materiality. OMG's *Primitive Hut* used decomposing materials (hemp, sawdust, and a proprietary bioresin[2]), modular systems and living trees to allow part of the pavilion to dissolve itself, part to be demountable (eliminating nails and adhesives), and part to be a natural growing system, which would significantly replace the wood used in construction.[3]

Rising Canes by Penda and the *ICD Aggregate Pavilion* both attempted similar feats of using modular systems that rely entirely on self-supporting geometries. These pavilions address a growing issue in Municipal Solid Waste separation: composites and adhesives. By eliminating any binding materials (nails, screws, adhesives, etc.) each pavilion is able to be disassembled and relocated, or have its materials repurposed as necessary.[4]

The above pavilions, and others like them, appear in architectural media, where they are, more often than not, considered to be 'playful' works. Palettes, bottles, beer kegs, paper, corrugated cardboard and cardboard tubing, shipping containers, barrels, car parts, kitchen sinks, and many more waste products make up a motley crew of materials used in the last decades by architects in the experimental pavilion genre. Touted as, "innovative" or "new and exciting," these projects are still considered outside the norm (Williams, 2016; Citimetric Staff, 2015).

Of course, environmental thinking is nothing new: the idea of engaging in issues of waste and architectural installations had its heyday in the late 1960s, as a result of a consumerist culture that was producing commodities and waste exponentially. By 1970, the annual household income had increased to 376% of its 1918 value (U.S. DOL, 2006). While the increase in household income grew at a relatively consistent pace, the percentage allotted for non-necessities saw a drastic increase due to the expansion of availability of material products. Consequently, the increase in the rate of refuse between 1920 and 1970 was five times the rate of population growth (National League, 1973). This dramatic increase of municipal waste pressured federal governmental agencies (and individuals) to address the issue of municipal waste management across national, regional, and local levels. It was not until 1965, when US legislation passed the Solid Waste Disposal Act that the US addressed the increasing demand for a better waste management system, and the EPA was founded in 1970 (Solid Waste Disposal Act of 1965, 2012).

Eccentric houses of plastic bottles and beer cans appeared. In 1966, the commune of Drop City arose, a collection of domes made from material scavenged from automobile junk yards (Grossman & McCourt, 2012). Spurred on by various environmental crises as well as the evocative image of the earth from space—and the image of fragility and finitude associated with that—many architects were motivated to rethink their wasteful practices. Pioneered by architect Michael Reynolds, the Earthship projects, beginning in the 1970s (but still being built today), proposed houses using both recycled and natural local materials, usually tires, aluminum cans, and rammed earth. Beyond material sustainability, these houses aimed to be self-sufficient in terms of energy production and human needs (Reynolds, 1990).

Manipulating Affordances

The effect of using these materials is, of course, more than pragmatic. The shifting meaning embodied in the transformation of a waste product into an aggregation or misused architectural element produces a mental calculation: the object is known, but the context and use-value are displaced. Like a mystery reader, the viewer must do some work in deducing the original material, must consider its original life and its new one, and as such a relationship of communication is developed between the viewer and the material object that makes for a more engaged experience.

This phenomenon has been described by James Gibson as a fundamental part of perception. According to Gibson (1986), the primary perception of an object is its function: "what it *offers* the animal, what it *provides* or *furnishes*, either for good or ill " (p.7). Affordances exist as "action possibilities" (Gibson, 1986, p.7) latent in the environment, existing independently of the animal's ability to recognize them, but always in relation to and dependent upon the animal's capabilities.

When a material is misused, as is the case for waste transformed into architecture, the affordance is obscured and the perceiver reassigns the affordance of the object. This perceptual play is a technique that has been used by numerous artists, in pieces in which the perception of the overall form is in conflict with the familiar and useful form of the individual component: Tony Cragg's amorphous forms made of dice, Tara Donovan's landscapes made of plastic straws, cups, and pencils, El Anatsui's tapestries made of ringpulls and bottletops all play this perceptual game between large and small scale in which the object and its original affordance are in conflict with the current role as aggregate (Cragg, 2011; Donovan, 2018; Anatsui, 2014).

Through this playful technique, the use of waste materials provokes the public to rethink the materials that they had previously assigned a single function, to think about the potential second lives of objects, and eventually, potentially, about their own behaviors and practices around waste.

Architectural Temporality

While the temporary pavilion genre has shifted in response to criticism, the architecture discipline as a whole has been less agile. Construction materials, including stone, gravel, and sand, comprise around three-quarters of total raw materials use in the U.S (Matos, 2012). In 2014, 484 million metric tons of Construction and Demolition (C&D) debris were generated in the country (more than twice the amount of generated municipal solid waste) and although some European countries are achieving construction waste recovery figures of 80%, the U.S. is lagging significantly behind at 48% (Bossink, & Brouwers, 1996; Earth 911).

While temporary pavilions have borne the brunt of the criticism, it is worth remembering that all buildings built today are temporary. A survey conducted in Japan, concluded that the average lifespan of an office building in 1990 was between 23 and 41 years and many individual components are replaced in a much shorter time-frame (Yashiro, Kato, Yoshida & Komatsu, 1990). That means that architects need to consider demountability when designing a building, just as they have begun to do in temporary buildings.

The engagement with sustainable materials has been slowly growing since the 1990s when the U.K. first introduced their sustainable design rating system, Building Research Establishment's Environmental Assessment Method (BREEAM) and since the USGBC introduced their version of a sustainable rating system, Leadership in Energy and Environmental Design (LEED) in 2000 (United States Green Building Council, 2018). Since its conception, LEED has seen a steady 12% increase in registered projects through 2013, with new 500 projects added each month (Shutters, 2016). This increase equates to a rise in sustainable buildings, checking off boxes in LEED's accreditation system, relating to material use, energy, transport and many other categories. For example, if construction materials, including furniture and furnishings, use recycled content that accounts for greater than 10% of the total value of the materials in the project LEED will award 1 point. Additionally, if these materials use more than 20% recycled content LEED will award 2 points (United States Green Building Council, 2016).

The less glamourous but perhaps more pressing question is that of the post-demo lives of the materials. After the materials have been used, where do they go? LEED has experimented with offering credit for C&D Waste Management through their pilot credit program. This pilot credit, although now closed, aimed to "reduce construction and demolition waste disposed of in landfills and incineration facilities by recovering, reusing and recycling materials". Users were given the option to either divert 50% of total construction and demolition material from three material streams, divert 75% of total construction and demolition material from four material streams, or generate less than 10 kilos of construction waste per square foot (United States Green Building Council, 2009). Whether LEED will adopt this pilot into their points system remains to be seen.

In the case of *Party Wall*, steel was returned to the manufacturer, smelted, and likely supports some new New York building for years to come. The skateboards were used as partitions, and sometimes as fuel, (which was, in fact, their original fate). Ground screw foundations and pools were returned to their users and the water-weight bags were donated to a farm to be used as water cisterns. As with other pavilions using recycled materials, the best case scenario is that the materials are sent back to the recycling loop. In many cases, such as the cases of metals and many plastics, these loops are infinite. Materials are melted, formed, remelted reformed and so on. However, many materials are *not* infinitely recyclable. Paper, for example, only has 5-7 loops before its fibers are too short. Wood and fabrics can only be downcycled. Concrete and composites are likely not recycled at all.

Beyond Recycling

In the same way that zero net energy is a dubious goal (we should be *producing* energy, not aiming for zero), an architectural intervention that simply returns materials to a linear cycle should not be our highest aspiration.

To do better, we might affect social practices and effectively change local behaviors. To do better we might engage in economics of the problems and understand how bundling or amassing might impact collection. To do better, we might find materials that are currently not being recycled, and through our intervention, ameliorate their predicament, and transform them, through use in architecture, into materials that can be recycled.[5] To do better we might intervene not only in the waste segment of the cycle, but in the design and manufacturing segments, so that the materials that are being produced will be fabricated in ways that can demount and detach.

It is in this latter capacity that John Habraken was a ground-breaker in his 1960s collaboration with Heineken. When he began to design a pavilion using glass bottles, understood that the design innovation was not in the use of a found object, but that the object's past could be delved into and reformulated: Habraken redesigned the bottle to make it more fitting for the life of a brick. This project was generated in response to a series of problems: a housing shortage in the Caribbean resulting from the lack of building materials available, and a surplus of beer bottles due to the economic unfeasibility of shipping empty bottles back to the Netherlands. While the project was ultimately unsuccessful due to Heineken's marketing reservations, the redesign of a product in order to address an unsolved social, political, or economic problem was a novel idea that has led to other innovations (Bokern, 2008).

The project laid the groundwork for the idea of a temporary pavilion to take on the role of temporary disaster relief using waste materials. The UNITED BOTTLE project by Switzerland based UNITED BOTTLE GROUP, addresses one of the most highly produced consumer goods in the world: the plastic bottle. With over 480 billion plastic bottles sold around the world per year, the plastic bottle is an easily accessible building material in any geographic location (Laville & Taylor, 2017). With small but effective changes to the form of these Polyethylene Terephthalate (PET) plastic objects, the project provides the opportunity for a second life-cycle as building units, without the use of adhesives. Through the bottle's inherent qualities of a vessel for containment, the project envisions these units filled with an array of different elements (local sand, earth, hair wool, paper, and textiles) all of which provide different architectural properties. Using these locally sourced or produced substances, UNITED BOTTLE can achieve all the architectural necessities of thermal, acoustical, structural, and even aesthetic characteristics found in conventional buildings (Hebel, Heisel & Wisniewska, 2014).

However, the use of plastic bottles is far from a commercial application. Acknowledging the need for more applicable systems, some of the UNITED BOTTLE architects, together with Werner Sobek, decided to test the idea of a fully recyclable building.

Built at the NEST research building on the campus of the Swiss Federal Laboratories for Materials Science and Technology (Empa), the *Urban Mining & Recycling (UMAR)* unit, created by Werner Sobek with Dirk E. Hebel and Felix Heisel, uses only resources that are reusable, recyclable, or compostable. According to Heisel (2018):

This places life-cycle thinking at the forefront of the design: instead of merely using and subsequently disposing of resources, they are borrowed from their technical and biological cycles for a certain amount of time before being put back into circulation once again. Such an approach makes reusing and repurposing materials just as important as recycling and upcycling them (both at a systemic and a molecular/biological level, e.g. via melting or composting. (p.4)

The project uses untreated wood, repurposed copper sheets, mycelium boards, innovative recycled bricks, repurposed insulation materials, leased floor coverings, and a range of other products that can be separated and sorted before being recirculated.

While these material choices are important, more crucial still are the connections between materials. All details were redesigned to eliminate glues and seals, and instead replace them with folds and screws.

A New Architecture

Change is inevitable. Returning to our original material shortages, such looming deadlines for raw material depletion, combined with rising consumption, means that a paradigm shift in domestic behavior, manufacturing, construction, and design cannot be avoided. It will be necessary, in the very near future, to reengage not only with our waste, but with the design and manufacturing of all products that will someday become waste.

As an industry that used 2.45 billion metric tons of material in 2008, one might conclude that a paradigm shift in the discipline, and in the industry, is overdue (Matos, 2012). However, there is an additional imperative here. As part of an ongoing debate between architecture and design, it has been noted that architecture is (at least in part, if not in all) an art, which goes beyond both the function and trendiness that design embraces (Eisenman, 2009). Architecture, as an art, has the ability to communicate. In the urgency of material finitude and resource management, such ideas can be diminished. However, we, as architects are uniquely placed to harness this ability, and to affect, though the legibility in our work, the behaviors and the policies that shape our futures.

REFERENCES

Anatsui, E. (2014). *Ascension.* Retrieved from http://el-anatsui.com/artworks/

Andersen Windows and Doors. (2012). *A-Series Double-Hung Windows (2008-Present).* Retrieved from https://www.andersenwindows.com/-/media/aw/files/technical-docs/parts-catalog/partscatalog-windowsincludingstormwatch-2008-present--a-series-hung.pdf

Baker-Brown, D. (2017). *The Re-Use Atlas.* London, UK: RIBA Publishing.

Bokern, A. (2008). Message in a Bottle. In A. Bahamon & M. C. Sanjines (Eds.), *Remateiral: from waste to architecture* (pp. 24–27). New York, NY: W.W. Norton & Company.

Bossink, B. A. G., & Brouwers, H. J. H. (1996). Construction waste: Quantification and source evaluation. *Journal of Construction Engineering and Management, 122*(1), 55–60. doi:10.1061/(ASCE)0733-9364(1996)122:1(55)

Bureau of Planning and Sustainability. (2016). *City of Portland 2016 Recycling Program Summary*. Portland, OR: Bureau of Planning and Sustainability.

Chan, K. (2012 May 8). *Pop-Up Populism: How the Temporary Architecture Craze is Changing Our Relationship to the Built Environment.* Retrieved from http://www.blouinartinfo.com/news/story/802841/pop-up-populism-how-the-temporary-architecture-craze-is

Citimetric Staff. (2015 April 15). *9 building materials made entirely from waste products. CitiMetric*. Retrieved from https://www.citymetric.com/skylines/9-building-materials-made-entirely-waste-products-932

Cragg, T. (2011). *Dice Sculptures*. FAIC 2011. Retrieved from http://www.thisiscolossal.com/2011/10/dice-sculptures-by-tony-cragg/

Croaker, T. (2013 March 30). Build It and They Will Go. *Sydney Morning Herald*. Retrieved from https://www.smh.com.au/entertainment/art-and-design/build-it-and-they-will-go-20130327-2gsvc.html

Donovan, T. (2018). *Hyperobjects*. Retrieved from https://www.pacegallery.com/news/3019/tara-donovan-in-hyperobjects-at-ballroom-marfa

Eisenman, P. (2009). RE: RE: Architecture or Design: Wither the Discipline? In C. O'Donnell (Ed.), *Journal of Architecture, Issue 8* (pp. 178–181). Ithaca, NY: AAP Publications/Actar.

Frondel, M., Grosche, P., Huchtemann, D., Oberheitmann, A., & Peters, J. (2007). *Trends der Angebots- und Nachfragesituation bei mineralischen Rohstoffen: Endbericht. Forschungsprojekt Nr. 09/05 des Bundesministeriums für Wirtschaft und Technologie (BMWi)*. Retrieved from https://www.econstor.eu/bitstream/10419/70880/1/737674393.pdf

Gadano, P. (2016). Architecture Activating Awareness. In C. O'Donnell & S. Chodoriwsky (Eds.), *This Is Not A Wall* (p. 256). Ithaca, NY: Cornell AAP Publications.

Gibson, J. J. (1986). *The Ecological Approach to Visual Perception*. Lawrence Erlbaum Assoc. Inc. Publishers.

Grossman, J., & McCourt, T. (Producers), & Grossman, J. (Director). (2012). Drop City [Documentary]. United States: Pinball Films.

Hebel, D. E., Heisel, F., & Wisniewska, M. H. (2014). *Building from Waste: Recovered Materials in Architecture and Construction*. Basel, Switzerland: Birkhauser. doi:10.1515/9783038213758

Heisel, F. (2018). Economics for a Circular Anthropogenic Environment. In C. O'Donnell & D. Pranger (Eds.), Cyclo: Architecture of Waste. Academic Press. (forthcoming)

HGTV. (2017). *The Benefits of Recycled Demin Insulation*. Retrieved from https://www.hgtv.com/remodel/mechanical-systems/the-benefits-of-recycled-denim-insulation

Humes, E. (2012 April 13). Garbage: A Costly American Addiction. *Forbes*. Retrieved from https://www.forbes.com/sites/kerryadolan/2012/04/13/garbage-a-costly-american-addiction/#3ae122a65b1c

Laville, S., & Taylor, M. (2017 June 28). A million bottles a minute: world's plastic binge 'as dangerous as climate change'. *The Guardian*. Retrieved from https://www.theguardian.com/environment/2017/jun/28/a-million-a-minute-worlds-plastic-bottle-binge-as-dangerous-as-climate-change

Mairs, J. (2016 June 6). Alejandro Aravena uses over 90 tonnes of recycled waste for entrance rooms of Venice Biennale 2016. *Dezeen*. Retrieved from https://www.dezeen.com/2016/06/02/venice-architecture-biennale-2016-recycled-waste-exhibition-entrances-alejandro-aravena

Matos, G. (2012). *Use of Raw Materials in the United States from 1900 Through 2010 (USGS)*. Washington, DC: United States Department of the Interior.

McDonough, W., & Braungart, M. (2002). *Cradle to Cradle*. New York, NY: North Point Press.

National League of Cities and United States Conference of Mayors, Solid Waste Management Task Force. (1973). *Cities and the Nations Disposal Crisis*. Washington, DC: U.S. Government Printing Office.

Nike. (2018). *Nike Grind*. Retrieved from https://www.nikegrind.com/about/

Philips. (2018). *Circular Lighting at Schiphol airport*. Retrieved from http://www.lighting.philips.com/main/cases/cases/airports/schiphol-airport

Reynolds, M. (1990). *Earthship: How to Build Your Own* (Vol. 1). Solar Survival Architecture.

Shutters, C., & Tufts, R. (2016, May 27). *LEED by the numbers: 16 years of steady growth*. Retrieved from https://www.usgbc.org/articles/leed-numbers-16-years-steady-growth

Simmons, A. (2016, April 22). The world's trash crisis, and why many Americans are oblivious. *Los Angeles Times*. Retrieved from http://www.latimes.com/world/global-development/ la-fg-global-trash-20160422-20160421-snap-htmlstory.html

Solid Waste Disposal Act of 1965, 42 U.S.C. §§ 6901-6908a (2012).

Solid waste: diversion, AB-341, 112th Cong. (2011).

Stahel, W. (2016). *The Circular Economy Package*. Retrieved from https://www.eesc.europa.eu/resources/docs/2016_01_26_presentation-stahel.pdf

Stahel, W., & Reday-Mulvey, G. (1981). *JOBS FOR TOMORROW: The Potential For Substituting Manpower For Energy*. New York, NY: Vantage Press, Inc.

T.E.A.M. (2017). *Clastic Order*. Retrieved from http://tpluseplusaplusm.us/clasticorder.html

The 7th Environment Action Programme (EAP), 1383/2013/EU. (2013).

The Ragdale Foundation. (2018, March 8) *Call for proposals: The Ragdale Ring*. Retrieved from http://ragdale.org/wp/wp-content/uploads/2017/03/2018-Ragdale-Ring-RFP.pdf

United States Department of Labor, Bureau of Labor Statistics. (2006). *100 Years of U.S. Consumer Spending: Data for the Nation, New York City, and Boston, BLS Report 991*. Washington, DC: U.S. Government Printing Office.

United States Environmental Protection Agency. (2016). *Advancing Sustainable Materials Management: 2014 Fact Sheet: Assessing Trends in Materials Generation, Recycling, Composting, Combustion with Energy Recovery and Landfilling in the United States*. Washington, DC: U.S. Government Printing Office.

United States Environmental Protection Agency. (2017). *EPA report: Sustainable Management of Construction and Demolition Materials*. Washington, DC: U.S. Government Printing Office.

United States Green Building Council. (2009). *Construction and demolition waste management*. Retrieved from https://www.usgbc.org/credits/new-construction-core-and-shell-schools-new-construction-retail-new-construction-healthca-38

United States Green Building Council. (2016). *Recycled Content*. Retrieved from https://www.usgbc.org/credits/commercial-interiors-retail-ci/v2009/mrc4

United States Green Building Council. (2018). *About USGBC.* Retrieved from https://new.usgbc.org/about

Warerkar, T. (2018, January 16). Governors Island's 2018 City of Dreams Pavilion finalists announced. *Curbed NY.* Retrieved from https://ny.curbed.com/2018/1/16/16897862/

Williams, A. (2016, September 1). Multi-use community center makes the most of recycled materials. *New Atlas*. Retrieved from https://newatlas.com/re-ainbow-hp-architects/39175/

Yashiro, T., Kato, H., Yoshida, T., & Komatsu, Y. (Eds.). (1990). *Proceedings from CIB 90 W55/65 Symposium: Survey on real life span of office buildings in Japan.* Academic Press.

ENDNOTES

1 2013 MoMA PS1 brief, as disseminated to the five competitors. Courtesy of CODA, 2012.

2 Materials were developed and produced by e2e Materials, Ithaca.

3 OMG, Primitive Hut, at Omi International Arts Center, Ghent, New York.

4 Rising Canes by Penda ICD Aggregate Pavilion 2015 by researchers and students from the University of Stuttgart.

5 This was the goal of the authors' 2017 Strauch Design Studio "Cyclo: Architectures of Waste" taught at Cornell University, Department of Architecture, with Dirk Hebel and Felix Heisel.

Chapter 6
Comprehensive Evaluation for Mortars and Concretes Incorporating Wastes

Alberto Marcelo Guzmán
Universidad Tecnológica Nacional, Argentina

Noemí Graciela Maldonado
Universidad Tecnológica Nacional, Argentina

Graciela Affranchino
Universidad Tecnológica Nacional, Argentina

ABSTRACT

Sustainability is concerned with the most efficient use of resources where the residues play an essential role. Trends in concrete technology include natural or artificial additions and additives in order to reduce the consumption of cement. The characterization of the wastes is of great importance with respect to the amount that must be incorporated into the matrices of construction materials both for its economic and engineering impacts (strength and durability). The authors study the impact in strength, durability, and sustainability of the use of finely ground waste of ferroalloys in concrete. The behavior of durability of sustainable concrete also is evaluated. The proportioning between traditional materials and these additions involves preliminary tests on pastes and mortars. Also, they study the impact of the use of different plastic wastes (polyethylene) in different percentages. They evaluated consistency, compressive strength, suction capability, and leaching.

DOI: 10.4018/978-1-5225-6995-4.ch006

INTRODUCTION

Modern society has developed based on the use of concrete in its different forms and applications, becoming the most consumed material after water. Concrete is the most used construction material in the world, a composite material consisting mainly of aggregates immersed in a cement matrix. Every year the global concrete industry uses 1.6 trillion tons of cement, 10 trillion tons of rock and sand and 1 trillion tons of water. Demand is expected to reach 18 trillion tons per year, starting in 2050.

These statistics indicate the impact that the manufacturing of concrete produces in nature. For this reason, achieving a sustainable concrete requires to moderate the degradation of the soil, energy consumption and pollution of the environment. Some of the measures that contribute to the concept of sustainable concrete are: to rationally dose the constituent materials of a concrete, to use added cements, to use recycled aggregates, to take advantage of the reuse of water, and to use alternative fuels in the manufacture of the constitutive materials.

The popularity of concrete is due to the many advantages offered by this material. It can be durable and of high resistance with a correct combination of cementitious materials and additions, additives, aggregates and water. Generally it can be elaborated with the raw materials available in each place. It also has an assortment of applications among which can be mention: a high reflectance value helps to reduce the heat island effect in built-up areas, it can be used without terminations and with the correct mixture it is resistant to the weather; it can be porous to help absorb infiltration water, it can be incorporate waste and recycled materials into the mix, reducing the consumption of raw materials (Chun, Claisse, Naik, & Ganjian, 2007).

The construction, use, repair, maintenance and demolition activities consume resources and generate waste. According to the report of the United Nations Environment Program (2014) and the Sustainable Building and Construction Initiative (SBCI), the construction industry consumes between 25% and 40% of global energy consumption, it generates between 30% and 40% of solid waste and 20% of liquid waste and emit between 30% and 40% of greenhouse gases. An alternative to help environmental sustainability is the reduction of the use of materials to make concrete, especially cement. The supplementary cementitious materials (SCM) allow to replace a part of the cement in a mixture for concrete.

Many of these materials come from industrial waste (Maldonado & Helene, 2007). The questions that are asked when a waste is incorporated into a material of great use such as concrete are: can it be incorporated into concrete? Can it be incorporated into reinforced concrete? How is its relationship with the environment regarding the reduction of solid waste? How much could be incorporated in the proportioning of the mixtures? How would the regulations apply in these cases? These questions are answered when there are results of research and laboratory and field work.

In this chapter two cases are studied: the incorporation of residues of siderurgical origin as a replacement for a certain part of the cement or fine aggregate, both in mortars and concretes, and the incorporation of plastic wastes in mortars as a replacement for a fraction of the aggregate. The incorporation of non-standard additions is evaluated by laboratory tests, where the use benefits for the desired characteristics of the concrete is demonstrated. The compatibility between cements and both admixtures is tested in pastes and the resistance and durability are evaluated in concretes. The experimental results are evaluated and they are oriented on the applications in relation to the current regulations on sustainable construction for dry semiarid climate.

BACKGROUND

The term "sustainability" is a noun that indicates "quality of sustainability"; often it is commonly used as sustainability, free translation of the English term "sustainability", since it is in the Anglo-Saxon countries where the term "sustainable" was used for the first time as "sustainable development"(Mimbacas, 2013).

Kathib (2009) points out that there are many books published on construction materials but very few are related to environmental and sustainability issues. Regarding the subject in concretes, it can be observed that the problem of sustainability has to do with the cementing materials and the environmental issue with the aggregates mainly.

The aggregates make up the part of the concrete due to its volumetric presence, which is why it is considered a great step to reduce its quantity or replace it with waste of compatible characteristics. An example of this are governmental actions to protect the environment, such as in Modena, Italy (Langer, Giusti, & Barelli, 2003) for a comprehensive plan of extractive activities or the quarry management plans required in Argentina (Clariá, Irassar, López, & Bonavetti, 2012).

The issue of global warming due to the industry's carbon dioxide emissions has forced the cement industry to decrease the manufacture of cement clinker and the incorporation of supplementary cementitious materials (SCM) originating in the treatment of industrial waste, both in the conformation of blended cements or compounds as in their use as fine aggregate.

In 1983 the commission chaired by the Norwegian Gro Harlem Brundtland reached a consensus document in which the controversial conflict that had occurred until then between development and environment is reconciled. In the so-called Brundtland Report, "sustainable development" is defined as "one that meets the needs of the present without compromising the possibility of future generations to meet their own needs". This is how it tries to reconcile developmentalism and environmentalism positions, understanding environmental conservation as a requirement for economic progress. It not only considers environmental aspects, but considers them as essential requirements for development.

The problem of the environment with respect to concrete has been incorporated since the beginning of the 21st century in international regulations, as a result of the problems of pathologies generated in concrete due to the aggressive conditions of the environment (presence of chlorides and sulfates, in particular). The European norms establish the environmental categories (Rostam, 1999) that with the local nuances have been incorporated into the regulations on reinforced concrete, as is the case of the Argentine Concrete Regulation, CIRSOC 201 (2005).

A major effort to include the issue of the sustainability of construction materials, especially the concrete that is the material manufactured in greater quantity has been the formation of the Center for Green Concrete in Denmark. In the period 1998-2002 they have researched to create new knowledge about sustainable concretes through the development of technological solutions available with new concretes with industrial waste products. They have also developed a guide to new materials in the production of concrete (Glavind, 2009).

The three components of sustainability are: the environment, the society and the economy. It could be extended to a fourth component as energy because of its growing importance worldwide (Kevern, 2010). These components must be balanced among themselves to achieve sustainable development. If the environment is combined with the economy, the theme is viable, if the environment is combined with society, the issue must be tolerable and if society is combined with the economy, the issue must be equitable. The balanced balance of the three components should result in sustainability (Figure 1).

Each of the components of sustainability has a number of variables involved that make its measurement complex. The cost and the initial environmental / social benefits of any product or system do not provide a real picture of all interactions since they should be considered from "the cradle to the grave" and more appropriately from "the cradle to the cradle".

In the case of a construction with a service life of 60 years, 90% of the environmental impact occurs during the use phase of the building, while the remaining 10% appears as built-in energy used for its construction, of which only between 2% and 3% correspond to construction materials (Figure 2).

When analyzing the feasibility of a construction, the following must be taken into account (Brunatti, 2014):

- Location of the work with a prior evaluation of the environmental impact it would produce
- Estimated energy consumption

Figure 1. The 3 principles of sustainability

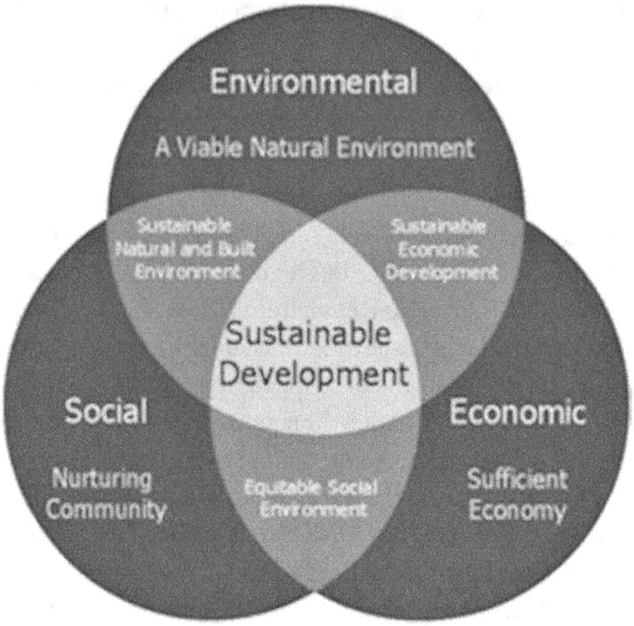

- CConstruction system
- Use of less polluting materials
- Construction waste
- Recycling of materials

To meet the objectives of sustainable development, the construction industry relies on three important bases (Brunatti, 2014):

Figure 2. Environmental impact of the building. Source: AFCP, 2010

- The durability of structures
- The recycling of materials
- The use of construction waste

But these objectives cannot be considered independently, but they must ensure a holistic approach integrating knowledge of all processes involved.

In relation to mitigation measures for these gases that cause alterations in the thermal equilibrium of the planet, the Intergovernmental Panel on Climate Change (IPCC, 2007) indicates that it is the building sector that has the greatest economic potential for mitigation by 2030 in comparison to other sectors such as transport, agriculture and forestry.

When is a concrete construction successful in terms of sustainable development? Among the different possible answers is the reduction of the use of resources since the concrete is manufactured using postindustrialized byproducts, where waste materials can be used and the same components of the concrete can be obtained by reusing it. Table 1 shows the differences between a traditional concrete and a sustainable concrete.

The incorporation of these supplementary cement materials (SCM) has updated the study of the methods of dosage of structural concretes due to the use of a third generation of additives known as superplasticizers, with retarders or with setting control that have made the use of mineral additions of different origin.

Also, the concepts of durability, service life and sustainable environment have introduced a new generation of codes and they are impacting concrete technology and environmental sustainability when it comes to industrial waste (Levy & Helene, 2004). A fundamental contribution of the new national regulations (CIRSOC 201, 2005) requires a concrete design for a service life of 50-year, therefore the environment is incorporated as a further action on the structure, which must be identified to establish the criteria of protection of the structure. Exposure conditions are classified as general and/or specific. The general conditions correspond to those that produce degradation in the structures by processes of corrosion and the specific ones correspond to actions of freezing and thawing, attack by substances contained in soils and waters of contact.

With respect to the incorporation of non-standard additions, as in this case, it must be demonstrated by laboratory tests that use the characteristics of the concrete and that does not produce unfavorable reactions, it does not alter the protection of the reinforcements and does not affect the volumetric stability of the hardened concrete.

Currently, the intensive use of plastics has generated a waste problem due to its lack of environmental degradability. The management of solid waste is one of the main environmental problems due to the scarcity of spaces for landfilling due to saturation and cost.

Table 1. Differences between traditional concrete and sustainable concrete. Source: Brunatti, 2014

Material	kg CO_2 / t of concrete	Standarized Concrete		Sustainable Concrete	
		kg/m3 concrete	kg CO_2 / t of concrete	kg/m3 concrete	kg CO_2 / t of concrete
Cement	1000	320	320	150	150
Granulate blast-furnace slag	630		0	90	57
Fly ash	0		0	70	0
Silica fume	0		0	10	0
Natural Coarse Agreggate	135	1100	149	770	104
Air-cooled slag	80		0	330	25
Natural Fine Sand	63	800	50	560	35
Sand of slag	80		0	240	19
Admixtures	0.21	2.5	0.0005	2.5	0.0005
Water	0	180	0	0	0
Recycled Water	0		0	180	0
Total		2402.5	519	2402.5	392

Concrete is a construction material that allows the incorporation of waste in its manufacture, in the case of plastics, a selective recycling of them allow to obtain construction materials with other properties, minimizing the volume of waste collection and risks of consequential pollution. Miller (2007) has proposed using plastic waste as an aggregate of concrete, improving its thermal insulation properties. Zainab et al (2008) of the University of Baghdad, Iraq participated in a study to determine the efficiency of the use of plastic waste in the production of concrete. Marzouk et al (2007) studied the use of PET plastic bottles as a substitution for sand, showing that up to 50% substitution of sand by the addition of granulated plastic does not affect the strength of the mixture.

Sanchez et al (2012) obtained a 43% energy saving when using plastic with respect to a conventional concrete wall in Argentina.

CASE 1: USE OF WASTE OF FERROALLOYS IN PASTES, MORTARS AND CONCRETES

Characterization of Materials

Portland-pozzolan cement (CPP40) is used and its properties comply with the Argentine national standard (IRAM 50001, 2017).

The industrial wastes originate in silica and calcium furnaces and of silica and iron, which generate calcium silicide (SC) and ferrosilicon (FS) as raw material respectively. Production wastes are kept in outdoor stacks without a certain destination of use. The chemical compositions do not qualify the material either as granulated blast-furnace slag according to national standard (IRAM 1667, 2016) or as silica fume as stated by ASTM standard C1240. The physical properties and chemical analysis of the cement and milled local industry residues are listed in Table 2. The diffraction test is used to identify different phases of calcium silicide but it is not possible to identify the phases due to their amorphous state in the ferrosilicon.

The coarse aggregates (AG) were extracted from alluvium bed of mountain rivers near the Andes Mountains in the West of Argentina, They are predominant by rounded gravel from basaltic and granite rock. The coarse aggregates have a maximum size of 19.5 mm and a fineness modulus of 6.16.

The specific gravity, water absorption, and fineness modulus of washed rounded river sand (F) (passing through 5 mm sieve opening size) are 2.57, 0.81 and 3.1, respectively.

A commonly used plasticizer (P) of lignosulfonic base (29% solid contents) and a high range water reducing superplasticizer (S) of melamine polymer (36% solid contents) were used with different combinations (Table 4).

Studies of Pastes

SCM replacements: 2, 5, 7 and 10% by weight of cement and at three water / cementitius materials ratios (w/cm) = 0.40; 0.45 and 0.50 were evaluated in cement pastes. The fluidity behavior of the pastes were studied by the Kantro mini-slump test method (Kantro, 1980). Figure 3 presents the results of the extended surface for the different w/cm ratios and different levels of replacement of the two residues. For the w/cm = 0.4 mixtures with a replacement of 10% FS, the extended surface test could not be performed due to the lack of workability of the paste.

Furthermore the percentage of additives used is evaluated by the Kantro mini-cone test, looking for the most economical combination between plasticizer (P) and superplasticizer (S) under laboratory conditions for three levels of cement replacement (2, 5 and 7% by weight of the cement) and for w/cm = 0.40. The dosages used are:

Table 2. Physical and chemical compositions of cement and milled local industry residues

Item	CPP40	SC	FS
Specific gravity	2.95	2.38	2.23
Coefficient Pozzolanic	0.68	---	---
Loss of ignition, %	1.53	6.01	4.25
SiO_2, %	28.4	56.2	73.38
Al_2O_3, %	4.23	0.71	2.44
CaO, %	54.6	21.1	6.83
MgO, %	2.2	5.9	5.9
SO_3, %	3.59	0.80	0.48
Na_2O, %	0.48	1.16	4.17
K_2O, %	0.59	3.32	3.41
MnO, %	1.04	5.86	1.01
Fe_2O_3, %	4.90	0.33	3.83

Figure 3. Extended surface results tests for different replacements of SCM and w/ cm ratio

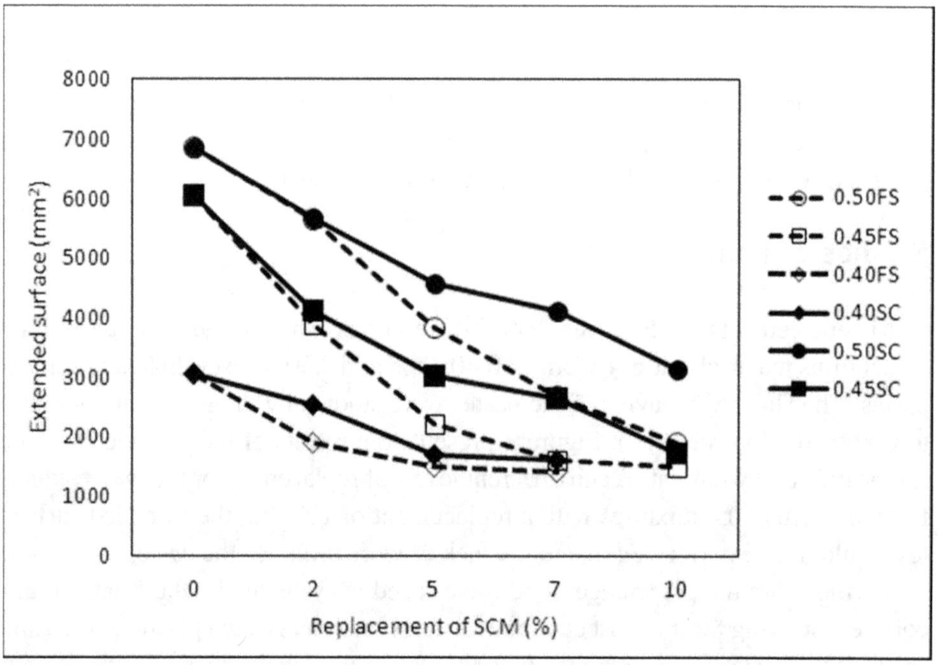

0.35% for plasticizer (A1), 0.35% plasticizer + 0.60% superplasticizer (A2) and 0.35% plasticizer + 1.0% superplasticizer (A3) to the weight of the cement. Figure 4 presents the results of the extended paste surface when using different combinations of additives.

Results of pastes studies: It is observed in Figure 3 that for the three w/cm considered (0.40, 0.45 and 0.50), the use of FS and SC lead to a loss of fluidity of the paste, resulting in increased as the percentage of SCM increases. In all cases, the calcium silicide (SC) gives rise to a paste which has less loss of flow than when ferrosilicon (FS) is added. In the case of the SC and for the three w/cm ratios, it is observed that the lines represented in Figure 4 have approximately the same slope. In the case of the ferrous silicon (FS), this situation is given for w/cm ratios of 0.45 and 0.50. In both residues, the saturation level in the paste is at the 20% of SCM, therefore, it is not manageable in the fresh state.

When additives are added to the paste in Figure 4 it is observed that the use of the SCM lead to a loss of the flowability of the paste for all the doses used, resulting in a greater loss as the percentage of incorporated SCM increases. When using

Figure 4. Results of extended surface tests on pastes for different combination of additives and replacements of SCM for ratio w/cm=0.40

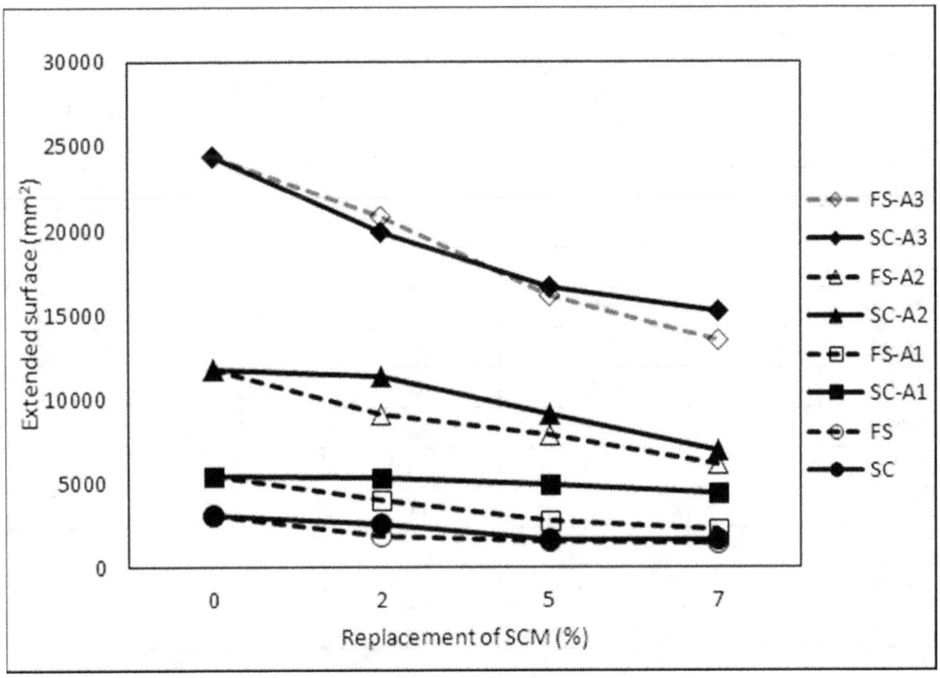

plasticizer (A1: 0.35% of plasticizer) the loss of fluidity with calcium silicide is approximately of the same order for each of the percentages of replacements used, being more accentuated when the ferrosilicon is used. This same situation is detected when doses (A2) are used. The loss of fluidity is more significant with increasing doses of superplasticizer (A3) but with minimal difference in the behavior of the two residues.

Morters Studies

The mortars are made with a w/cm = 0.50 and they are evaluated with different levels of replacement of Portland cement (C) and fine aggregate (F): 5, 10 and 20%.

The consistency of the mortar is evaluated by the standard (IRAM 1570, 1994) method for mortars in the fresh state. Table 3 presents the results of the tests of consistency for 70 mm diameter of mould and Figure 5 presents the results of the consistency evaluation for different levels of SCM and fine aggregate.

Table 3 presents the results of mechanical resistance in mortars for the different levels of SCM and Figure 6 and Figure 7 show the results of the compression and flexural tests at 28 days for the different SCM levels studied respectively.

Table 3. Results of tests of consistency and mortar strengths

Mortar	ø withspread (mm)	Consistency (%)	Flexural strength (MPa)	Compressive strength (MPa)
CPP	80.0	14,29	8.2	37.8
SC05C	76.7	9,57	9.0	50.2
SC10C	73.3	4,71	9.6	49.6
SC20C	72.0	2,86	13.8	52.3
SC05F	80,0	14,29	12.0	57.7
SC10F	80.0	14,29	15.2	80.2
SC20F	74.5	6,43	16.0	79.6
FS05C	7.5	7,14	9.42	47.66
FS10C	7.4	5,71	10.35	36.33
FS20C	7.07	1,00	6.55	30.7
FS05F	79.0	12,86	10.8	58.44
FS10F	75.5	7,86	12.5	63.98
FS20F	71.0	1,43	12.1	60.55

Figure 5. Consistency results according to standard IRAM 1570 for different levels of replacement

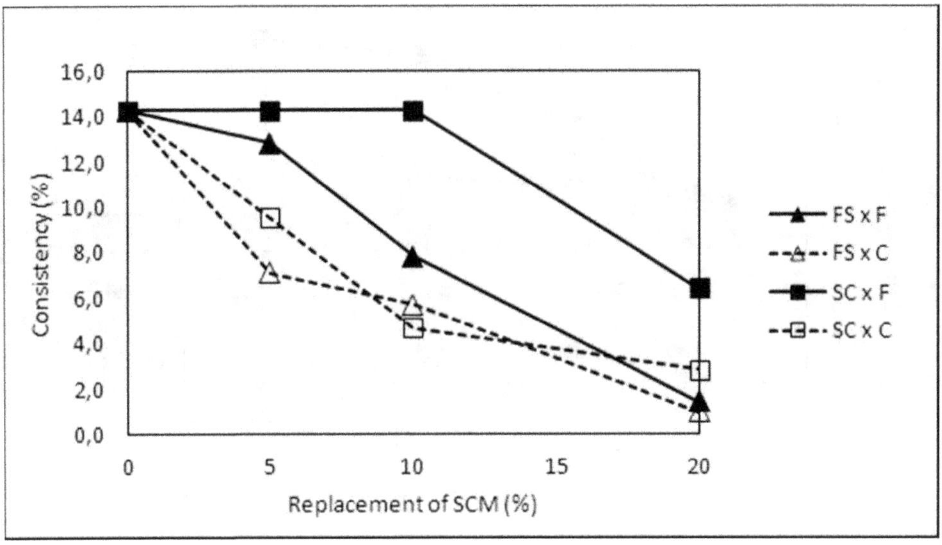

Figure 6. Results of 28-day mortar compressive strength for different replacement levels

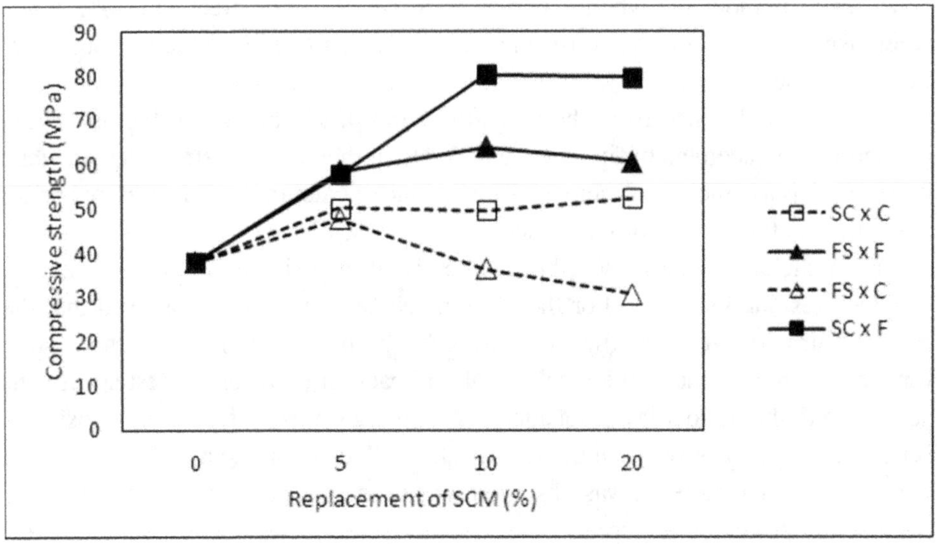

Figure 7. Results of 28-day mortar flexural strength for different replacement levels

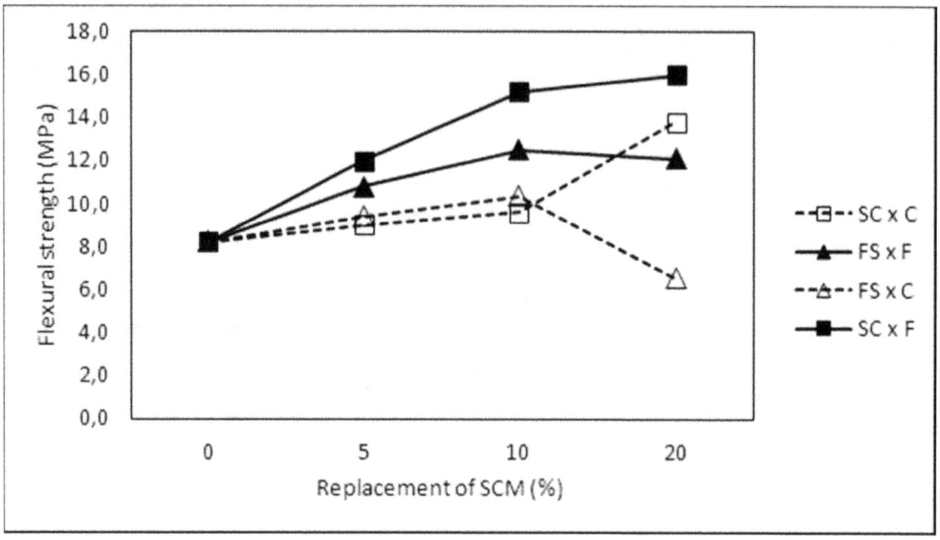

Results of studies on mortars: Regarding the fresh state of mortars, it is observed, that for the use of residue as a replacement of fine aggregate, the consistency decreases as the percentage of SCM increases. In the case of calcium silicide, the consistency is modified when it comes to SCM values greater than 10%, while when using ferrosilicon the consistency decreases uniformly as the percentage of SCM increases.

The results obtained from the strength of mortars indicate that the use of the residue as a replacement of fine aggregate led to higher levels of strength than when it was used as a cement replacement. The residue of calcium silicide presents better strength than the residue of ferrosilicon.

When calcium silicide is used, the compressive strength increases as the percentage of added residue increases. For the 5% replacement level, it is observed that the strength turns out to be of the same order for both uses. It should be noted that for the 20% replacement, the mechanical compaction used for the test specimens according to the standard was not adequate, leaving some voids. When ferrosilicon is used, the compressive strength increases for both uses only to a 10% replacement percentage and a decrease was observed when the 20% percentage is used due to compaction problems (Figure 5).

When calcium silicide is used the flexural strength increases as the percentage of added residue increases. In the replacement level corresponding to 5%, it is observed for both cases of use, that the strength turns out to be of the same order. For 20% replacement, there were gaps that produce less resistant level. When using ferrosilicon, the flexural strength increases for both uses only up to 10% replacement and a decrease was observed when using the percentage of 20% due to problems of compaction (Figure 6).

Concrete Studies

For concretes, 5% by weight of cement replacement and 10% of replacement by weight of the fine aggregate were chosen based on the results of mortars and pastes. Three values of the weight ratio of aggregate / mass dry mass of cementitious material (cm) are used for each concrete family: 3.5 / 1, 5.5 / 1 and 6.5 / 1 to make the nomogram of mixing design (Maldonado & Helene, 2005).

Table 4. Proportions of materials used for concrete and fresh test results

Mixture	CPP (kg)	C	w/cm	F	AG	SC	FS	% addition	Slump (mm)	
									Without additive	With additive
R-3.5	488	1	0.42	1.4	2.1	0	0	0	40	40
R-5.5	338	1	0.61	2.2	3.3	0	0	0	120	120
R-6.5	293	1	0.70	2.6	3.9	0	0	0	190	190
SC05C-3.5	464	0.95	0.42	1.4	2.1	0.05	0	0.34	20	40
SC05C-5.5	321	0.95	0.61	2.2	3.3	0.05	0	0.48	35	115
SC05C-6.5	278	0.95	0.70	2.6	3.9	0.05	0	0.48	75	190
FS05C-3.5	464	0.95	0.42	1.4	2.1	0	0.05	0	40	40
FS05C-5.5	321	0.95	0.61	2.2	3.3	0	0.05	0.21	70	150
FS05C-6.5	278	0.95	0.70	2.6	2.9	0	0.05	0.40	90	180
SC10F-3.5	488	0.86	0.37	1.4	2.1	0.14	0	0.64	1	40
SC10F-5.5	398	0.78	0.50	2.2	3.3	0.22	0	1.45	10	135
SC10F-6.5	293	0.74	0.56	2.6	3.9	0.26	0	1.37	25	170
FS10F-3.5	488	0.86	0.37	1.4	2.1	0	0.14	0.46	1	40
FS10F-5.5	398	0.78	0.50	2.2	3.3	0	0.22	1.25	10	110
FS10F-6.5	293	0.74	0.56	2.6	3.9	0	0.26	1.37	30	190

References: CPP (Pozzolanic Portland Cement), C (cement), w/cm (ratio water/cementing materials), F (fine aggregate), AG (coarse aggregate), SC (calcium siliciure), FS (ferrosilicon), R (reference without SCM)

The tested proportions and the results of the slump cone settlement measured in fresh condition are presented in Table 4. The results of mechanical and durability tests are presented in Table 5.

Results of the tests in concrete: The dosage method chosen allows to show the different aspects evaluated in the behavior of the concrete. Figure 8 corresponds to the nomogram with the tested proportions for pozzolanic portland cement, where the quadrants corresponding to the compressive strength, workability, amount of cementitious material and durability evaluated by capillary suction for the age of 28 days are represented (Maldonado & Helene, 2005, 2007).

Evaluating the workability of fresh concrete, it is observed that the main difference between the mixtures (FS) with the mixtures (R) and (SC) is due to the fact that the ground residue acts as a fine aggregate modifying the workability, as observed in mortars (Table 3) and in concrete (Table 4 and Figure 8).

In Table 5 the value of compressive strength represents the average of three tests at the same age. The statistical adjustment of the obtained results confirms that the compressive strength verifies Abrams' Law for each family of concrete (Figure 8). The effects of the SCM studied on the compressive strength are observed after 28 days, always exceeding the levels reached by the reference concrete and gaining more strength with higher levels of SCM (Table 6).

Figure 8. Nomogram of proportion of 28-day concrete mixtures

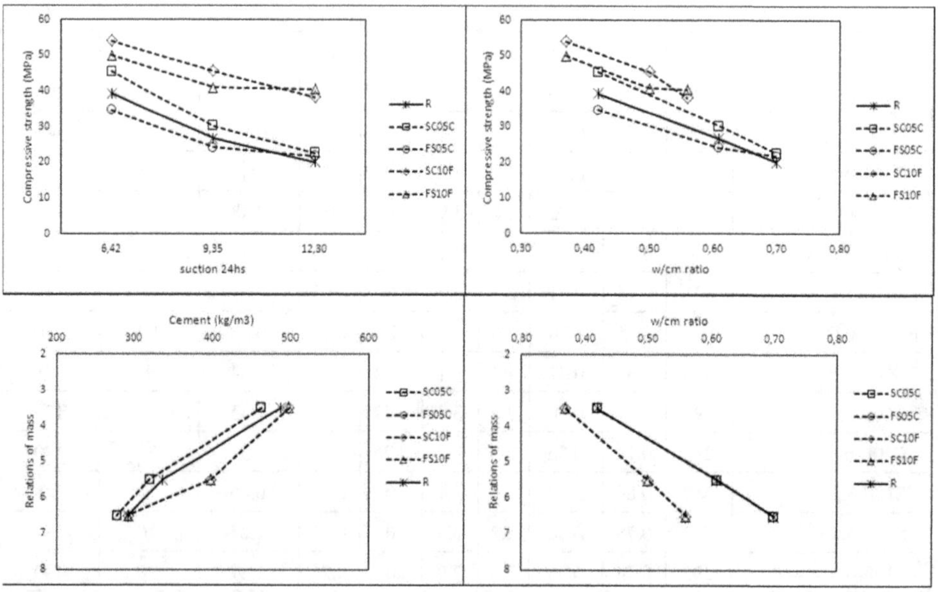

The compressive strength gain in the long term (180 days) is verified when 5% of SC is used as replacement of Portland cement with the three proportions studied, with a higher SC contribution (110 and 113%) than ferrosilicon (112 and 109%) for a 5% cement replacement.

The compressive strength gain is verified when 10% FS is used in replacement of fine aggregate in the poorest proportions (1:6.5) in the long term, while in the richest proportions (1:3.5) the values of maximum strength are for 10% of SC in replacement of fine aggregate.

To identify the effect of the residues, the capillary absorption test according to standard (IRAM 1871, 2004) was performed by the increase in weight of the specimen by capillary water absorption. The obtained 24-hour suction capacity is presented in Table 5.

The capillary suction test indicates that the industrial waste used does not reach the normative limit for the impermeability level lower than 4 $g/m^2 s^{1/2}$ for samples dried at 50ºC for water capillary suction requirements (CIRSOC 201, 2005), but the suction capacity of the mixtures with residue is lower than those of the reference mixtures. Therefore the capillary suction test indicates that the proportion of the materials and the level of replacement of SCM influence the result. For the leaner mixtures (1: 6.5) the capillary suction value at 24-hour is greater for the 5% cement replacement by SC, while for the other replacements, the value is significantly lower than the reference. For the richer mixtures (1:3.5) the capillary suction value at 24-hour is equal or less for all replacements, becoming significantly lower for the FS (Table 4).

Resistance to leaching is the most important parameter for the evaluation of matrix immobilization because water could be the first liquid available for potential ion dispersion. In this case, the proposed leach tank method was used (Sugiyama et al, 2007). Hardened concrete slices are used, extracted from the same samples sectioned for the suction test, and placed in contact with distilled water as a leaching liquid for measuring the pH. In Figure 9, the variation of the pH measurements during the leaching test are shown at 1, 2, 3, 4, 7, 14 and 28 days for the reference mixtures R3.5 and R6.5 with the different combinations of the SCM tested.

The behavior of the concrete with FS and SC residues can be observed with the electrical conductivity of the leachate where the differences can be compared to the reference concrete (R) in Figure 10.

The leachates through-out the testing period with pH values greater than 10 indicate that the interstitial pore fluid in contact with hydrated cementitious materials is buffered by the presence of alkaline ions. So far as the local condition of the leachant remains unchanged, the leaching remains very low.

Table 5. Results of compressive strength and sorptivity of hardened concrete

Mixture	f´c 7	f´c 28	f´c 56	f´c 91	f´c 180	24-hour suction	m^3 concrete cost
	(MPa)					($g/m^2\ s^{1/2}$)	(us$)
R-3.5	31.5	39.1	41.0	42.9	49.9	6.42	74.86
R-5.5	17.3	26.6	29.3	30.4	32.3	9.35	57.12
R-6.5	14.1	20.0	23.3	25.6	25.8	12.30	53.03
SC05C-3.5	33.2	45.2	48.8	52.0	55.0	6.05	70.92
SC05C-5.5	19.1	30.0	35.2	33.9	37.9	10.08	57.81
SC05C-6.5	11.9	22.5	26.6	26.4	31.2	12.83	53.40
FS05C-3.5	28.7	34.5	46.0	49.5	55.5	4.60	69.53
FS05C-5.5	15.0	24.1	27.9	29.3	31.8	7.64	54.27
FS05C-6.5	12.7	21.7	25.0	26.4	30.0	8.20	49.55
SC10F-3.5	35.8	53.7	58.9	58.7	59.7	5.25	77.97
SC10F-5.5	24.8	45.4	48.1	48.9	48.5	7.00	80.59
SC10F-6.5	18.7	38.1	41.4	39.1	43.5	7.66	62.31
FS10F-3.5	35.8	49.7	52.0	50.5	57.7	4.21	75.95
FS10F-5.5	27.3	40.8	42.5	43.7	45.4	6.19	78.75
FS10F-6.5	23.7	40.4	41.4	45.6	48.9	6.45	62.31

Figure 9. Results of pH variation during the leaching test

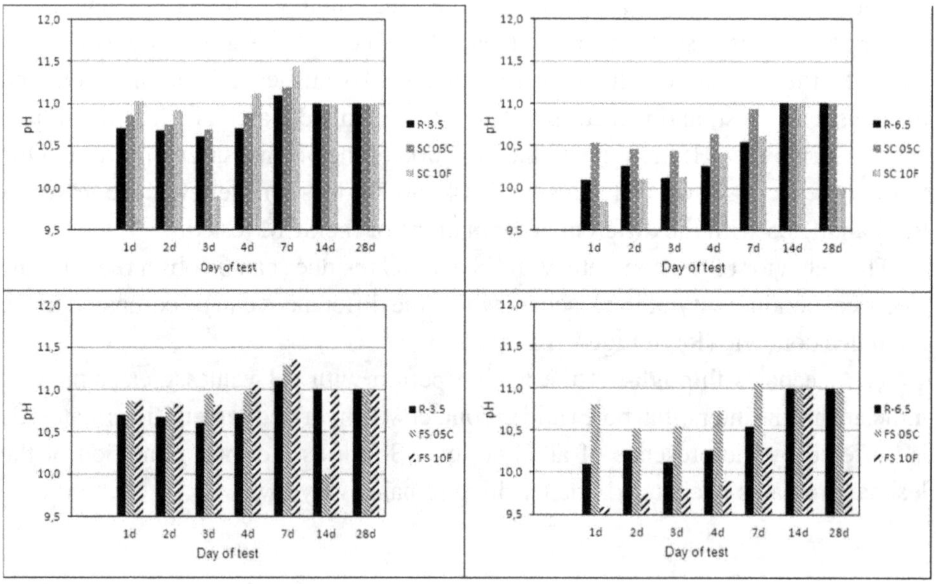

Figure 10. Results of electrical conductivity in leaching test

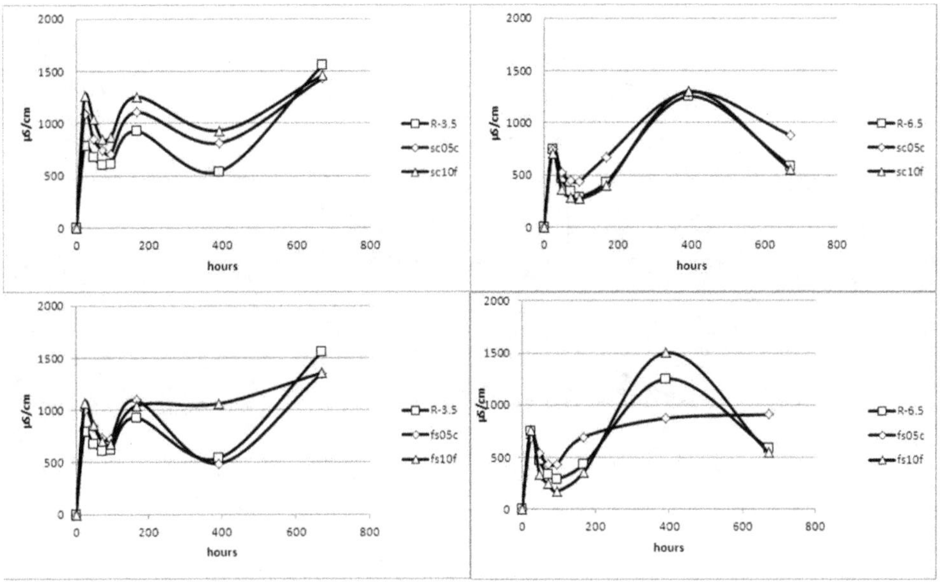

Table 6. Alkali content of the concretes studied according to CAS Standard A 23.2-27 A expressed in kg of Na2Oeq / m3

Mixture	Reference mixtures	Cement replacement	Fine replacement
3.5	4,22	4,77	6,35
5.5	2,92	3,30	5,24
6.5	2,53	2,83	4,90

Figure 11 shows the amount of alkalis released during the leaching process, the amount being higher for the case of ferrosilicon. Therefore, if the probability of alkali-aggregate reaction is studied, given that the local aggregates are potentially reactive, and the one that can be presented under certain environmental conditions for an evaluation of the application of the Canadian standard CSA A 23.2-27A (Milanesi, Pappalardo, & Violini, 2008) would indicate a slight level of reactivity (Table 6) for the poorer mix with cement replacement (limit 3 kg of Na2Oeq / m3), which warrants further study of the problem with local aggregates and the remaining mixtures.

Figure 11. Total content of alkalies released during the leaching test

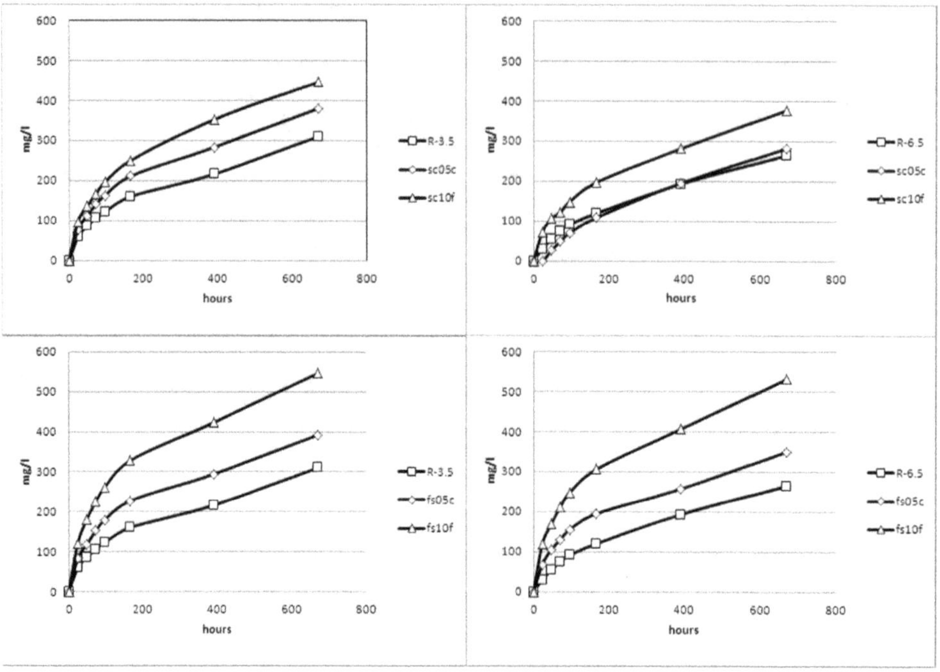

Applications

The results of concrete performance in relation to the cost of a m3 of the studied concrete with respect to 28 day strength and the capillary suction at 24 hours are presented in Figure 12.

Regarding the strength, the need to incorporate additives increases its cost, however, the presence of the SCM improved the capillary suction behaviour, the result being more significant when FS is used.

The studied industrial milled residues modify the fresh concrete properties because they act as fine aggregate. Up to 10% it modifies the workability but not the mechanical properties.

The levels of resistance achieved by the concretes made with the different mixtures in study indicate that these residues allow to reach the maximum levels of resistance for Mendoza, Argentina with rounded natural aggregates and locally produced cements.

The results obtained from the compressive strength of concretes made with different SC and FS replacement levels indicate that both finely ground residues and in the amounts of 5% cement weight replacement and 10% weight replacement of fine aggregate present a comparable action with the pozzolanic action by the gain of resistance to compression in the time.

In the case of the SC, a fast pozzolanic reaction between 7 and 28 days is not detected, but the filling effect improves the levels of permeability. In the case FS its behavior can be associated to the operation of the slag rather than to pozzolan. In all cases it is important to keep the amount of calcium sulphate generated during the production process due to the incompatibilities that can cause between additives and cementing material controlled.

The replacement levels used are not sufficient to obtain a highly impermeable concrete according to the regulatory requirements, but the values obtained of capillary suction capacity indicate that they are smaller than those reached by the standard concrete, so that their incorporation can be considered beneficial for the purpose of improving impermeability.

The incorporation of the residue of the local industry ground has not obtained a better efficiency with respect to the costs of materials, since it requires an increase in the dose of additives with respect to the standard concrete, hence the importance of the evaluation of pastes and mortars to adjust the most appropriate dose to incorporate.

The selected hardened concrete leaching test allows to establish the differences in the release of chemical compounds that can affect the behavior of concrete, and in this case chemical analyzes such as alkali content indicate that additional studies are required to ensure the durability life with the local aggregates when used in foundations or pavements.

Figure 12. Strength and suction efficiency of tested concretes

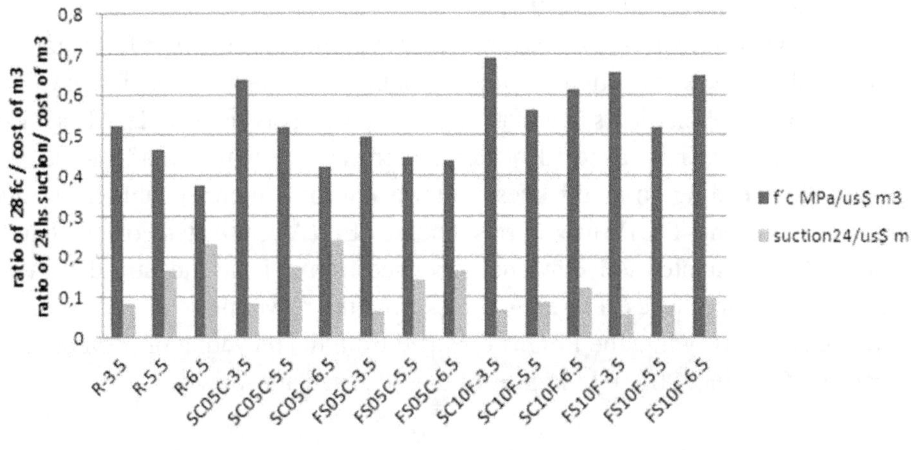

CASE 2: USE OF PLASTIC AND SIDERURGICAL RESIDUES IN MORTARS

Characterization of Materials

A Portland limestone blended cement (CPF-40) is used according to national standard (IRAM 50000, 2017).

As a fine aggregate, naturally fragmented rounded sand of sedimentary origin is used. This type of fine aggregate is locally available in nearby quarries.

As plastic waste have been used three forms of it: PET, PEA and PEG, which are detailed:

PET (Polyethylene Terephthalate) is a strong and lightweight material, used for beverage containers, juices, water, alcoholic beverages, edible oils, among others. Chemically, PET is a polymer obtained by a polycondensation reaction between terephthalic acid and ethyleneglycol. It belongs to the group of polyesters called synthetic materials. PET was developed primarily for the production of synthetic fibers by the British Calico Printers in 1941. The patent rights were then sold to DuPont and ICI, which in turn sold the regional rights to many other companies. Although originally produced for fibers, PET then began to be used as films to manufacture packaging elements. In the mid-1960s and early 1970s, the technique for expanding biaxially oriented bottles was commercially developed. The current bottles are the most significant use of PET resins. The PET presents high resistance to wear and corrosion, as well as good chemical and thermal resistance. It is a material particularly resistant to biodegradation due to its high crystallinity and the aromatic nature of its molecules, which is why it is considered non-biodegradable. It can be produced in the form of fibers, strips and sheets.

In this study, the used PET was of the recycling disused plastic containers, giving rise to what were called flakes (sheets) with a size of the order of 5 mm x 5 mm. Figure 13 shows the recycling of PET.

PE (Low Density Polyethylene) is chemically the simplest polymer. It is obtained from the polymerization of ethylene. It is one of the most common plastics due to its low price and simplicity in its manufacture. The low density polyethylene has good thermal and chemical resistance, adequate impact resistance, whitish, and can become transparent depending on its thickness. It turns out to be more flexible than high density polyethylene. The PE used corresponds to PEA (Agglutinated Polyethylene), and to PEG (Granulated Polyethylene), obtained from a local industry dedicated to the recycling of plastics. The PEA comes from the recycling of plastics used as wrapping elements, while the PEG is obtained from the recycling of polyethylene containers (Gaggino, 2012). In Figure 14 the PEs used are shown.

A byproduct derived from the manufacture of calcium silicide SiCa (SC) was used as a siderurgical waste. The SC is manufactured in electric arc furnaces submerged, then used in the ferroalloy industry. These additions are finely ground, but turn out to be thicker than cement. Figure 14 shows the waste used.

Morters Studies

A reference mortar (R) was prepared, the results of which allowed the comparison with the results obtained from the added mortars. Table 7 indicates the designation used for the mortars evaluated.

The proportioning of the mortars was established in parts with respect to the weight of the cement. In the case of plastic waste and due to its low density, the replacements with respect to a certain part of the sand were considered with respect to the volume of this. Table 8 shows the used proportions.

Figure 13. Flakes of PET, PEA and PEG recycling

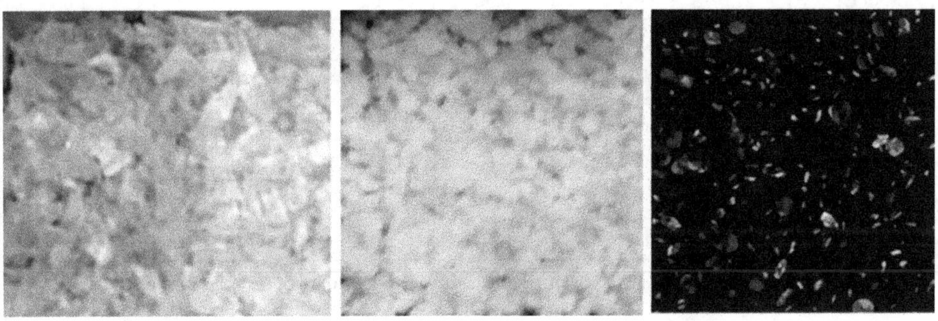

Figure 14. Industrial waste from the manufacture of calcium silicide

Table 7. Designation of mortars

Mortar	Plastic waste	% Replacement	Siderurgical waste	% Replacement
R	-	0	-	0
50PET	PET	50	-	-
50PEA	PEA	50	-	-
50PEG	PEG	50	-	-
10SC	-	0	SC	10
50PET10SC	PET	50	SC	10
50PEA10SC	PEA	50	SC	10
50PEG10SC	PEG	50	SC	10

Table 8. Proportions of mortars

Mortar	Cement [part]	Sand [part]	Water [part]	*Plastic waste [part]	Siderurgical waste [part]	Cost per unit of cement [us$]
R	1.0	3.0	0.60	0.0	0.0	0.3
50PET	1.0	1.5	0.50	0.5	0.0	1.6
50PEA	1.0	1.5	0.45	0.5	0.0	1.6
50PEG	1.0	1.5	0.50	0.5	0.0	1.6
10SC	0.9	3.0	0.60	0.0	0.1	0.3
50PET10SC	0.9	1.5	0.50	0.5	0.1	1.6
50PEA10SC	0.9	1.5	0.45	0.5	0.1	1.6
50PEG10SC	0.9	1.5	0.50	0.5	0.1	1.6

* The replacement of plastic waste has been established in volume with respect to that of the sand

Consistency

The consistency of the mortars was evaluated from the runoff test according to what is established in the standard (IRAM 1570, 1994). Qualitatively in this evaluation, the premise was to obtain mortars with adequate consistencies, such that they would allow their manageability, achieving a good mixing of the materials, an easy placement of the mortar in the molds, and a convenient compaction and completion of the same, without the occurrence of segregation phenomena. Table 9 shows the results obtained from the runoff test, while Figure 15 shows some of the consistencies of the different mortars evaluated.

Table 9. Runoff measured

Mortar	ø [mm]	ø withspread [mm]	Runoff [%]	[%] respect to R
R	70	136	94	100
50PET	70	79	13	14
50PEA	70	87	24	26
50PEG	70	116	66	70
10SC	70	110	58	62
50PET10SC	70	82	18	19
50PEA10SC	70	81	16	17
50PEG10SC	70	97	38	40

Figure 15. Runoff measured: a) Reference b) 50PEA c) 10SC d) 50PET10SC

Unit Weight

Prior to the mechanical strength test, each set of three specimens corresponding to each of the evaluated mortars was weighed and measured. From the average obtained from these measurements, the unit weight of each mortar was quantified. Table 10 shows the results obtained.

Mechanical Strengths

The 28-day mechanical resistance to bending and compression of the obtained mortars was evaluated, in accordance with the provisions of the standard (IRAM 1622, 2006). Table 11 shows the average results obtained for each of the mortars. In Figure 16 presented breakages are shown.

Table 10. Unit weights

Mortar	Unit weight [N/m³]	Relation with R[%]
R	21393	100
50PET	19155	90
50PEA	17302	81
50PEG	18724	88
10SC	20851	97
50PET10SC	18750	88
50PEA10SC	18446	86
50PEG10SC	17115	80

Table 11. Flexural strength and compressive strength

Mortar	Flexural strength [MPa]	[%] respect to R	Compressive strength [MPa]	[%] respect to R
R	6.2	100	20.2	100
50PET	6.5	105	22.2	110
50PEA	4.7	76	14.8	73
50PEG	6.2	100	22.2	110
10SC	6.0	97	22.5	111
50PET10SC	5.7	92	20.8	103
50PEA10SC	5.7	92	17.4	86
50PEG10SC	4.9	79	14.5	72

Figure 16: Flexural breakages a) 50PEG b) 10SC c) 50PET10SC

Applications

The replacements in volume of plastic waste by a certain part of fine aggregate establish that the corresponding 50% replacement is the most convenient, giving rise to an adequate mortar consistency.

The three plastic wastes considered, PET, PEA and PEG, in general give rise to mortars with an adequate behavior. Particularly with the PEG not only are very good results obtained both in fresh and hardened state, but also, it turns out to be the lighter added mortar.

As for the weight replacements of the SC steel residue for a certain part of cement, it could be established that the replacement corresponding to 10% is the most convenient.

Regarding those mortars added with incorporation of plastic and iron and siderurgical waste, it was possible to observe from the comparison of the results with the reference mortar R, that the mortar referred to as 50PET10SC presents an adequate handling in fresh state, with only a slight decrease of the resistance levels.

From the point of view of the economy, the challenge is to lower the cost of transport and energy for the recycling of plastic, since it has an important impact on the final market value but it is a viable alternative from the point of view of sustainability.

These results are promising to contribute to the sustainability of the environment.

FUTURE RESEARCH DIRECTIONS

It is desirable to test specimens at the age of 20 or more years of local exposure, and efforts should be made to find out the actual durability of concrete with industrial residues.

CONCLUSION

Environmental sustainability requires the use of industrial waste having cementitious properties comprehensively evaluating the possible combinations for design and proportioning but in its relationship with the safety, durability and interaction with the environment when used in situ.

To use siderurgical residues, it is necessary to evaluate the chemical composition of the residues and their probable interaction with the environment, requiring diffraction studies to understand the durability behavior in hardened state (case alkali-aggregate reaction).

Regarding the percentages of siderurgical residues, it is necessary to carry out tests in the fresh state to evaluate their behavior, since the fineness of the grind is a condition for the use of additives, which increase the cost of the concrete.

The same applies to plastic waste, which requires testing in the fresh state prior to the adoption of proportions to evaluate mechanical resistance.

Regarding the use of waste originated in plastics, it is necessary to incorporate the evaluation of the available market, since it may not be sufficient for a large-scale application, especially for the costs of transformation energy in aggregate and transport.

ACKNOWLEDGMENT

This research was supported by the Universidad Tecnológica Nacional Facultad Regional Mendoza [grant PICT2010/21].

REFERENCES

Asociación de Fabricantes de Cemento Portland. (Ed.). (2010). *La industria del cemento y la sostenibilidad*. Author.

Brunatti, C. A. (2014). *La industria del cemento Portland y la sostenibilidad. Ciudad Autónoma de Buenos Aires*. Asociación de Fabricantes del Cemento Portland.

Chun, Y.-M., Claisse, P., Naik, T. R., & Ganjian, E. (2007). *Sustainable Construction Materials and Technologies*. London: Taylor & Francis.

Clariá, M., Irassar, E. F., López, R., & Bonavetti, V. (2012). Cements. In N. Maldonado & M. F. Carrasco (Eds.), *Ese material llamado hormigón* (pp. 19–78). Buenos Aires: Asociación Argentina de Tecnología del Hormigón.

Gaggino, R. (2012). Water-resistant panels made from recycled plastics and resin. *Construction & Building Materials*, *35*, 468–482. doi:10.1016/j.conbuildmat.2012.04.125

Glavind, M. (2009). Sustainability of cement, concrete and cement replacement materials in construction. In Sustainability of construction materials. Woodhead Publishing Limited.

INTI. (2005). CIRSOC 201 Reglamento Argentino de Estructuras de Hormigón. Buenos Aires, Argentina: INTI.

IPCC. (2007). Cambio climático 2007: Informe de síntesis. Contribución de los Grupos de trabajo I, II y III al Cuarto Informe de evaluación del Grupo Intergubernamental de Expertos sobre el Cambio Climático. Ginebra, Suiza.

IRAM. (2017). *Argentine Standards Catalogue*. IRAM.

Kantro, D. L. (1980). Influence of water reducing admixture on properties of cement-paste –a miniature test. *Cement, Concrete and Aggregates*, *2*(2), 95–108. doi:10.1520/CCA10190J

Kathib, J. (2009). Sustainability of construction materials. Woodhead Publishing Limited.

Kevern, J. T. (2010, Spring). *How Do We Teach the Next Generation of Engineers Green Building?* Paper presented at the ACI Convention Spring 2010, Chicago, IL.

Langer, W. H., Giusti, C., & Barelli, G. (2003). Sustainable development of natural aggregate, with examples from Modena Province, Italy. *SME Transactions*, *314*, 138–144.

Levy, S., & Helene, P. (2004). Durability of recycled aggregates concrete: A safe way to sustainable development. *Cement and Concrete Research*, *34*(11), 1975–1980. doi:10.1016/j.cemconres.2004.02.009

Maldonado, N. G., & Helene, P. (2005). ¿Hay diferencias entre métodos de dosificar hormigones de alta performance para evaluar durabilidad. In *Proceedings of Structural Concrete and Time, fib Symposium* (*vol. 1,* pp.163-170). La Plata, Argentina: AATH, AAHE & LEMIT.

Maldonado, N. G., & Helene, P. (2007). The importance of mixture proportioning in sustainable construction. In Y-M. Chun, P. Claisse, T.R. Naik, & E. Ganjian (Eds.), Sustainable Construction Materials and Technologies (pp. 223-230). Taylor & Francis.

Marzouk, Y. O., Dheilly, R.M., Queneudec, M. (2007). Valorization of postconsumer waste plastic in cementitious concrete composites. *Waste Management,* (27), 310–318.

Milanesi, C., Pappalardo, M., & Violini, D. (2008). Consideraciones sobre la aplicación del criterio canadiense para inhibir la reacción álcali-sílice. In *Memoria Los áridos como factor de desarrollo* (Vol. 1, pp. 475–481). Cámara Argentina de la Piedra.

Miller, H. (2007). *Tesis de maestría.* Facultad de Ingeniería y Medio Ambiente Construido, Universidad de Wolverhampton, Reino Unido.

Mimbacas, A. (2013). *Construction Sustentability. Boletín técnico Nº 7.* ALCONPAT Internacional.

Rostam, S. (1999). Durability. In Structural concrete, textbook on Behaviour, design and performance update knowledge of the CEB/FIP Model Code 1990. fib Bulletin 3.

Sánchez, S., Oshiro, A., & Positieri, M. (2012). *Cálculo de la huella ecológica del hormigón. V Congreso Internacional.* 19ª Reunión Técnica de la Asociación Argentina de Tecnología del Hormigón, Bahía Blanca, Argentina.

Sugiyama, T., Takahashi, S., Honda, M., & Sakai, E. (2007). Current State of the JSCE Standard on Test Method for Leaching of Trace Elements from Hardened Concrete. *Proceedings of Special Sessions Sustainable Construction Materials and Technologies, 1*, 197–203.

Téllez Martínez, L.A., Villarreal, U., Armenta Menchaca, C., Porsen Oveergard, R., & Bremer Bremer, M.H. (2014). *State of Play of Sustainable Building in Latin America.* UNEP – United Nations Enviroment Programme. Sustainable Buildings and Climate Initiative.

Zainab, Z., Ismail, E., & Hashmi, A. (2008). *Use of waste plastic in concrete mixture as aggregate replacement.* Waste Management.

Chapter 7
Energetic Forms of Matter

Aletheia Ida
University of Arizona, USA

ABSTRACT

One of the challenges that architects and designers are confronted with in contemporary contexts is the need to address an ethical responsibility towards the health of the environment through understanding the energetic processes embedded in materials and their compositions. A scientific explanation of material fundamentals, including chemistry, physical structure, and embodied energy, provides the greatest insight to material property performance values and relative environmental impacts. This information aids architects in making informed decisions about building materials in the design process. This chapter addresses the book topic of reusable and sustainable building materials through the position that all matter is a form of energy, just as living systems are the transmutation of matter and energy. The seven major material groups, which include natural materials, non-technical ceramics, technical ceramics, metals, polymers, foams and elastomers, and composites, are presented with examples and applications discussed.

INTRODUCTION

The materials employed in architecture are selected through a complex decision-making framework, including aspects of aesthetics, function, durability, cost, availability, and embedded cultural norms and industrial standards. The challenge that architects and designers are confronted with in contemporary contexts is the need to address an ethical responsibility towards the health of the environment through the energetic processes embedded in such material choices and compositions. A scientific understanding of material fundamentals, including chemistry, physical

DOI: 10.4018/978-1-5225-6995-4.ch007

structure, and embodied energy, provides the greatest insight to material property performance values and relative environmental impacts, and to making informed decisions about building materials in the design process. For future architects, it is imperative that knowledge of materials and energy flows extend beyond accepted best practices and current building technologies, as the expertise that is needed for making a shift towards reducing our environmental impact requires this enhanced depth in realms not traditionally central to our practice (Brownell, 2010).

This chapter addresses the book topic of reusable and sustainable building materials through the position that all matter is a form of energy, just as living systems are the transmutation of matter and energy. The seven major material groups, which include natural materials, non-technical ceramics, technical ceramics, metals, polymers, foams and elastomers, and composites, are presented and studied in comparative modes for various characteristics. Stereomicroscopy images are shown for each material group, revealing microstructures that express the potential for energy transfer and other performance properties. Embodied energy values are presented for each material group, revealing the potential environmental impact upon construction implementation. In this way, the significance of a correlation between the micro-scale structure of materials with the macro-scale availability of material resource also begins to reveal the natural environmental characteristics for regional appropriateness through an energy metric (i.e. evolution of plant systems, soil hardiness, and response to climatic adversity). Ultimately, the information presented in this chapter contributes to fundamental material considerations in the architecture design process for realizing an ideal *energetic form of matter* in the artifact that manifests.

BACKGROUND

Energy and Matter

The primary constituents of our physical world can essentially be distilled into two distinct aspects – photons that comprise energy and molecules that comprise matter. Without the energy provided in the form of photon particle-waves from the sun or other residual magma, life on earth would be nonexistent. Matter would still exist, but most likely in a lifeless state. The beauty of natural living systems lies at the intersection between matter and energy. We observe the trees and plants that endure seasonal changes evident in their physical appearance. Animals in the wild provide innate cues to environmental response based on habitat choices and built-in physiognomic mechanisms. While it is the differences in energy (either forms of energy or intensities) that instigate the responsiveness of living systems, it is the material compositions of these living entities that allow for such effects to manifest. Furthermore, there tends to be an efficiency in nature between energy expended and material affect. For instance, the branching growth mechanism of trees is based on factors such as nutrients absorbed through root systems by capillary action into new growth that favors certain amounts of natural light. The behavior of animal choices for how they construct their habitats, such as termites, birds, and beavers, provide examples of site specific material compositions (dirt mounds, woven biopolymers, and wood dams). Each conveys unique characteristics for succinct survival needs. Aside from chemical energy expended in the form of calories due to the physical labor endured by each animal, no other equipment or power tool requiring energy is used in the construction process of these habitats. The embodied energy of such habitats could basically be considered the total caloric output required to locate and extract the materials, plus the caloric output for forming the constructions. The materials used also have an energy history as a result of other biogeochemical processes external to the life of the animal's construction (i.e. wood growth, soil sedimentation, etc.). In buildings, the integration of energy as a primary source from the sun intersects with materials around and through the enclosure systems. More than one-third of energy expenditures in industrialized countries is attributed to maintaining interior building environments for human thermal comfort and lighting needs (Davies, 2004), thus the significance of the material compositions and respective fundamental heat transfer functions through the solids, gases, and liquids in the building enclosure is critical.

Chemistry Fundamentals

Basic chemistry is an important premise of knowledge in order to develop an intuitive understanding of different material behaviors and also of energy. Thermodynamics provides the theorems and proofs for how energy behaves in the forms, modes, and transformations that it might undergo. The insights of thermodynamic laws also govern how atoms generally behave and inform the structures and organizational changes that might be incurred by molecules. Chemistry exposes these essential physical laws for the underlying explanations while it also extends into biology for its living applications. Biological systems are composed of atoms and molecules, whose structures are defined by chemistry and whose environmental responses constitute changes in molecular structures. The basic building blocks of matter include atoms, which are composed of a nucleus at the center containing protons and neutrons that is surrounded by layers of electron clouds. Atoms are electrically neutral because their protons (positive charge) and electrons (negative charge) cancel each other out. Atoms can lose or gain electrons, which results in an ion that is an electrically charged atom. If an atom has gained electrons it is an anion and is negatively charged, but if it has lost electrons it is a cation and is positively charged. Chemical bonding between atoms tends to take place readily between anions (likely to lose their electrons) and cations (likely to gain electrons) in the form of ionic bonding. Ionic bonding tends to result in rigid solid structures. Covalent bonding, on the other hand, occurs when atoms bond together and share electrons in their outer layers. Covalent bonds typically result in soft squishy structures, but not exclusively. The bonding of atoms results in molecules, such as the ionic bond that forms salt ($NaCl$) or the covalent bond that forms water (H_2O). A third form of bonding is metallic, which occurs between cation metal atoms. The metal atoms lose their additional electrons from the outer layer creating a sea of electrons surrounding the cations, which is what gives metals the properties of malleability, ductility, and conductivity (Atkins, 2015).

Energy and Entropy

Some of the fundamental laws of energy provide us with insight as to the energy content embedded in different materials based on the energy exchanged to create certain chemical bonds. The First Law of thermodynamics explains that there is a given amount of total energy content within the universe, which can neither be created or destroyed. The Second Law of thermodynamics explains that energy will inherently transform from higher to lower grade states under any natural process, thus increasing the entropy or the disorder of a system and degrading the quality of energy. The valence of elements describes the typical number of covalent bonds that can be formed and infers a certain amount of energy that is released in the forming of such bonds. Energy may also be released when bonds between atoms are broken, such as combustion reactions, but such bond-breaking requires an initial energy input to make this happen. Exothermic reactions are those that release energy in the form of heat, while endothermic reactions are those which absorb energy in the form of heat. Enthalpy describes the total heat content of a system and tends to decrease with exothermic reactions and increase with endothermic reactions. Although enthalpy may increase or decrease in a given system, the entropy will always be increasing and thus the quality of energy or heat moving from higher to lower grade states. Reactions between atoms and molecules can occur when the rate of molecular movements is increased by raising the temperature to encourage collisions between molecules. Chemical reactions can also be enhanced with the introduction of a catalyst that provides for a reduced activation energy in order for atoms to be rearranged and bonds to form. Natural catalysts in human bodies are protein molecules known as enzymes and provide for the continuum of life with constant catalysis. For a period of time during stages of growth, living entities experience an internal decrease in entropy, since the reproduction of cells and the processes of mitosis and meiosis occur towards higher grade energy states and increased order. However, the total entropy between the living system and its environment continues to increase towards greater disorder and lower grade energy states.

MATERIAL GROUPS

The materials that are used in buildings, and for other modern products and technologies, are grouped according to characteristics defining the typical behaviors. These defining characteristics include aspects such as chemistry, strength, density, and thermal conductivity. The resulting material groups include natural materials, non-technical ceramics, technical ceramics, metals, foams and elastomers, polymers, and composites (Ashby, Shercliff, & Cebon, 2007). An introduction to each material group is provided here to convey the primary properties and chemistry while highlighting the thermal performance of each material group and showing select stereomicroscopy examples.

The material examples demonstrated in each section are primarily from recent research projects by the author and students that are inspired by *eotechnic* concepts in building technology. The term *eotechnic* (early technique) is derived from Lewis Mumford, which identifies the priorities in built environment technologies with water and wind forces for energy production through interactions with wood and glass materials (Mumford, 1934). The *eotechnic* mode is considered one of clean environmental processes with heightened human senses and phenomenological effects. The *eotechnic* period in history preceded the formal industrial revolution when metals and alloys became the prevalent technology in addition to coal and fossil-fuel energy processes. The emphasis in material relationship to building compositions depicted throughout is generally focused on the building enclosure system, which mediates between the exterior climate fluctuations and interior programmatic conditions (Yu, 2013). The impact of energy content embedded in materials through sourcing and production processes provides a rough indication to the environmental impact and also is somewhat correlated to material costs. The energy content described for each material group does not include transportation costs to a project site for implementation in buildings, which would be a part of the product life cycle calculation. The subsequent section describes some comparisons and trade-offs between embodied energy, thermal performance, and recyclability of the various building materials.

Natural Materials

The natural materials are primarily biopolymers and wood and tend to be relatively low density and lightweight. The range of density in this material group fall between 100 kg/m^3 - 1,000 kg/m^3, with a tension yield strength ranging between 0.5 MPa – 95 MPa. Biopolymers are comprised of cellulose, hemicelluloses, lignin, and pectin. Cellulose, a polysaccharide made of numerous glucose monomers, comprises the primary structural fiber of wood through formation of microfibrils that have both crystalline and non-crystalline regions (Gibson, 2012). The microfibrils are bound together in fibril aggregates by a matrix of hemicellulose and lignin. Wood and wood products are relatively low in embodied energy and cost when in a fairly pure and untreated state (paper is an exception for instance) (Ramage et al., 2017). The recyclability of wood products is dependent on the purity and quality of the material at its end life, with composites such as medium density fiberboard (MDF) being hardly recyclable (Abott, 2013). Due to the sponge-like covalently bonded structures of natural materials, they have a fairly low range of thermal conductivity, from about 0.1 W/mK – 0.5 W/mK. An example of the typical wood growth pattern of the cross-grain rings is seen in Figure 1a and an enlarged image of the macrofibril bundles is seen in Figure 1b. When wood is processed and manufactured into building products the techniques of sawing and milling affect the surface of the material as seen in Figure 1c and Figure 1d.

Emerging Applications

Wood has been implemented in building construction since the earliest examples of dwellings from around the world. Natural materials that are readily available close to building sites, such as straw bale, have also been used for enclosure systems in small and medium scale buildings for high thermal resistance properties. However, it is not until the 21st century that wood is now being implemented as the primary structure in large-scale buildings. The recent development of cross-laminated timber (CLT) has resulted in a number of multi-story buildings to be constructed with wood framing systems rather than steel or concrete (Dickson & Parker, 2014; Guo et al, 2017). Bamboo construction is also more prevalent in contemporary architecture and is a renewable resource that has a rapid and prolific growth cycle (Chele, 2012). Bamboo is extremely light weight with very high strength in relation to the material density and is therefore popular for both construction scaffolding and building structures. Beyond the building application, broader environmental issues can be addressed through selection of natural materials, such as the use of drought-ridden pinyon pine in Figure 1 to enable enhanced forestry management practices and alleviate forest-fire hazards.

Figure 1. a) Pinyon pine wood along cross-grain at 7x magnification (top left); b) Pinyon pine wood along cross-grain at 32x magnification (top right); c) Pinyon pine wood with saw-blade cutting at 7x magnification (bottom left); and d) Pinyon pine wood with saw-blade marks at 32x magnification (bottom right).

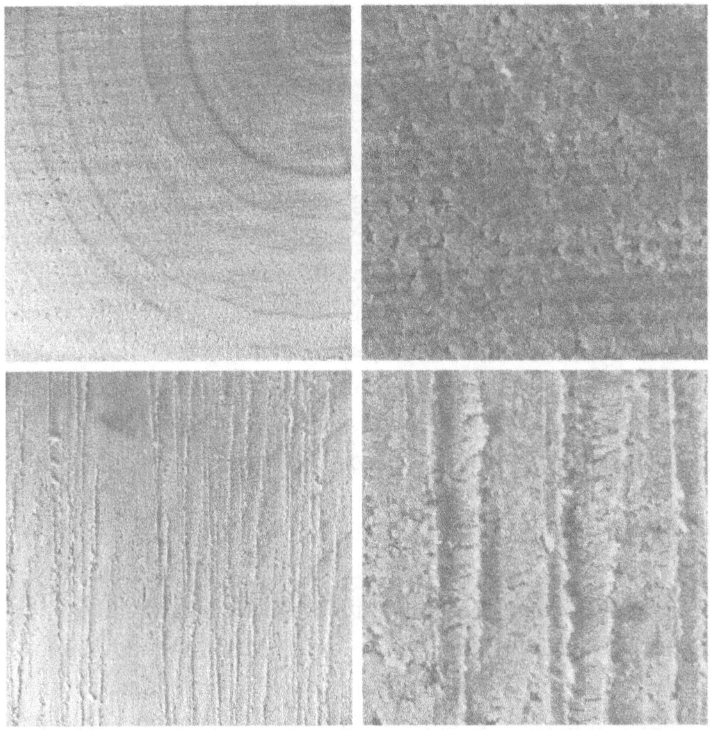

Non-Technical Ceramics

Non-technical ceramics are comprised of naturally occurring geological materials, such as stone, and also various forms of masonry and concrete made from clays and soil mixtures. While the density of ceramics is typically in the mid-range from 1,000 kg/m^3 – 10,000 kg/m^3, they can include some of the highest compressive strengths ranging between 10 MPa – 10,000 MPa. The chemical compositions of stone fall into three basic categories: siliceous, argillaceous, or calcareous. The main element of siliceous stone is silica and includes types like sandstone or quartz. The main element of argillaceous stone is alumina, which become clay-like in types of slate. Calcareous stone is comprised primarily of calcium carbonates or lime and includes types such as marble. Most stone materials require pressurization or compressed sedimentation for millennia before their identified state is formed. The formation of rocks also results in three major types: igneous (solidification and cooling of magma), sedimentary (low temperature sediment accumulation), and metamorphic (application of temperature and pressure on pre-existing rock surfaces) (Mibei, 2014). Masonry and concrete are a combination of soil aggregates mixed with clay or cement and cured by processes of either heating or hydration. Non-technical ceramics comprise some of the lowest energy content and most affordable building materials available (Morel et al., 2001; Reddy & Jagadish, 2003; Estokava & Porhincak, 2015) but can also range up to high energy content and cost due to rarity for ceramics not typically utilized in buildings. Non-technical ceramics are challenging to recycle due to the various compositions of mixes required in the processing as well as the large force required for crushing power (Zhang, 2008), but some stones and pure porous ceramics will be more easily recyclable at end of life. Because of the close packing structure of ionic bonds in most nontechnical ceramics, the thermal conductivity range moves slightly higher than other natural materials starting around 0.2 W/mK and up to about 0.6 W/mK. However, it should be noted that the thermal capacitance of many ceramic and stone materials in buildings is relatively high, so the transfer of heat through these materials tends to move at a relatively slower rate since thermal diffusivity is inversely proportional to the heat capacitance (Cobirzan et al., 2016).

Emerging Applications

Stone and masonry have long been implemented in building construction, in part because of the readily available resource near construction sites and in part because of the durability and longevity these materials offer. In some cases, ancient techniques and uses of soils and minerals in construction are re-emerging in contemporary architecture. Minerals such as salt provide unique possibilities for building materials that have added health benefits due to the ionization effects that are known to improve air quality (Chervinskaya & Zilber, 1995; Hedman et al., 2006; Horowitz, 2010). Salt construction methods include treating the material similar to stone and masonry units, as seen in Figure 2, by cutting the quarried salt blocks into modules. Salt can also be carved, or printed with binding agents, into unique shapes and forms. Salts have the unique benefits of high heat capacitance and phase change properties, which enhance the energy conservation of buildings when integrated in the enclosure system. Rammed earth construction techniques are also utilized in contemporary architecture by making use of the soils found directly at a project site and compressing and stamping them into formwork to create thick building envelopes with high heat capacitance (Ciancio, 2015). Other techniques that are emerging include making concrete-like materials with methods that engage biological reactions in the curing process to eliminate environmentally harmful carbon dioxide emissions (Soleimani, 2017).

Figure 2. a) Himalayan salt block pieces cut to modules for adaptive salt enclosure (top left); b) Adaptive salt block module with backlighting (top right); c) Himalayan salt block at 10x magnification (bottom left); and d) Himalayan salt block at 32x magnification (bottom right).

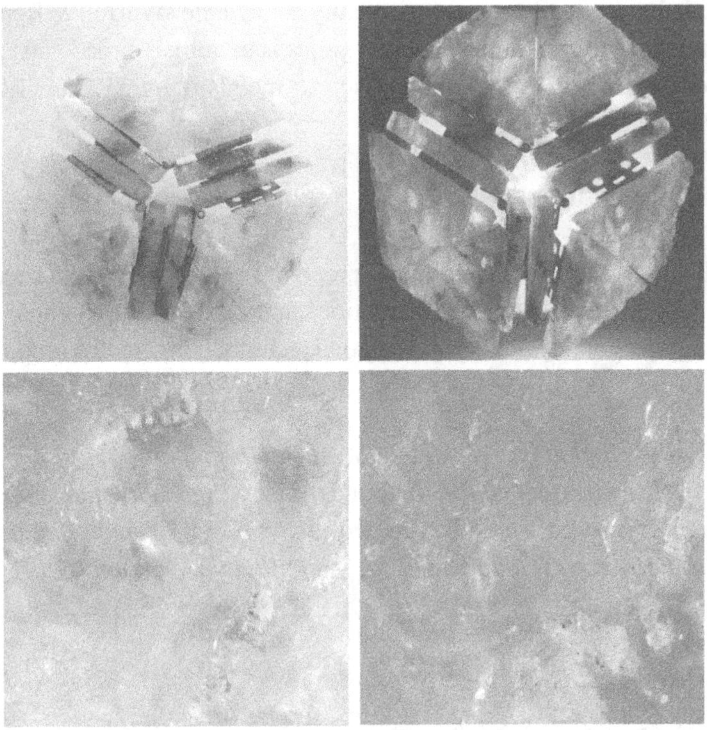

Technical Ceramics

The technical ceramics are primarily glass materials and have similar properties and characteristics with other non-technical ceramics, except that the process of heating molten sand with other oxides to make the product causes some of their unique amorphous properties. The density of glass tends to fall around 3,000 kg/m^3 with compressive strength around 500 MPa. Most glass materials are composed of silica with modifying oxides such as soda or lime. The formation of most glass requires a three-part chemical formula consisting of formers, fluxes, and stabilizers. The formers make the highest percentage of each blend with silicon dioxide as the most common. Fluxes reduce the melting temperature of the formers and may include sodium carbonate or potassium carbonate. Stabilizers, such as calcium carbonate, are required to make the finished glass water resistant (Vogel, 1994). Many different combinations of formers, fluxes and stabilizers constitute the myriad array of glass materials that are possible. Glass tends to be higher in energy content and cost than most typical ceramic building materials since more heat energy is required in the production process (Wallenberger & Bingham, 2010). Glass is mostly recyclable and the processes for capturing glass waste streams and downgrading or re-melting the materials into reusable contents is readily achievable (Worrell & Reuter, 2014). The thermal conductivity of technical ceramics is extremely broad and can be quite high exceeding such values in comparison to metals, ranging from 1.0 W/mK – 1,200 W/mK.

Emerging Applications

Early civilizations explored techniques for replicating the material phenomena found in quartz and other pure glass-like formations in the earth. One example of an early glass making method is *faience*, an Egyptian technique in which a mixture of silica

Figure 3. a) Recycled glass melted in a low temperature kiln and sandblasted (top left); b) Recycled glass melted in a low temperature kiln resulting in amorphous and porous samples (top right); c) Porous recycled glass at 7x magnification (bottom left); and d) Porous recycled glass at 32x magnification (bottom right).

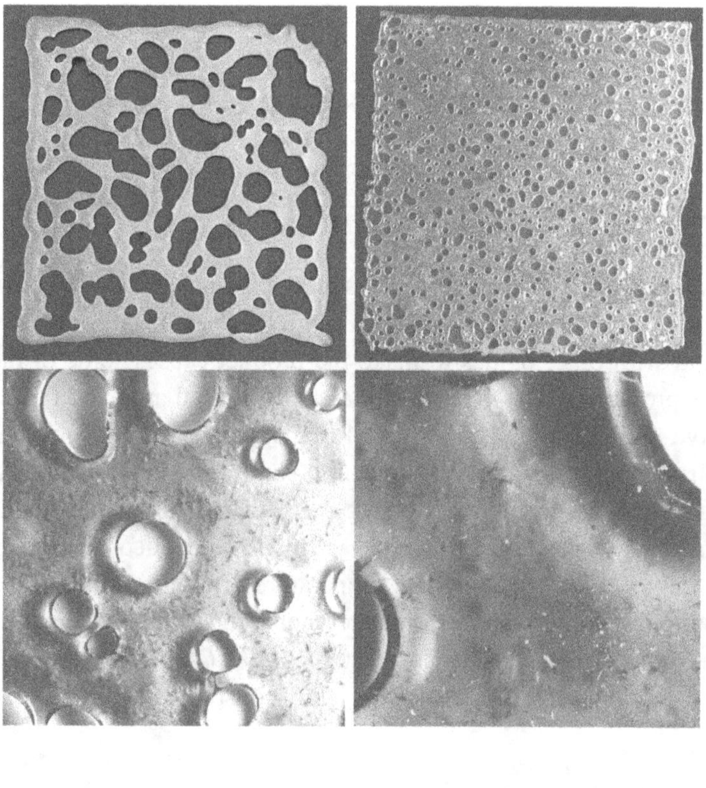

with other elements such as sodium carbonates, chalk, and kaolin are blended into a paste with water and then heated in low-temperature kilns with fine sands or ashes used for glazes (Nicholson, 2012). Because of the intensive amount of energy input and heat required to melt silica and glass constituents, larger scale production of glass for building materials did not pick up until the late nineteenth century. Recent research is addressing the processing energy required to make glass, and notions of returning to early techniques such as *faience* are being considered. Some examples of low-temperature kiln melting of recycled glass parts are shown in Figure 3. Because glass is highly conductive yet offers transparency, it is considered one of the most problematic and beneficial materials in building envelope systems respectively. Emerging concepts on utilizing the conductive properties of glass to embrace the capture and transfer of heat in useful ways to building systems are being explored. Specialized glass materials that integrate spectrally selective coatings with photo- or thermo-responsive properties have emerged in the past few decades and aid in the reduction of heat transfer while improving natural daylighting into buildings (Costanzo et al., 2016; Long & Ye, 2014; Saeli et al., 2010; Hee et al., 2015). Double-skin envelope systems that utilize an air chamber between two or three layers of glazing provide unique benefits in a broad range of climates for improved energy conservation, natural ventilation, and natural daylighting (Ghaffarianhoseini et al., 2016; Shameri et al., 2011; Chou et al., 2009).

Metals

There are numerous naturally occurring metals and alloys that endure various modes of processing to create products used in buildings. The metals include some of the heaviest, densest materials available (2,000 kg/m^3 – 25,000 kg/m^3) and are especially strong in tensile yield (up to 5,000 MPa). Metals comprise almost seventy-percent of the periodic table of elements and exhibit metallic bonds with free electrons in the outer valence bands and crystalline chemical structure. In most practical applications, different metals are combined together in mixtures as alloys to modify the properties of the pure metals in order to produce more desirable characteristics. Common metals in building materials include aluminum and iron alloys such as steel. Other metals are also used in select building products including copper and zinc. Base metals are easily oxidized and corroded while noble metals are resistant to these processes. Ferrous metals contain iron and tend to be magnetic.

The extraction methods for metals requires an intensive amount of energy and results in this material group being one of the highest in energy content with mid to high end costs (Reddy & Jagadish, 2003). Metals are mostly recyclable with the processes widely established for separating and melting existing metal products into reusable materials (Morawski, 2006). As a result of the dense metallic bonding

Figure 4. a) Aluminum at 10x magnification (left); and b) Aluminum at 32x magnification (right).

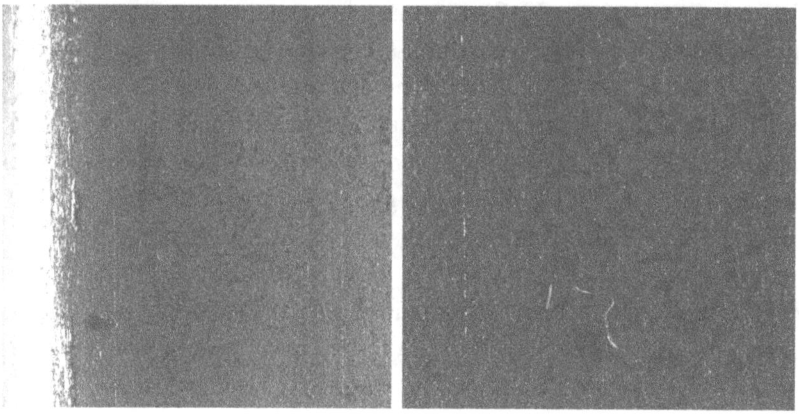

of atoms in metals with the free electron fields, most metals have a high thermal conductivity with values from 8.0 W/mK – 800 W/mK. Figure 4 depicts the surface of aluminum struts and demonstrates the density and opacity of the solid material.

Emerging Applications

Metals emerged as building materials with the industrial revolution in the nineteenth century. Contemporary construction processes emphasize the integrated design of metal systems for future deconstruction and recycling (Kieran Timberlake, 2008). Appropriation of metal constructions from other industries, such as shipping containers, are also a mode of recyclability and reuse in contemporary architecture projects (Bowley & Mukhopadhaya, 2017; Vijayalaxmi, 2010). Metal panel designs in rain-screen systems in contemporary architecture also make use of unique fabrication techniques, such as laser-cutting, three-dimensional forming, and additive manufacturing to accommodate solar shading geometries and for material conservation (Klammt et al., 2012; McGee & Ponce de Leon, 2014; Gebler et al., 2014). These emerging fabrication methods allow for more specific control of building surface geometries in relationship to environmental forces for energy conservation benefit, and also minimize material waste in the fabrication process. Some of the environmental challenges of integrating certain metals in building construction include the negative effects on soil and water systems when corrosive chemical discharge is incurred. Protective coatings to make metal environmentally benign provide some improvements in this aspect (Ahmad, 2006). There are also metal building skins that integrate functionalized surface characteristics, such as titanium dioxide for photocatalysis of nitrogen oxides and air-cleaning properties, that are becoming more prevalent in urban environments (Hot et al., 2017). In addition, emerging applications of shape-memory-alloys (SMA) on building skins to responsively mediate the temporal variances in environmental conditions and improve building energy conservation are being developed (Formentini & Lenci, 2018).

Polymers

The polymers group includes plastics and polymeric materials that are synthetically produced with chemical processes. The strength and density are comparable to rubbers, with weights ranging between 900 kg/m^3 – 3,000 kg/m^3 and tensile yield strengths of 8 MPa – 90 MPa. Polymers are made of long chains of repeating monomers and typically include links by carbon atoms. Thermoplastics are easily reversible back to a liquid state while thermosets cannot easily be reversed from their solid form. Synthetic polymers require a decent amount of heat energy in the production process but are mass produced at scales that allow for mid-range costs to be achieved (Biron, 2014). Many polymers are not recyclable because they are thermosets that cannot be melted and separated. The thermal conductivity of polymers ranges from 0.09 W/mK – 1.1 W/mK, which is comparable to other natural wood materials.

Emerging Applications

Waxes and phase change materials (PCMs) as shown in Figure 5a are also types of polymers and may be synthetically produced in the form of paraffin, which is a petroleum derivative. Such materials have especially high heat capacitance and may be used to improve the storage of heat within building materials for thermal lag effects that may be desired. Current research on methods for integrating PCMs into building systems and materials includes techniques of micro-encapsulation and mixing with concrete, or fabrication of sandwich panels for building envelopes since some translucency can also be achieved (Vigna et al., 2018; Cabeza et al., 2007). Another emerging application for polymers in buildings is the introduction of polyacrylamide hydrogels, as seen in Figure 5b, to improve multifunctional environmental management of moisture, light, and heat (Wu et al., 2016; Smith, 2017). Other areas of emerging research in polymers is the study of processing polysaccharides, which come from renewable biopolymer resources and waste streams such as shrimp shells or pecan shells (Habibi & Lucian, 2012). Though the material properties are derivatives of the natural materials group, chemical synthesis and processing is required to make such materials and so these are included with the polymers group. Polysaccharide biopolymers include chitin and agar, materials that come from the skin-like shells of crustaceans or from seaweed respectively. Polysaccharides are promising environmentally friendly alternatives to carbon-based polymer materials used in buildings, though ongoing research is required for understanding methods to prevent natural microbial processes and decay. Figure 5c depicts a polylactic acid sample processed with a foaming technique to encourage air entrainment and improve sound absorptance properties. Another example presented in Figure 5d is a chitosan biopolymer processed with low temperature synthesis and room temperature curing.

Foams and Elastomers

The lightest weight material group, which also has the lowest strength, is foams. These lightweight materials are between 100 kg/m^3 - 300 kg/m^3 density, and between 0.1 MPa – 10 MPa tensile yield strength. Elastomers, or rubbers have slightly higher density and strength, ranging between 1,000 kg/m^3 – 3,000 kg/m^3 and 5 MPa – 50 MPa respectively. Foams and elastomers are used in buildings typically in the form

Figure 5. a) Paraffin wax (top left); b) Polyacrylamide hydrogel at x magnification (top right); c) Polylactic acid in cured form after foaming process during heating (bottom left); and d) Chitosan polysaccharide biopolymer in cured form at 7x magnification (bottom right).

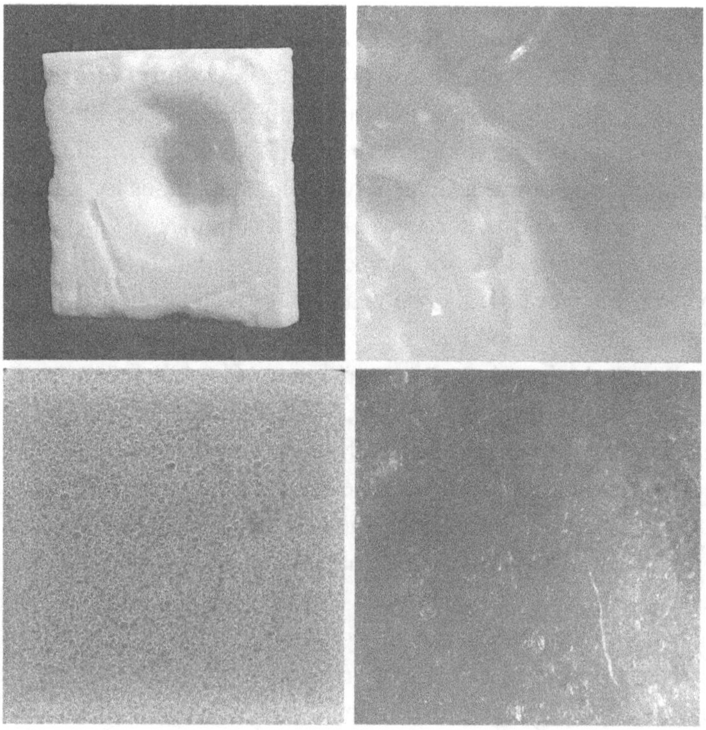

of insulation, thermal breaks, and gaskets and seals between dissimilar materials. Solid foams consist of polymer chains formed around numerous air pockets and characterized by a large internal surface area. While a number of solid foams and foam-like structures are found in nature, such as sea sponges, cork, honeycomb, and radiolara bone, common industrial solid foams are synthetically produced with polyurethanes or other polymers (Cantat, 2013). Rubbers are elastomers that are able to undergo reversible extension and retraction cycles, and thus embody high potential energy in the expanded or stretched state and high kinetic energy in the transition stage returning to rest. Natural rubber comes from rubber trees and is a polymer of isoprene. Industrial rubbers are comprised of covalently bonded chloroprene. Rubbers range from low to high energy content while most foams reside at the higher embodied energy spectrum due to intensive processing steps (Chanchaichujit & Saavedra-Rosas, 2018). The costs, which are based on material weight, is mid-range for foams and rubbers. Many foams and rubbers are not recyclable because they are thermosets that cannot be melted and separated. Of all material groups, foams have the lowest thermal conductivity and thus the highest thermal resistance, largely due to the amount of air entrainment within their structures. The thermal conductivity of foams falls between 0.025 W/mK – 0.1 W/mK. This makes foams the most ideal thermal insulators in building construction and historically is targeted as the primary means for energy conserving measures for building enclosures.

Emerging Applications

Due to the indoor air quality challenges resulting from the off-gassing of volatile organic compounds (VOCs) from spray-foam and polyurethane insulation materials (Ginley & Cahen, 2012; Spengler et al., 2001), development of non-environmentally harmful insulation materials is emerging. Some examples include foam-like polysaccharides, such as lyophilized agarose or chitosan shown in Figure 6. These enlarged stereomicroscopy images depict the sponge-like characteristics of the foams indicating a barrier to heat transfer because of the expansive internal surface area surrounded by air. In addition, these polysaccharide-based foams come from renewable resources or capture otherwise waste streams in the environment and also are easily recyclable. Super-insulator materials such as aerogels are extremely lightweight and can also provide some transparency and daylight transmittance through exterior building walls with excellent energy conservation benefits (Buratti & Moretti, 2012; Cotana et al., 2014), though initial costs for this emerging material are proven very high for typical building applications. Further development of aerogel studies includes utilizing biomaterials and transfer of the technology from aerospace into buildings with lowering production costs (Jiménez-Saelices et al., 2017).

Figure 6. a) Lyophilized agarose (top left); b) Lyophilized agarose at 32x magnification (top right); c) Lyophilized chitosan at 10x magnification (bottom left); and d) Lyophilized chitosan at 32x magnification (bottom right).

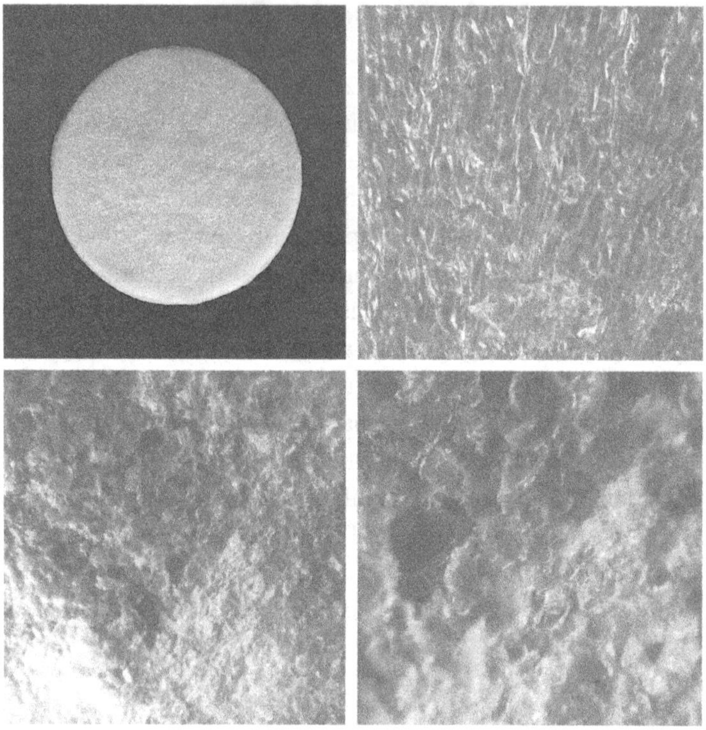

Composites

Composite materials are made with combinations of two or more different materials and include an area of material science with the greatest potential for new and emerging developments. Often times composites are invented to achieve a unique combination of properties that is not otherwise available in a single material. Some examples of composites include Glass Fiber Reinforced Concrete (GFRC) and Carbon Fiber Reinforced Polymers (CFRP). The density of composites is in a pretty narrow range of about 1,500 kg/m^3 – 2,000 kg/m^3, while the tensile yield strengths are relatively high from around 100 MPa – 1,000 MPa. Because composites are specialized materials, both the costs and energy content tend to be on the high end of the spectrum, though alternative composites with natural materials can be quite affordable. Most composites are not recyclable because the combination of materials cannot be easily separated and returned to their raw form (Zhang, 2008). The thermal conductivity of most composite materials lies somewhere between 0.5 W/mK and 5.0 W/mK, higher than wood and natural materials but lower than metals. A regional composite material example shown in Figure 7 combines Opunita (prickly-pear) cactus skeleton meat within a resin to provide improved thermal and acoustic properties with the transparency of the clear polymer.

Emerging Applications

Since composite materials include an almost indefinite possibility of combinations, new applications for composites is constantly emerging. Most recent contemporary examples include the integration of carbon fibers for reinforcing with polymers and enhancement of strength with minimal weight (Saeed et al., 2016). CFRPs enable lightweight long-span structures with minimal materials. Cellulosic composites are also developing with the integration of nanocellulose techniques such as electrospinning (Vallejos et al., 2012), and also some based on early compositing techniques of heating pressed boards and blocks with natural materials and muds (Wei et al., 2016).

CONCLUSION

The concept of sustainability for building materials is complex and multi-faceted due to the varied impacts on environmental, social, or economic systems. Different societies have developed different material fabrication and building construction techniques, resulting in variations of how materials are implemented in buildings in different cultures. The sustainability of regional construction practices and expertise of those labor communities is another consideration for material selection. The scientific content presented here emphasizes the environmental benefits through energy conservation metrics, which consequently have direct impact on economic metrics as well.

Each material group presented in the prior section provides specific characteristics and properties that are useful in the design of buildings. As shown, some materials provide enhanced retention of energy or resistance to heat transfer while others might provide high strength to light weight, transparency, or acoustic benefits. All forms of materials have an embedded energy value from the chemical bonding mode and microstructure to the regional sourcing and application method.

Figure 7. a) Composite of clear resin and prickly-pear cactus skeleton (top left); b) Opunita cactus branch and resin composite at 10x magnification (top right); c) Opunita cactus branch and resin composite at 32x magnification (bottom left); and d) Opunita cactus branch and resin composite at 32x magnification (bottom right).

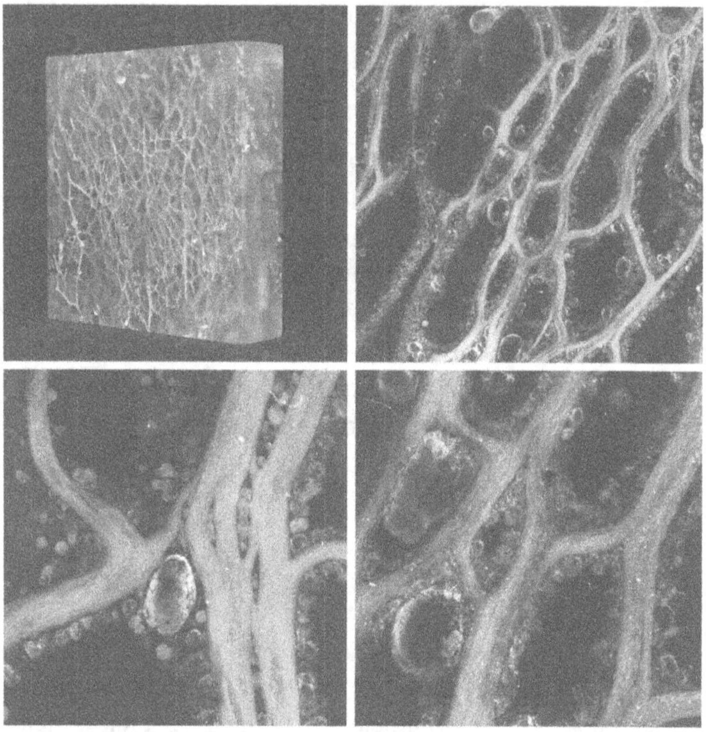

Material Chemistry

The chemical bonding structure of materials explains energy transfer at a nano-scale as shown in the comparison of material group chemical notation examples in Figure 8. The natural materials, such as wood and polysaccharides of agarose and chitin in the foams and composites, present chemical structures void of carbon and made possible with covalent bonds. The ceramics present the ionic bonding structure that results in dense hard matter. Glass chemical notation explains the amorphous semi-crystalline ionic bonding that forms when the atoms are heated and effectively melted together. The metals depict the metallic bonding that allows for free electrons to move between atoms under different temperature conditions. The synthetic polymers that are carbon-based present denser but soft material structures through covalent bonding. The chemicals that comprise different materials are resources that vary in availability dependent on geographic regions and in some cases economic wealth.

Figure 8. Material groups and examples of chemistry for each, including (from top to bottom) cellulose, sodium chloride, alumino-silicate, steel, polyacrylamide, agarose, and a polyacrylamide-chitosan composite.

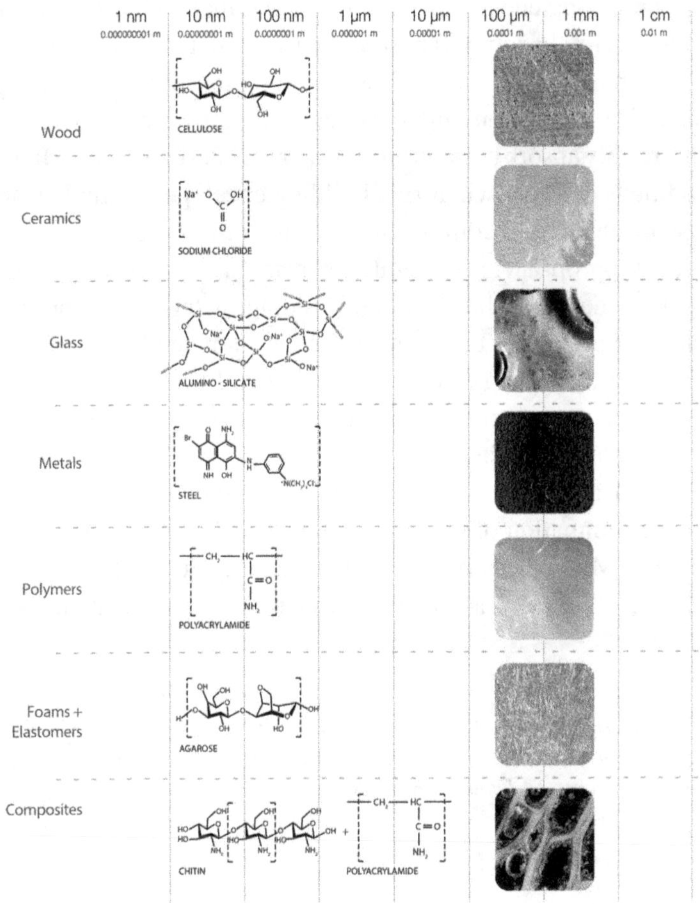

Material Response to Climate

The regional proximity of material sourcing and production to a project site impacts the embodied energy content of the materials employed in buildings. Very early vernacular building techniques depended entirely on the material resources immediately surrounding construction sites. The materials that tend to be readily available in any given geographic region have an inherent relationship to that climate condition. Climate types are defined based on the characteristics of soils and plant materials that have naturally occurred and evolved in different regions. The Koppen-Geiger climate classification system (Koppen & Geiger, 2018) identified more than nineteen different climate zones in the world. These zones can be roughly grouped into four major types as shown in Figure 9, including hot-humid, hot-arid, mixed temperate, and cold. Different construction types are preferable in different climate regions for purposes of conserving energy and improving thermal comfort. Hot-arid and cold climate conditions imply compact building forms with thick exterior walls and high insulation or high heat capacitance. Hot-humid regions imply lightweight exterior walls to maximize ventilation opportunities and high pitch roofs of biopolymeric materials that readily absorb humidity and heat from interior spaces. Mixed temperate climate conditions call for switch-rich building envelope assemblies and adaptive material systems that can accommodate seasonal differences.

The regional appropriateness of building material selection is a significant factor in the embodied energy of the building. Building materials and products that are produced in close proximity to a project site have a much lower embodied energy since the distance and energy of transport is greatly reduced. Embodied energy accounts for the energy required for material sourcing, processing and production into the material technologies to be employed in buildings. Accounting for the regional origin of material resources extends to the concept of *emergy,* which is an all-inclusive summation of the available energy inputs consumed by any given material system (Moe & Srinivasan, 2015). Emergy accounts for the originating source of solar, geothermal, and biochemical energy required for materials to develop

Figure 9. Global climate map depicting four basic climate zones for varying construction typologies including: hot-humid, hot-arid, mixed-temperate, and cold.
Source: Simplified and redrawn by author based on Koppen-Geiger Climate Map http://koeppen-geiger.vu-wien.ac.at/ (accessed January 18, 2018).

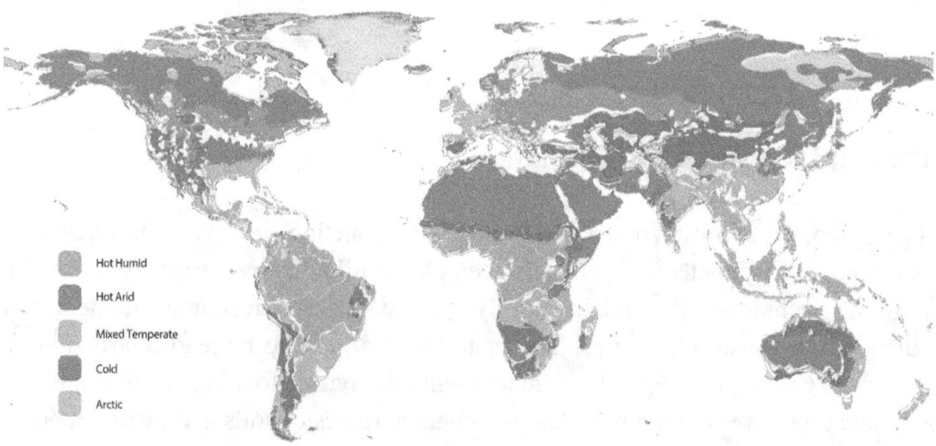

**For a more accurate representation see the electronic version.*

into their current state at sites of extraction. Life cycle of material technologies includes the embodied energy in addition to the energy required for transport of building materials to the project site, lifespan of the material system during building use, and energy required to deconstruct or recycle the material products at their end life (McDonough & Braungart, 2013). Emergy analysis, life cycle analysis, and the basic concept of embodied energy originated in the disciplines of systems ecology and industrial ecology (Odum, 1983). These disciplines shed light on how all of our material and energy systems are interrelated to each other beyond the autonomous building itself, and the challenges of considering how our building designs impact the broader environment.

Energy Conservation in Buildings

The selection and composition of materials in the design of buildings has long-term implications on the energy that will be consumed in the operation and maintenance during the building's lifespan. A great amount of the energetic processes in building operations is attributed to the thermal conditioning of the interior spaces (Davies, 2004). Due to the differentials in temperature and climate conditions between outdoors and thermal comfort needs of indoors, a vast amount of energy exchange is apt to occur at the building enclosure system (Sadineni et al., 2011). The composition of material assemblies for the building envelope and its geometries holds great implications on the *energetic form* in response to the climatic conditions and human needs.

Different trade-offs and benefits can be realized with different material selections. Comparisons of thermal and strength properties of material groups with the embodied energy, recycle fraction, and potential energy conservation in buildings are shown in Figure 10. Since wood is a renewable resource, the long-term environmental impact is much lower than using steel or concrete as primary building structure materials. Both steel and concrete, in comparison, have relatively high environmental impact due to the amount of carbon dioxides released into the atmosphere as a result of the manufacturing processes. However, steel is more readily recyclable and reusable as a building material in comparison with wood, which tends to lose its strength and degrade over time. The use of non-technical ceramics in buildings often has very low environmental impact especially if the materials utilized are taken directly from or close to the building site, such as rammed earth or salt construction. Such materials also tend to conserve energy for buildings by thermal regulation. The material considerations in an architecture design process will inherently have implications on the *energetic form* of the building. The *energetic form* implies the embedded energy value of a building made visible through the composition of its materials.

Figure 10. Energy conservation potential of different material groups in buildings compared with embodied energy, thermal conductivity, recycle fraction, and strength properties.

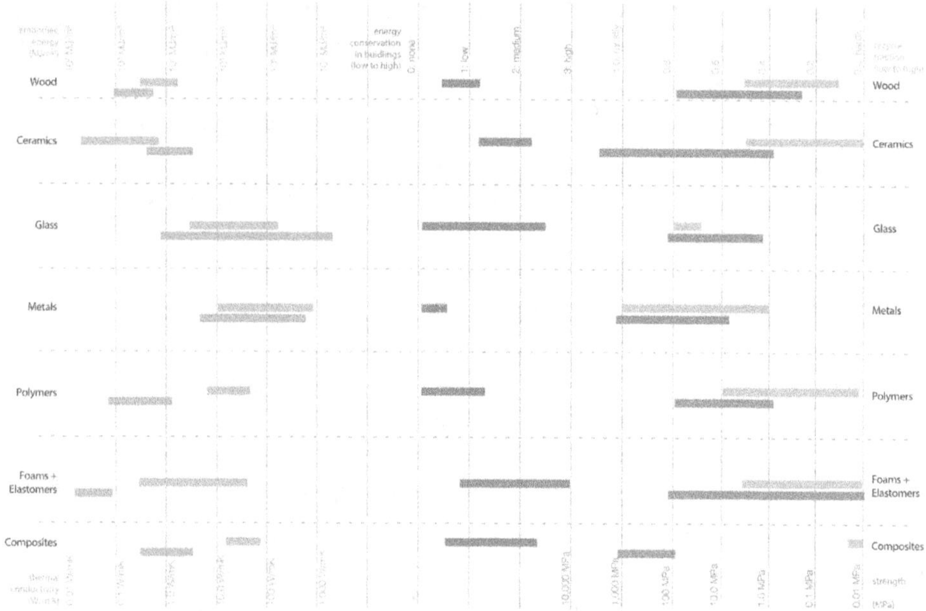

ACKNOWLEDGMENT

The author is grateful to Patty Jansma, Facility Manager of the University of Arizona (UA) Imaging Core for providing guidance with the stereomicroscopy imaging. Gratitude is also extended to Paulus Musters, Manager of the UA School of Architecture (SoA) material fabrication laboratories for providing guidance and assistance with students for producing emerging material compositions. Some of the material examples depicted in this chapter were developed by architecture students enrolled in the author's Material Properties and Tests course in Spring 2017 at the UA SoA, including the following: Fatemeh Sharef-Zadeh (Himalayan salt-block modules), Barrett Miesfeld (recycled porous glass), Alexander Villa (paraffin wax studies), Rahil Zarpoush (polylactic acid acoustic panels), Derek Runge (lyophilized agarose and chitosan foams), and Andy Kraut (Opunita cactus and resin composite).

REFERENCES

Abott, A. (2013, November 8). Novel process to manufacture starch-based recyclable medium density fiberboards. *Industrial Processing*.

Ahmad, Z. (2006). *Principles of Corrosion Engineering and Corrosion Control*. New York, NY: Elsevier, Ltd.

Ashby, M., Shercliff, H., & Cebon, D. (2007). *Materials Engineering, Science, Processing and Design*. Burlington, MA: Elsevier Ltd.

Atkins, P. (2015). *Chemistry: A Very Short Introduction*. Oxford, UK: Oxford University Press. doi:10.1093/actrade/9780199683970.001.0001

Bowley, W., & Mukhopadhaya, P. (2017). A sustainable design for an off-grid passive container house. *International Review of Applied Sciences and Engineering*, *8*(2), 145–152. doi:10.1556/1848.2017.8.2.7

Brownell, B. (2010). Material Ecologies in Architecture. In L. Tilder & B. Blostein (Eds.), *Design Ecologies: Essays on the Nature of Design*. New York, NY: Princeton Architectural Press.

Buratti, C., & Moretti, E. (2012). Glazing systems with silica aerogel for energy savings in buildings. *Applied Energy*, *98*, 396–403. doi:10.1016/j.apenergy.2012.03.062

Cabeza, L. F., Castellon, C., Nogues, M., Medrano, M., Leppers, R., & Zubillaga, O. (2007). Use of microencapsulated PCM in concrete walls for energy savings. *Energy and Building*, *39*(2), 113–119. doi:10.1016/j.enbuild.2006.03.030

Cantat, I. (2013). Uses of Foam. In *Foams: Structure and Dynamics*. Oxford, UK: Oxford University Press. doi:10.1093/acprof:oso/9780199662890.003.0001

Chanchaichujit, J., & Saavedra-Rosas, J. F. (2018). *Using Simulation Tools to Model Renewable Resources: The Case of the Thai Rubber Industry*. Palgrave Macmillan, Springer Nature. doi:10.1007/978-3-319-55816-5

Chele, E.-S., Ricardo, M.-C., Ana, P.-M., & Teresa, M.-R. (2012). Bamboo, from Traditional Crafts to Contemporary Design and Architecture. *Procedia: Social and Behavioral Sciences*, *51*, 777–781. doi:10.1016/j.sbspro.2012.08.239

Chervinskaya, A. V., & Zilber, N. A. (1995). Halotherapy for treatment of respiratory diseases. *Aerosol Medicine*, *8*(3), 221–232. doi:10.1089/jam.1995.8.221 PMID:10161255

Chou, S. K., Chua, K. J., & Ho, J. C. (2009). A study on the effects of double skin facades on the energy management in buildings. *Energy Conversion and Management, 50*(9), 2275–2281. doi:10.1016/j.enconman.2009.05.003

Ciancio, D. (2015). *Rammed Earth Construction: Cutting-Edge Research on Traditional and Modern Rammed Earth.* London: Taylor & Francis, Ltd. doi:10.1201/b18046

Cobirzan, N., Balog, A. A., Belean, B., Borodi, G., Dadarlat, D., & Streza, M. (2016). Thermophysical properties of masonry units: Accurate characterization by means of photothermal techniques and relationship to porosity and mineral composition. *Construction & Building Materials, 105*, 297–306. doi:10.1016/j.conbuildmat.2015.12.056

Costanzo, V., Evola, G., & Marletta, L. (2016). Thermal and visual performance of real and theoretical thermochromic glazing solutions for office buildings. *Solar Energy Materials and Solar Cells, 149*, 110–120. doi:10.1016/j.solmat.2016.01.008

Cotana, F., Pisello, A.L., Moretti, E., & Buratti, C. (2014). *Multipurpose characterization of glazing systems with silica aerogel: In-field experimental analysis of thermal-energy, lighting and acoustic performance.* Academic Press.

Davies, M. G. (2004). *Building Heat Transfer.* West Sussex, UK: John Wiley & Sons, Ltd. doi:10.1002/0470020555

Dickson, M., & Parker, D. (2014). *Sustainable timber design.* New York, NY: Routledge.

Estokova, A., & Porhincak, M. (2015). Environmental analysis of two building material alternatives in structures with the aim of sustainable construction. *Clean Technologies and Environmental Policy, 17*(1), 75–83. doi:10.100710098-014-0758-z

Formentini, M., & Lenci, S. (2018). An innovative building envelope (kinetic façade) with Shape Memory Alloys used as actuators and sensors. *Automation in Construction, 85*, 220–231. doi:10.1016/j.autcon.2017.10.006

Gebler, M., Uiterkamp, A. J. M. S., & Visser, C. (2014). A global sustainability perspective on 3D printing technologies. *Energy Policy, 74*, 158–167. doi:10.1016/j.enpol.2014.08.033

Ghaffarianhoseini, A., Ghaffarianhoseini, A., Berardi, U., Tookey, J., Li, D. H. W., & Kariminia, S. (2016). Exploring the advantages and challenges of double-skin facades (DSFs). *Renewable & Sustainable Energy Reviews, 60*, 1052–1065. doi:10.1016/j.rser.2016.01.130

Gibson, L. J. (2012). The hierarchical structure and mechanics of plant materials. *Journal of the Royal Society, Interface*, *9*(76), 2749–2766. doi:10.1098/rsif.2012.0341 PMID:22874093

Ginley, D. S., & Cahen, D. (Eds.). (2012). *Fundamentals of Materials for Energy and Environmental Sustainability*. New York, NY: Cambridge University Press.

Guo, H., Liu, Y., Wen-Shao, C., Shao, Y., & Sun, C. (2017). Energy saving and carbon reduction in the operation stage of cross laminated timber residential buildings in china. *Sustainability*, *9*(2), 292. doi:10.3390u9020292

Habibi, Y., & Lucian, A. L. (Eds.). (2012). *Polysaccharide Building Blocks: A Sustainable Approach to the Development of Renewable Biomaterials*. Hoboken, NJ: Wiley. doi:10.1002/9781118229484

Hedman, J. T., Hugg, J., Sandell, T. H., & Haahtela, T. (2006). The effect of salt chamber treatment on bronchial hyperresponsiveness in asthmatics. *Allergy*, *61*(5), 605–610. doi:10.1111/j.1398-9995.2006.01073.x PMID:16629791

Hee, W. J., Alghoul, M. A., Bakhtyar, B., Elayeb, O., Shameri, M. A., Alrubaih, M. S., & Sopian, K. (2015). The role of window glazing on daylighting and energy saving in buildings. *Renewable & Sustainable Energy Reviews*, *42*, 323–343. doi:10.1016/j.rser.2014.09.020

Horowitz, S. (2010). Salt cave therapy: Rediscovering the benefits of an old preservative. *Alternative and Complementary Therapies*, *16*(3), 158–162. doi:10.1089/act.2010.16302

Hot, J., Topalov, J., Ringot, E., & Bertron, A. (2017). Investigation on parameters affecting the effectiveness of photocatalytic functional coatings to degrade NO: TiO_2 amount on surface, illumination, and substrate roughness. *International Journal of Photoenergy*, 14.

Jiménez-Saelices, C., Seantier, B., Cathala, B., & Grohens, Y. (2017). Effect of freeze-drying parameters on the microstructure and thermal insulating properties of nanofibrillated cellulose aerogels. *Journal of Sol-Gel Science and Technology*, *84*(3), 475–485. doi:10.100710971-017-4451-7

Klammt, S., Neyer, A., & Muller, H. F. C. (2012). Redirection of sunlight by microstructured components – Simulation, fabrication and experimental results. *Solar Energy*, *86*(5), 1660–1666. doi:10.1016/j.solener.2012.02.034

Koppen, W., & Geiger, R. (2018). *World Maps of Koppen-Geiger Climate Classification*. Retrieved January 18, 2018 from http://koeppen-geiger.vu-wien.ac.at/

Long, L., & Ye, H. (2014). Discussion of the performance improvement of thermochromic glazing applied in passive buildings. *Solar Energy, 107*, 236–244. doi:10.1016/j.solener.2014.05.014

McDonough, W., & Braungart, M. (2013). *The Upcycle: Beyond Sustainability – Designing for Abundance*. New York, NY: North Point Press.

McGee, W., & Ponce de Leon, M. (Eds.). (2014). *Robotic Fabrication in Architecture, Art and Design 2014*. Springer International Publishing. doi:10.1007/978-3-319-04663-1

Mibei, G. (2014). Introduction to types and classification of rocks. *Short Course IX on Exploration for Geothermal Resources by UNU-GTP, GDC and KenGen.*

Moe, K., & Srinivasan, R. (2015). *The Hierarchy of Energy in Architecture: Emergy Analysis*. New York, NY: Routledge.

Morawski, C. (2006). Embodied energy. *Alternatives Journal, 32*(1), 19.

Morel, J. C., Mesbah, A., Oggero, M., & Walker, P. (2001). Building houses with local materials: Means to drastically reduce the environmental impact of construction. *Building and Environment, 36*(10), 1119–1126. doi:10.1016/S0360-1323(00)00054-8

Mumford, L. (1934). The Eotechnic Phase. In *Technics and Civilization*. San Diego, CA: Harcourt, Brace and Co.

Nicholson, P. T. (2012). Stone… That Flows: Faience and Glass as Man-Made Stones in Egypt. *Journal of Glass Studies, 54*, 11–23.

Odum, H. T. (1983). *Systems Ecology: An Introduction*. New York, NY: John Wiley.

Ramage, M. H., Burridge, H., Busse-Wicher, M., Fereday, G., Reynolds, T., Shah, D. U., ... Scherman, O. (2017). The wood from the trees: The use of timber in construction. *Renewable & Sustainable Energy Reviews, 68*, 333–359. doi:10.1016/j.rser.2016.09.107

Reddy, B. V. V., & Jagadish, K. S. (2003). Embodied energy of common and alternative building materials and technologies. *Energy and Building, 35*(2), 129–137. doi:10.1016/S0378-7788(01)00141-4

Sadineni, S. B., Madala, S., & Boehm, R. F. (2011). Passive building energy savings: A review of building envelope components. *Renewable & Sustainable Energy Reviews, 15*(8), 3617–3631. doi:10.1016/j.rser.2011.07.014

Saeed, H. Z., Khan, Q. Z., Khan, H. A., & Farooq, R. (2016). Experimental investigation of stress-strain behavior of CFRP confined Low Strength Concrete (LSC) cylinders. *Construction & Building Materials*, *104*, 208–215. doi:10.1016/j.conbuildmat.2015.12.061

Saeli, M., Piccirillo, C., Parkin, I. P., Binions, R., & Ridley, I. (2010). Energy modelling studies of thermochromic glazing. *Energy and Building*, *42*(10), 1666–1673. doi:10.1016/j.enbuild.2010.04.010

Shameri, M. A., Alghoul, M. A., Sopian, K., Zain, M. F. M., & Elayeb, O. (2011). Perspectives of double skin façade systems in building and energy saving. *Renewable & Sustainable Energy Reviews*, *15*(3), 1468–1475. doi:10.1016/j.rser.2010.10.016

Smith, S. I. (2017). *Superporous Intelligent Hydrogels for Environmentally Adaptive Building Skins*. MRS Advances. doi:10.1557/adv.2017.429

Soleimani, M., & Shahandashti, M. (2017). Comparative process-based life-cycle assessment of bioconcrete and conventional concrete. *Journal of Engineering, Design and Technology*, *15*(5), 667–688. doi:10.1108/JEDT-04-2017-0033

Spengler, J. D., Samet, J. M., & McCarthy, J. F. (Eds.). (2001). *Indoor Air Quality Handbook*. New York, NY: McGraw-Hill.

Timberlake, K. (2008). *Cellophane House | Prefabricated Architecture & Design for Disassembly*. Retrieved May 6, 2018, from https://kierantimberlake.com/pages/view/14/cellophane-house/parent:3

Vallejos, M. E., Peresin, M. S., & Rojas, O. J. (2012). All-cellulose composite fibers obtained by electrospinning dispersions of cellulose acetate and cellulose nanocrystals. *Journal of Polymers and the Environment*, *20*(4), 1075–1083. doi:10.100710924-012-0499-1

Vigna, I., Bianco, L., Goia, F., & Serra, V. (2018). Phase Change Materials in Transparent Building Envelopes: A Strengths, Weaknesses, Opportunities and Threats (SWOT) Analysis. *Energies*, *11*(1), 111. doi:10.3390/en11010111

Vijayalaxmi, J. (2010). Towards sustainable architecture – a case with Greentainer. *Local Environment: The International Journal of Justice and Sustainability*, *15*(3), 245–259. doi:10.1080/13549830903575596

Vogel, W. (1994). Historical Development of Glass Chemistry. In *Glass Chemistry*. New York, NY: Springer-Verlag. doi:10.1007/978-3-642-78723-2_1

Wallenberger, F. T., & Bingham, P. A. (Eds.). (2010). *Fiberglass and Glass Technology: Energy-Friendly Compositions and Applications*. New York, NY: Springer. doi:10.1007/978-1-4419-0736-3

Wei, P., Rao, X., Yang, J., Guo, Y., Chen, H., Zhang, Y., & Wang, Z. (2016). Hot pressing of wood-based composites: A review. *Forest Products Journal, 66*(7), 419–427. doi:10.13073/FPJ-D-15-00047

Worrell, E., & Reuter, M. (Eds.). (2014). *Handbook of Recycling: State-of-the-Art for Practitioners, Analysts, and Scientists*. New York, NY: Elsevier.

Wu, Y., Connelly, K., Liu, Y., Gu, X., Gao, Y., & Chen, G. Z. (2016). Smart solar concentrators for building integrated photovoltaic facades. *Solar Energy, 133*, 111–118. doi:10.1016/j.solener.2016.03.046

Yu, M. L. (2013). *Skins, envelopes, and enclosures: concepts for designing building exteriors*. New York, NY: Routledge.

Zhang, F. (2008). Framework for building design recyclability. *ProQuest Dissertations & Theses Global,* 304617969.

ADDITIONAL READING

Addington, M., & Schodek, D. (2005). *Smart Materials and Technologies for the Architecture and Design Professions*. Oxford, UK: Architectural Press.

Bateson, G. (2000). Form, Substance, and Difference. In *Steps to an Ecology of Mind*. Chicago, IL: University of Chicago Press.

Borden, G. P., & Meredith, M. (Eds.). (2012). *Matter: Material Processes in Architectural Production*. New York, NY: Routledge.

Dyson, A. (2002). Recombinant Assemblies. *Architectural Design*, *72*(5), 60–66.

Fletcher, B. F. (1897). *The Influence of Material on Architecture*. London: B.T. Batsford.

Ginley, D. S., & Cahen, D. (Eds.). (2012). *Fundamentals of Materials for Energy and Environmental Sustainability*. Cambridge, UK: Cambridge University Press.

Hardy, S. (Ed.). (2008). *Environmental Tectonics: Forming Climatic Change*. London: AA Publications.

Hosford, W. F. (2014). *Materials for Engineers*. New York, NY: Cambridge University Press.

Princen, T. (2010). *Treading Softly: Paths to Ecological Order*. Cambridge, MA: The MIT Press.

Thompson, D. W. (1992). *On Growth and Form: The Complete* (Revised Edition). New York, NY: Dover Publications. doi:10.1017/CBO9781107325852

KEY TERMS AND DEFINITIONS

Embodied Energy: The energy content of a building material or building product that includes the energy to extract original material resources, the energy to process and make the product for its intended use, as well as the energy to transport the material product to the building site.

Emergy: The total available energy consumed by the matter and content of a building product inclusive of the biogeochemical energy required to create the original material resources (i.e., sun, water, lightning, etc.), the energy for material sourcing, processing, and fabrication, as well as the energy to transport the product to the site for its intended use.

Energetic Form: The embedded energy value of a building design made visible in the composition of its materials.

Entropy: Related to the second law of thermodynamics principle that all forms of energy transform from higher to lower grade states in natural circumstances as a general trend of increasing disorder in the universe.

Eotechnic: Early techniques of industry and energy production with the utilization of wood and glass materials and water and wind forces.

Life-Cycle: The total lifespan of a building material or product and its energy value for that duration including the energy it conserves for the building during implementation.

Lyophilize: A processing method that sublimates ice from a material with vacuum pressure or high-speed centrifugal spinning techniques resulting in a porous foam-like structure; freeze-drying.

Microstructure: The structure of materials at the micron scale.

Recycle Fraction: The fraction of a material that can be recycled cost effectively.

Stereomicroscopy: A microscopy technique that makes use of an instrument capable of assimilating a three-dimensional imaging effect to enhance the surface interface detail of the sample being viewed.

Chapter 8
Building Relationships:
Changing Technology and Society

Jennifer Loy
University of Technology Sydney, Australia

Tim Schork
University of Technology Sydney, Australia

ABSTRACT

This chapter describes how digital immersion, changing social values, and environmental and economic pressures have the potential to create a paradigm shift in relationships between people and their built environment with the growing sustainability imperative. It responds to emerging opportunities provided by digital technologies for the construction, maintenance, and heritage curation of the life of buildings, and draws on aligned changes in thinking apparent in manufacturing, healthcare, business, and education in the 21st century. The ideas that shape this chapter are relevant to architects and educators, but also to scholars and practitioners across disciplines because they provide an innovative approach in responding to the types of changes currently impacting societies worldwide.

INTRODUCTION

The emergence of rapid unsustainable growth – in population; cities; resource consumption; depletion of topsoil; freshwater supplies and living species; pollution flows; and economic output that is measured and guided by an absurd and distorted set of universally accepted metrics that blinds us to the destructive consequences of the self-deceiving choice we are routinely making. (Gore, 2013, introduction)

DOI: 10.4018/978-1-5225-6995-4.ch008

The twentieth century created a legacy of practice driven primarily by profit over environmental responsibility. With the growing world population increasing demand on resources tempered by the evidence of human impact on the biosphere, destructive social, environmental and economic values and behaviors need to change. In architecture and construction this creates drivers for rethinking the relationships that people have with their built environments. This chapter speculates how new architectural and construction practices based on digital construction, digital immersion and ubiquitous computing can drive a shift in thinking about communication, monitoring, maintenance and material use as a response to the growing sustainability imperatives of the twenty-first century. The approach is based on an integration of digital technologies with a view of buildings as evolving, fluid structures that are designed, built and maintained with the aim of maximizing material resources.

The first decades of the twenty-first century appeared to herald in an all-encompassing globalisation with the dissolution of country borders, the development of complex economic interdependency and a gradual homogenisation of populations, as discussed by Ross (2016) in *The Industries of the Future*. However, the results of the Brexit vote in 2016, and the US Presidential election shortly afterwards, have suggested otherwise. Similarly, though discussion on Industry 4.0 is characterised by rhetoric on dehumanisation based on increased automation, robotics and artificial intelligence, arguably digital technologies that support communication, monitoring, analysis and bespoke implementation of services are actually allowing for a greater personalisation of provision across sectors. For example, digital immersion and personalised experiences, identified by Hajkowicz (2015) as global megatrends, are providing the basis for a potential paradigm shift in healthcare. In this shift, the role of the individual as a passive recipient of services is replaced with that of an active participant in a holistic, preventative, proactive healthcare regime. With the growth in accessible monitoring, from fitness trackers to home use blood pressure monitors, blood analysis services, bowel cancer monitoring, and apps for tracking general health and conditions, such as chronic fatigue, plus online services such as DNA analysis, and physiotherapy provided by video link, the engaged individual is empowered in the monitoring of their own physiology. Preventative healthcare and early intervention are becoming enabled through these digital technologies, that could lead to a subsequent shift in the responsibility of the individual in terms of their level of engagement with their own maintenance and treatment. Australia provides an example of where this shift is starting to occur, with medical records moving online, and patients able to choose their doctors, including selecting different practitioners for different aspects of their care.

At the same time, there are emerging medical practices enabled by digital technologies creating changes in how hospital services, such as surgery, could be personalised in their delivery in the near future. For example, clinicians can now use multi-material 3D prints, based on actual patient scans, to plan surgical procedures and create bespoke parts prior to operations. Paul D'Urso at 'Anatomics' in Melbourne, Australia is a leader in the use of additive manufacturing (3D printing) for customised surgical implants and the development of individual medical instruments specific to each operation. This approach contrasts with conventional surgery where the surgeon is provided with a vast array of sterilised items that may or may not be required and has to respond to the operational requirements in situ. Rather than digital technology alienating the individual and reducing the quality of the experience, D'Urso argues it is supporting the empowerment of the individual and the development of personalised care and targeted spending as a new approach for the twenty-first century.

This empowerment is echoed in other industry sectors, for example with flexible learning tools enabled by digital technology, such as MOOCs (massive open online courses), changing the relationships between people and their learning experiences. It also fosters accessible entrepreneurship, for example through internet funding platforms, such as Kickstarter. Digital manufacturing, supported by the ability to transfer digital files online and the development of distributed manufacturing fabrication, rather than centralised manufacturing, has the potential to support a democratisation in making and, at its extreme, challenge the urbanization caused by the first industrial revolution and resultant work pattern imperatives. For single employer or industry sector dependent cities, such as Detroit, and cities where massive communication hubs, for example call centers, have been set up and then dismantled, the proposition of a future where flexible working patterns reduce a dependence on changes to permanent infrastructure is attractive. Even as an anti-globalization reaction sets in, internet freedoms remains relatively unchecked, with bottom-up movements, such as Uber and Air B&B, challenging traditional work practices and regulations. With the benefit of improved data visualization, these population changes can be mapped, highlighting and supporting the fluidity of movement. Populations are also forced to move because of natural disasters, such as the earth quake in Christchurch, New Zealand where ground liquefaction prevented rebuilding in some sections of the city, and significant flooding, such as in Brisbane in 2011. At the same time, political and economic pressures and changing demographics within populations affect the configuration of cities, for example where ageing populations change accommodation requirements or economic factors drive densification or displacement. Equally, open communication means that there is an awareness of opportunities arising elsewhere, as well as difficulties impacting particular zones that are causing population migrations, such as impending weather events (e.g. cyclones in Northern Territory, Australia) and armed conflict (e.g. Africa and the Middle East). The implications for city planning, architecture and building practices are immense and difficult to track. However, with the development of the Smart City agenda, it is timely to consider how the emerging opportunities provided by digital technology, combined with a view of the technology aligned with the approach and social values expressed by the healthcare example, could create increased individualisation, personal expression and personal responsibility, added value and engagement. This chapter considers these developments in relation to architecture, material use and construction practices, and specifically possibilities provided by digital technologies in supporting sustainability strategies and innovation for the twenty-first century.

BACKGROUND LITERATURE

Architectural innovation and transformations in practice over the last century have frequently been closely linked to technological change. Seminal publications, tracking cultural and critical discipline responses to the implications of technological change, arguably began with the work of Walter Benjamin (1936) in the nineteen thirties, who was concerned with the impact of mechanical reproduction on the representation of architectural concepts. Donald Schön was a philosopher and urban planner at MIT whose work in the nineteen sixties focussed on the adaptation of social systems to technological change (Schon, 1967). In the book *Architectural Representation and the Perspective Hinge*, Alberto Pérez Gómez and Louise Pelletier underline by means of the perspective drawing how tools of representation have a direct influence on the conceptual development of design and the generation of form (Pérez, Gómez & Pelletier, 1997). Similarly, Thomas Kvan argues that modes of representation change the way we understand and design a building (Kvan 2004). At the turn of the century, William Mitchell led the integration of architectural practice with the computer (Mitchell, 1999; Mitchell, 2001). More recently, the advances in digital modelling work for building simulations by William Braham and Jonathan Hale (Braham & Hale, 2007) significantly transformed practice. Emeritus Professor of Architecture, Yehuda Kalay (Kalay, 2006), argues that computing and telecommunication have become the new media of architecture. Inventions and innovations, from the compass and perspective drawing to projective geometry, have led to the computer and its integration into architectural practice. This integration is manifest in a new conceptual vocabulary, changed understandings of space and form, new practices of design, representation and making that inform the educational curriculum. Researchers discuss the relationship between the virtual and the physical, the transition and intended product outcome in terms of an interdependency between technological tools, techniques and media and design (Cook, 2008; Evans, 1995, 1997; Gänshirt, 2007). Recent developments in computational design methods and digital fabrication by Gramazio et al (2014) and Menges et al (2017) have demonstrated how an integrated approach and close engagement with computation and robotic fabrication can expand not only the range of possibilities beyond the automation of traditional fabrication techniques in unforeseen directions, but also contribute to more sustainable design strategies. The research informing this chapter draws on this work and considers the transformative potential it provides to support a re-evaluation of the role of the individual and communities in relation to sustainability strategies enabled by digital technologies.

BUILDING RELATIONSHIPS

The difference between a house and a home is the sense of connection the terms communicate. Yet even where a home may have been built by the subsequent occupier, the connection with the fabric of the building is generally limited to superficial maintenance. Equally, whilst the layout of the building for domestic occupation may be influenced by the owner, the structure of the building itself rarely is. The relationship between the occupiers and the building remains passive during its useful life. However, if the approach discussed in relation to healthcare is applied to buildings, the paradigm shift would move the occupier from passive recipient of the service of housing, to an informed, proactive occupier who is engaged in monitoring and responding to the life of the building itself. The research at the Massachusetts Institute of Technology (MIT) Media lab demonstrates the possibility of monitoring a building to an advanced level, including the movement of users, the temperature, humidity and even the sounds within the building. Their work at Tidmarsh Living Observatory takes this further, suggesting that digital technology could enhance the human experience of a space. The potential to develop a widespread web of sensors and communication tools to provide the occupier with a comprehensive understanding of the complex, ongoing performance of the house is emerging, but not in use. This approach would require not only technological development but also education on understanding the information provided by the sensors and the implications of the results. Support systems to respond to problems anticipated by a monitoring regime would need to be developed. The overall life of buildings, based on their performance and evolution would be available for analysis, informing building practices in the future. Other emerging practices enabled by digital technology that can be discussed as applied to architecture and construction, include the use of advanced digital building information modelling techniques and bespoke digital fabrication. In conventional industrial manufacturing, for example, products are designed for mass production. Because traditional manufacturing processes, such as injection moulding, require expensive moulds for each part, the aim for designers is to create generic products that will be suitable for as many customers as possible. For this reason, designs are built to take stresses at the extreme end of the load spectrum, resulting in products that are necessarily over engineered for their actual in-use requirements. However, new digital fabrication techniques are changing the mindset that drives traditional manufacturing. Additive manufacturing is an example. This term refers to a range of manufacturing processes that create objects in layers without the need to build an initial mould to shape the object. This means a component can be built as a single, one-off piece, rather than as part of a large run (Gibson, Rosen, & Stucker, 2014). This changes what it is geometrically possible to build, but more significantly, it challenges established business models used in manufacturing. Companies, such as Digital Forming and Materialise, are providing online platforms for designers to work professionally with their clients using additive manufacturing and parametric versions of their designs (with the structural limitations built in), that can be adjusted in collaboration with the customer and tailored to their needs. From a sustainability point of view, this is very significant as it has the potential to break the hold mass production currently has on manufacturing. In conventional manufacturing, a company is forced to develop generic designs that, once launched, cannot be adjusted in response to subsequent customer feedback, even if problems with the design emerge, and sell as many of the same item as possible irrespective of the customer or situation or demand. This inevitably leads to overproduction, discounting and landfill. Allwood and Cullen (2012) argue that "ruthlessly pursuing standardisation" in the twentieth century led to the standardisation of parts that were generally heavier than optimised ones (p.169). Materials are lost in the process and by design. It is a wasteful approach to meeting customer needs.

If additive manufacturing becomes widely used in business models and manufacturing, it should lead to personalised designs specific to an individual's needs and the situation the product will be used in. The material used will be minimised to the specific requirements and pre and post-consumer waste reduced. Applying this approach to architecture and construction, the aim would be to stop building generic, over engineered houses expected to stay the same and stand indefinitely, but instead to build to the specific needs of the initial occupants and the situation, and no more. In this strategy, just as with additive manufacturing of customised products, there is an upfront investment in greater planning and research into understanding the customers and site, its challenges and opportunities, as well as a more realistic assessment of the use of the building than is generally currently practice. This approach has the potential of fostering the development of lightweight, temporary structures, rather than ones that are built for the extremes of enduring use, as if the building needs to remain standing in its current form indefinitely and accommodate a myriad of possible utilities. The view of the house instead as a fluid, adaptable, temporary and ultimately recyclable dwelling is dependent on the occupant developing a very different relationship with their home. At the moment, people tend to be fairly passive recipients of the service of housing they purchase. Maintenance is generally superficial and in response to external indicators of problems, rather than built on an in-depth knowledge of the internal behaviours of the structure of the building during its use and the durability of materials within it over time and under particular stress factors. In the approach suggested here, the emphasis shifts to developing a proactive prevention of problems in home maintenance based on in-depth information gathered by ongoing monitoring, in the same way as healthcare. This would require a paradigm shift in thinking. When the building was first designed, there would need to be more planning in anticipation of the actual use, access to monitoring, and development requirements for the occupants over an agreed lifespan. Materials could then be used more sparingly as building structures would be embedded with sensors to measure and monitor internal stresses. Sensors and the ability to collect and analyse information has become increasingly accessible:

Information processing has become inexpensive and widely available, much closer to the building blocks in a monument than a precious jewel at the top. It is nearly as easy to incorporate information processing into a mass-produced object as it is to create a custom injection-moulded plastic part. The capability to collect, organise, and manipulate information has become a component instead of the goal or a digital product design. (Kuniavsky, 2010, p. 43)

According to Kuniavsky, sensors can be used to convert multiple external phenomena into electrical information, such as "light, pressure, location, heat, sound, movement, chemicals and many other things" (p.50). The sensors would allow the occupants to track the health of the building, providing a fail-safe to alert the occupants to any issues and anticipate problems that emerge through use or unexpected pressures (for example unusual weather activities). Smart materials that are fed by information sensors are under development that respond intelligently to their environment themselves (Kuniavsky 2010), suggesting a future where the fabric of the building could potentially alter in response to external influences, for example temperature change, to protect the materials. In the meantime, buildings could be designed for easier access to critical parts of a structure for the replacement of parts for maintenance, as it would be anticipated that this would be more necessary with the strategy of reducing generic building over engineering. The underlying premise would be based on the recognition that the majority of buildings, and in particular houses, are temporary structures.

In the US, the average time a first home owner occupies a house is eleven years (Beekman & deBoer, 2014), and the expected life of a house in Australia is forty years. Buildings are demolished for a range of reasons. These include the inability to adapt the floorplan to meet changing requirements, for example additional bathrooms and changing economic demand for housing in an area, leading to the vacant buildings or to densification. These changes suggest that buildings should be designed so that they can be adapted or disassembled. This means the structures need to be accessible and able to be dismantled or built on. Buildings in the twenty-first century are rarely likely to be occupied by the same people for their entire lives. Even if they are, then the facilities that the occupiers would want are likely to change as occupancy rates grow or decline depending on personal circumstance. In addition, changes to the financial situation of those occupy the building are likely to change as the occupants age, either up during the height of a career, or down due to illness or redundancy, or old age. It should be not only possible, but normal, to add or take away modules to a building as circumstances changed, with all buildings designed on that basis.

Essentially a whole-systems thinking approach needs to be introduced based on the realities of current population movements and demographics. According to the Rocky Mountain Institute (RMI) "whole-systems thinking is a process through which the interconnectedness between systems are actively considered" (Kibert, 2005, p.13). From a sustainability point of view in designing, constructing, maintaining, adapting and dismantling buildings, the core principle of valuing and retaining materials is served by the responsible adoption of whole-systems thinking. This is based on maximizing materials by value adding, such as through the use of technology to extend the effectiveness of the use of the materials (see digital fabrication and façade design example later in this chapter), the minimum specification of materials based on monitored actual use and a proactive regime of replacement of critical or failing components, rather than over specification as an uninformed, precautionary strategy, and the retaining of the value of materials through design for disassembly and the protection of that value.

MAXIMIZING MATERIALS AND VALUE ADDING

Digital technologies are increasingly used in architecture, planning and construction, and their history, use and implications widely researched and discussed. Dunn (2012), for example, described the evolution of digital fabrication and generative design in architecture and their roles in changing relationships between humans and their environment, particularly in public architecture and work spaces. The

MIT Media lab was founded in 1967 by Nicholas Negroponte to explore the human machine interface and continues to drive architectural innovation through digital technology. Burry and Burry (2010) explore digital technologies at a more fundamental level for architecture and their potential to support responsive and reconfigurable environments. The dominant use of digital technology to maximize building architecture is illustrated in the projects highlighted by Kara and Bosia (2016) such as Al Fayah Park, the Merchant Square Bridge and the Central Bank of Iraq, for advanced planning models, with parametric modelling where, for example, as surface forms evolved, structural truss geometry is updated. Multiple scenarios are now tested to analyze the behaviors of the structures under different stress factors, supporting a bespoke building approach.

According to Kibert (2005), approximately ninety percent of all extracted resources in the United States are used to create the built environment. In the UK, the government estimates that fifteen percent of construction emissions are due to the embodied energy in the materials used (Allwood & Cullen, 2012). Where material use is evaluated in relation to labor and additional processing based only on financial considerations, then the over specification of materials is often the most economic approach to meeting performance requirements.

We use so much metal that we've designed and optimized our production processes to make it with great efficiency. However, a feature of that efficiency is that it is much cheaper to make a large volume of material of the same shape than to make each piece of metal from a different shape – there are significant economies of scale related to tooling costs, and the speed of continuous as opposed to discrete processes. As a result, it's almost always cheaper to make components with simple geometries, than to use less metal. (Allwood, 2012, p.169)

However, if triple-bottom line accounting is enforced by changed regulations, then this will change. If the driver for architectural design could be shifted to maximizing material resources, then, just as the use of tool-free additive manufacturing could potentially change production value from mass to customized, so the use of digital technologies in value-added construction techniques would become more economically viable.

Example of Research: Robotic Incremental Sheet Forming (RISF)

This project example is taken from a workshop led by Tim Schork (University of Technology Sydney) and Paul Nicholas (Centre for Information Technology and Architecture). It demonstrates how the utility of material resources can be maximized through the use of digital fabrication techniques. By using digital technology to meet performance characteristics beyond those the basic form of the material could achieve, the research demonstrates value-adding as a strategy to reduce the over engineering of architectural components. The context is façade design. As building envelopes get thicker, due to the changing regulatory frameworks/ building codes for thermal insulation, they also get heavier. As a negative consequence the required substructures and connections to support these façade systems also increase in size and weight. In this example digital deformation is used to add rigidity to the sheet material, in this case copper, thereby reducing the thickness of material required for the prescribed performance characteristics, as well as adding aesthetic value. For this project, a large-scale robot arm was used to deform the flat sheets of copper to predetermined ratios. The aim was to add to the structural strength of the thin, large sheets and to explore which forms created the maximum benefit and what the constraints, challenges and opportunities would arise.

As illustrated in Figure 1, the technique of rigidizing thin metal sheet can be used to create standard connection details, such as a standing seam. In addition, however, it can be used to create lightweight structures that integrate expression by adding structural integrity based on extending the surface of the construction product – in this case a copper building façade as shown in Figure 2. This construction method can be achieved with a single robot and pre-prepared mold. This has limitations as building the mold is a material intensive process and therefore is more viable as part of mass production. However, advanced computational design and mold-less Robotic Incremental Sheet Forming (RISF) using two robot arms working in tandem to deform the metal between them does not require pre-prepared molds. This therefore is an effective way to fabricate a highly differentiated thin wall façade system that maximizes the material value.

Figure 1. Research into creating deformities using robotic incremental sheet forming

Figure 2. Built sections of digitally deformed thin copper sheeting

The results of the project shown provided data on the potential to increase the strength of thin sheet material through different deformations and the limitations of the RISF, demonstrating the boundaries at which the machine could effectively operate and how the different strategies could be quantified for uniform bending, folding and deforming outcomes.

Architectural additive manufacturing technologies make it possible to control the target deformability behavior of global geometry by varying the material deposition of microstructures. This offers new ways to fabricate cellular materials with differentiated material behaviors that go beyond mimicking the structure and mechanical properties of materials found in nature. Recent advancements in the fields of computer-aided architectural design and robotic fabrication have enabled a closed

Figure 3. Optimized deformation

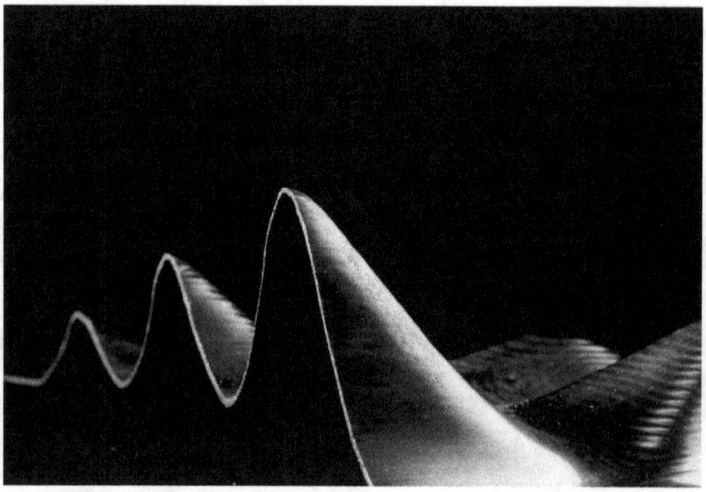

loop between the produced digital design model and the required information for fabrication. These advancements in computer-controlled fabrication, especially in robotic fabrication, afford high precision and differentiation of building components without necessarily resulting in an increase of labour and material costs. In Amsterdam DUS Architects are building the Canal House, a 3D printed structure printed from pellets of recycled plastic using a large frame fused deposition modeller. Whilst the printer itself is straight forward, the structures the architectural practice are building are innovative in that they are adaptable within the single structure. This means, for example, that walls can be designed for site specific needs on one side of the house that are different to the other.

The issue of resource-conscious design is central to sustainable construction, which ultimately aims to minimize natural resource consumption and the resulting impact on ecological systems. Sustainable construction considers the role and potential interface of ecosystems to provide services in a synergistic fashion. With respect to material selection, closing material loops and eliminating solid, liquid, and gaseous emissions are key sustainability objectives. (Kibert, 2005, p.9)

Cooling channels, voids that can be filled to increase thermal mass or bespoke lattice work to respond to dangers from impact in high risk zones for example, can be designed and built into the individual structures. 3D printing allows architects to break existing rules on the symmetry of building structures and allow for the entire house to be designed as a varied landscape. A good example is the work of DUS 3D Print pavilion KamerMaker, print small rooms using PLA, a corn-based bio plastic. The printer creates forms up to two by two by three and a half metres high.

The KamerMaker raises greater discussion concerning similar architectural techniques of the future, including the use of additive manufacturing technology and biodegradable plastics in on-demand architecture. (Beekmans & de Boer, 2014, p.53)

Organisations such as Emerging Objects are also demonstrating how 3D printing is allowing for innovations in architecture and construction, using build farms to create structures using distributed manufacturing. The geometrical complexities that digital manufacturing allows means that there is a potential for design for movement in response to earthquakes and high winds, and also disassembly that has not been possible until now.

Materials have also entered into a new realm of distinction with this onset of advancement in engineering and technology. We are at a point of history when technology allows for the "design" of specific materials to fit the unique needs of a building. Frank Gehry's signature metal panels are a great example: each is individually engineered for its precise position in the building. Such technology has introduced a period of new expressionism in the glory of materials and their qualities. (Ballard Bell & Rand, 2006, p. 10)

From a sustainability point of view, the use of new materials in digital manufacturing is also opening up new possibilities. For example, instead of a product manufacturer having to commit to a fully resolved design to launch on the market, the ability to develop short runs, or even individual test pieces of the actual finished product allows companies to develop a lower risk strategy of iterative development based on customer feedback. In the same way, buildings could potentially be adapted to use based on flexible-use building pods, such as created by UDS. However, just as over-moulding in manufacturing creates recycling problems, the mixing of materials in multi-head printers has the potential to restrict what can be reclaimed or recycled at the end of the building's lifespan. Research into the democratisation of manufacturing, and distributed manufacturing becomes possible in architecture too, with 'build farms', as demonstrated by the architects Emerging Objects in the construction of their quake column, are used to create large structures. Education, personal responsibility and changes to legislation will be needed as this fabrication revolution develops.

In the future, architectural designs will be freely available on the Internet and can be produced using local laser cutters or 3-D printers. Additive manufacturing is emerging as a new way of construction in the built environment...technological advancements are gradually making house printing a more realistic endeavour. Ready-to-print design concepts are serving to democratise a sector of city-making that once had high knowledge barriers to entry. (Beekmans & de Boer, 2014, p.55)

DESIGN FOR CHANGE

In the United States, the 140 million tons of construction and demolition waste produced annually comprise about one third of the total solid waste stream, consuming scarce landfill space, threatening water supplies, and driving up the costs of construction. (Kibert, 2005, p.10)

Rather than designing buildings to be engineered to that exact situation and specific building requirements, and monitored to ensure no unanticipated consequences, another approach it to view all buildings as for temporary occupancy. In this approach, they are specified materially to be temporary, and designed to be dismantled at the end of life and the materials reclaimed. One aspect of adding digital tagging and monitoring to materials and structures is that a chain of custody could be more effectively established and monitored. This is a new area of research for materials and sustainability. The approach is aligned to the introduction of extended producer responsibility in product manufacturing that has impacted practice in Europe. Ford Europe, for example, have run sustainability workshops on their response to extended producer responsibility legislation, where their vehicles are returned to the manufacturer at the end of their useful life. Whereas in the past, the responsibility would belong to the car owner in terms of disposal, car manufacturers in Europe have to take back the cars, so the cost and responsibility of their disposal is also theirs. In response, Ford, and other manufacturers, have redesigned their products so that they can be as far as possible disassembled at the end of life, all useful components removed from the chassis, and the remaining material reclaimed through separation and sorting processes once the vehicle is essentially put through a shredding machine.

Closed loop describes a process of keeping materials in productive use by recycling rather than disposing of them as waste. Products in closed loops are easily disassembled, and the constituent materials capable and worthy of recycling. (Kibert, 2005, p 10)

If extended produce responsibility was applied to buildings, so construction companies, along with major architectural firms were responsible for buildings once they were no longer occupied, there is a good chance that they would be designed very differently. In product design, there are good example of designing for disassembly. The Aeron chair by Herman Miller is one, where the office chair is designed for easy access to wearable parts for replacement, and the overall chair designed to be disassembled and materials reclaimed at the end of its useful life. In New Zealand, a high-profile furniture designer, David Trubridge, demonstrates taking this approach further, by designing lights from small pieces, computer numerically control routered out of plywood, that can be joined into different configurations to build lights that can be reshaped as the user wishes. Equally, a damaged component can be easily replaced, extending the usable life of the light. Arup Engineering demonstrates the potential future of designing buildings with topologically optimised connectors that would allow for bespoke construction for a specific site, but also potentially allow for easier disassembly, using metal 3D printing. This approach forms the basis of a rethinking of construction to be adaptable, rather than structurally fixed. Essentially, the proposition is that cities and buildings need to be constructed on fluid principles. This is not about adapting current practices but, much as in the paradigm shift discussed in relation to digitally enabled healthcare, rethinking architecture and construction completely as biospheres for example, completely different ways of living, different values in material use, and embracing the temporary.

The way contemporary cities are made no longer fits the dynamic of the age. Why does a society in which people become exponentially more flexible and mobile by the minute fail to make cities more adaptable to change? Cities are intense sites of activity and innovation, and yet most urban planning departments appear to be stuck in a post war mode that fails to address the needs of new activities and new users. (Beekmans & de Boer, 2014, p.16)

In architecture, there are increasing numbers of examples of structures that are built from modules that allow for flexible accommodation as well as more examples of temporary living emerging that utilise materials in different ways - embracing impermanence and an ecosystem approach. Temporary dwellings in refugee camps can be coated to provide waterproofing for a fifteen-year lifespan. What if all buildings were designed to only last for fifteen years? How would that change societies ideas of permanence, of community, relationships, work-related behaviors and patterns - the ownership of all forms of goods? As relationships are becoming more fluid, perhaps it would reduce social pressures if living spaces were not as invested – financially and emotionally, perhaps with populations not connecting with a single building structure, but constantly on the move: "Real freedom begins when our material burden is minimised" (Mitchell, Tiny Life, p.48).

What if all buildings were designed to grow or disintegrate on a twenty-year life cycle at most, preferably as an ongoing cycle of renewal? What material choices would be made with this strategy? One illustrative example of this idea is the City Egg. This is made of sacking, bamboo, wood chippings and grass seed. Just as mycelium is receiving attention in packaging, and salt and coffee grounds in 3D printing by architectural form Emerging Objects, so could different material choices be considered if buildings were designed for impermanence.

In reality, most traditional construction materials are able to be down-cycled, for instance used as a lower-value material in fill or road subbase, rather than recycled (Kibert, 2005). The aim therefore is to reduce the use of materials where possible, by reducing over engineering, and avoid materials reaching the point of reclamation by designing for disassembly and adaptability. Basically, what is required is a new, whole systems approach, not hampered by the constraints and habits of the past.

In the past, the workplace was a massive determining factor of where to live, ensuring that your daily commute would not consume hours each day. What happens, though, when work has turned fluid and spreads everywhere and nowhere? Never before in history have we had access to more international contacts in order to make ends meet. Even interpersonal relations are becoming more flexible, with divorce rates at their highest in human history. Home ownership is declining amongst younger generations, due to the barriers to entry into the housing market as well as a fear of spatial commitment. Despite all this flexibility, houses remain static physical objects – beacons of peace and rest in a fluid world – but, at the same time, it is dragging us down. (Beekmans & de Boer, 2014, p.19)

The advantage of this approach is that the availability of accommodation can be altered depending on the demand. This provides an arguably more realistic response to the densification of a city, for example, where the influx may be dependent on a major employer or is in response to a weather event. Beekmans and de Boer (2014) argue for temporary and supplementary accommodation, with examples such as the 'Fold Inn' by Dutch designer Lieke Jildou de Jong, the 'Loft Cube' by Aisslinger, airlifted onto an apartment block by helicopter, the 'Parasite' units by Malka Architecture, vertically-secured onto the side of a railway viaduct and the 'Sleepboxe's' installed at Moscow's Sheremetyevo International Airport, aligning to the small space 'Capsule Hotel' recently opened in Sydney, Australia. Whether because of changing work patterns or the significant extension to lifespans predicted by actuarial science, densification of living or a growth in fluid city living is a possibility.

MATERIAL CONNECTIONS IN A DIGITAL ERA

We are physical beings so it perhaps makes sense for us to identify and express our values using physical objects, which we like to touch and smell as well as read. (Miodwonik, 2015, p.45)

Alongside the approach of living lightly on the land, and maximising materials through the use of digital technologies, and reclaiming them at the end of life, there is still an argument for creating buildings that have emotional investment, both for individuals and communities. Yet in a digital era, with a more nomadic lifestyle, place and objects are less tangible. Rather than temporary products, Stuart Walker (2006) supports creating invested designs as a sustainability approach. Walker argues that the engagement in all objects is less about definitions of beauty than it is about connection:

A sustainable solution can be understood as one that possesses enduring value in terms of its meaning and characteristics. (Walker, 2006, p. 39)

This can be applied to buildings where the engagement level is heightened through strategies such as building in individual and community connection, for example through building into the fabric of the building memories of people, events and place. Currently buildings are not embedded with any history. Once occupants move on, the history of most buildings is lost. With increased population fluidity, this would happen more quickly, with implications for societies. Digital technology has the potential to contribute to this sustainability strategy too, by helping to create new layers of meaning within a building. Tom Dixon demonstrated this strategy for products with his work with Artek. Dixon advertised for the return of iconic stools manufactured by Artek, which he bought back and then resold with the story of where the chair had been since its construction digitally embedded in the stool, accessible by a smart phone app. By similarly embedding stories within the fabric of a building as it ages, layers of connection and meaning to individuals and communities would be built up that would mean it was more likely to be retained and repaired over time – and visited more if a public building. It would also act as a curator to community knowledge currently lost. This approach could be extended during the disassembly process, where reclaimed timber for example, could be embedded with its history to add value not only for the user but also the community.

The game changing aspect of the internet is that the community of sharers is not necessarily place based. That is, people can be communicating and trading between neighbourhoods or across cities and continents. (Legge, 2012, p.15)

Community is no longer fixed to a physical space. Online communities are well established, and the generations coming through now have grown-up within a digital world. Whilst older generations may have difficulty imagining the opportunities and individual responsibilities that would come from digital immersion inside a connected building and a web of connected products, for the next generation it may be easier to imagine a virtually connected city built on the fluid movements of nomadic people moving in response to weather patterns, work opportunities or just a need for change. If this is the case, then it may also be easier for the next generation to proactively take control of their use of resources and impact on the environment: "What might seem radical to one generation is considered normal by the next" (Legge, 2012 p.7).

Life-cycle assessment is a complex challenge, but as discussed by Hendrickson et al (2006), awareness, tracking and control are essential. For a child born into a future society where, for example, digital tracking is routinely and visibly linked energy use in the home on a daily basis or used to trace irresponsibly disposed of waste back to its source, or where perhaps micro-donations to combat climate change linked to pollution creation, for example because of a car journey, are common practice, there may be a more direct awareness of an individual's environmental footprints. When through apps or augmented reality, a timber building can tell stories of events and experiences that have happened within its walls and show images and even sensory experiences from where the trees it was built from were grown, relationships between people and the built environment will be a very different experience to the static, voiceless, engagements with materials and structures that exist now.

SOLUTIONS AND RECOMMENDATIONS

The results to date of the research informing this chapter and its recommendations suggest that technological change and in particular the use of integrated digital technologies in architecture and building, from planning, through construction techniques, to monitoring, analysis and re-purposing need to be strategically rethought for their significance in relation to social change, and its potential impact on sustainability.

FUTURE RESEARCH DIRECTIONS

Future research will explore the further development of digitally enabled architectural and construction practices and the potential benefits of social adaptations they could engender for a twenty-first century response to environmental and ethical imperatives.

CONCLUSION

Since the beginning of the twenty-first century, the dominant global paradigm has been one of disruption. From political to social, environmental to economic, established models of practice have been fractured, distorted or redirected. In the last fifty years of the twentieth century, change was predominantly incremental, with projections of increased globalization and homogeneity, convergence and contraction and collective environmental responsibility. Yet accepted attitudes and patterns of behaviour, values and strategies have been challenged by world events, conflict and emerging technologies.

For the construction industry and all that occupy static buildings, a total rethink of the relationship of the building to its environment, to the planet's environment and natural forces more broadly and to the occupants within them. Hawken, in the 2017 book Drawdown: The most comprehensive plan ever proposed to reverse global warming, argues that a completely new view of the construction and operation of buildings needs to change:

Buildings can function more like a forest, generating a net surplus of positives in function and form and exhaling value into the world. Buildings, in other words, can do more than simply be less bad. They can contribute to the greater good. (Hawken, 2017, p.188)

If that is going to happen, then city planning and the development of the built environment for the twenty-first century needs to be much more self-aware, and each individual equally aware of the incremental impact they contribute to on a daily basis. The patterns of living from the last century no longer fit. Expectations and behavior needs to change.

New structures for decision-making, financing, governing, managing, and producing had to be thought of in order to prevent western society as a whole from driving over the edge of a cliff. (Beekmans & de Boer, 2014, p.14)

Digital fabrication technologies, such as additive manufacturing, and digital monitoring and communication systems are providing new opportunities for customized construction that can be built more accurately to specific requirements and managed over time. Digital technologies are also creating new interactivity. The emerging suite of technologies operating with this time of change can act as a driver for social change based on innovations that addressing the challenges involved in utilizing building materials, designing for change, and developing viable, emotionally connected systems for material use and individual responsibility and awareness for long term sustainability.

REFERENCES

Allwood, J., & Cullen, J. (2012). *Sustainable materials: With both eyes open.* Cambridge, UK: UIT Cambridge.

Ballard Bell, V., & Rand, P. (2006). *Materials for architectural design.* London: Laurence King.

Beekmans, J., & de Boer, J. (2014). *Pop-up city: City making in a fluid world.* Amsterdam: BIS.

Benjamin, W. (1936). The Work of Art in the Age of Mechanical Reproduction. In H. Arendt (Ed.), *Illuminations* (pp. 219–253). New York: Schocken Books.

Braham, W., & Hale, J. (2007). *Rethinking technology: A reader in architectural theory.* Abingdon, UK: Routledge.

Burry, J., & Burry, M. (2010). *The new mathematics of architecture.* London: Thames & Hudson.

Cook, P. (2008). *Drawing: the motive force of architecture.* Chichester, UK: Architectural Design Primers, John Wiley & Sons.

Dunn, N. (2012). *Digital fabrication in architecture.* London: Laurence King.

Evans, R. (1995). *The projective cast: Architecture and its three geometries.* Cambridge, MA: MIT Press.

Gibson, I., Rosen, D., & Stucker, B. (2014). *Additive manufacturing technologies: Rapid prototyping to direct digital manufacturing* (2nd ed.). New York: Springer.

Gore, A. (2013). *The future.* New York: WH Allen.

Gramazio, F., Kohler, M., & Willmann, J. (2014). *The robotic touch: How robots change architecture.* Zurich, Switzerland: Park Books.

Hajkowicz, S. (2015). *Global megatrends: Seven patterns of change shaping our future.* Melbourne: CSIRO.

Hawken, P. (2017). *Drawdown: The most comprehensive plan ever proposed to reverse global warming.* London: Penguin Books.

Hendrickson, C., Lave, L., & Matthews, H. (2006). *Environmental life cycle assessment of goods and services: An input-output approach.* Washington, DC: Washington State, Resources of the Future.

Kalay, Y. (2006). The impact of information technology on design methods, products and practices. *Design Studies, 27*(3), 357–380. doi:10.1016/j.destud.2005.11.001

Kara, H., & Bosia, D. (2016). *Design engineering refocused.* Chichester, UK: John Wiley & Son. doi:10.1002/9781119164838

Kibert, C. (2005). *Sustainable construction: Green building delivery and design.* Hoboken, NJ: John Wiley & Son.

Kuniavsky, M. (2010). *Smart things: Ubiquitous computing user experience design.* Burlington, VT: Elsevier.

Kvan, T. (2004). *Modes of Representation. In Architectural Review Australia* (Vol. 90, pp. 82–83). Digital Architecture.

Legge, K. (2012). *Doing it differently.* Sydney: Place Partners.

Menges, A., Schwinn, T., & Krieg, O. D. (Eds.). (2017). *Advancing wood architecture: A computational approach.* London: Routledge.

Miodwonik, M. (2015). *Stuff matters: Exploring marvellous materials that shape our man-made world.* Seattle, WA: Mariner Books.

Mitchell, R. (2014). *Tiny house living: Ideas for building and living well in less than 400 square feet.* Sacramento, CA: Betterway Home.

Mitchell, W. (1999). A Tale of Two Cities: Architecture and the Digital Revolution. *Science, 285*(5429), 839–841. doi:10.1126cience.285.5429.839

Mitchell, W. (2001). Roll over Euclid: How Frank Gehry designs and builds. In J. F. Ragheb (Ed.), Frank Gehry, Architect (pp. 352- 64). New York: Solomon R. Guggenheim Foundation.

Pérez Gómez, A., & Pelletier, L. (1997). *Architectural Representation and the Perspective Hinge.* Cambridge, MA: The MIT Press.

Ross, A. (2016). *The Industries of the Future.* New York: Simon & Schuster.

Walker, S. (2006). *Sustainable by Design: Explorations in theory and practice.* Oxford, UK: Earthscan.

KEY TERMS AND DEFINITIONS

Additive Manufacturing: A range of digital manufacturing technologies that build objects in layers from a computer model without the use of molds.

Megatrends: Global patterns influencing behaviors and attitudes.

MOOCs: Massive open online courses were created to allow large numbers of people open access to education. The first MOOC was led by Stephen Downes of the Canadian National Research Council.

Robotic Incremental Sheet Forming: A process whereby robot arms are programmed to deform sheet metal. This can either by a single robot arm working to deform metal sheet into pre-prepared forms, produced, for example, by computer numerically controlled milling, or alternatively by two robot arms working in conjunction, deforming the material between the two heads in free space.

Sustainability: The ability to sustain life/lifestyle without impacting the ability of future generations to do the same.

Chapter 9
Urban Quality Assessment at the Neighborhood Scale:
An Experimental Approach

Valentina Puglisi
Politecnico di Milano, Italy

Andrea Ciaramella
Politecnico di Milano, Italy

ABSTRACT

This chapter describes the approach adopted within the framework of a multi-destination development project; the goal of which is to promote innovative technologies and methods to evaluate the environmental quality of an urban district under construction. This method of analysis has been tested on an area located in the former historic district of the Fiera di Milano, where a series of typical urban functions are inserted within a large public park. The success of the work is represented by indicators (air quality, acoustic, microclimate) that relate to the finished district and that can be compared with average values in the same city. The system may constitute a protocol capable of bringing benefits to local authorities. This type of assessment could be requested of developers/builders for complex projects, resulting in changes to the initial plan if the assessment identifies critical issues related to the design choices (orientation of buildings, green areas, traffic emissions, etc.) with the ultimate goal of creating neighborhoods with better environmental conditions.

DOI: 10.4018/978-1-5225-6995-4.ch009

INTRODUCTION

The assessment of urban quality in the European current practice is carried out following legislative requirements or international standards: the Strategic Environmental Assessment (SEA) and the Environmental Impact Assessment (EIA) are the most widespread environmental impact assessment systems at European and Italian level.

SEA is a decision making support process whose main goal is to estimate the environmental effects of plans and programs before their approval (ex-ante), during their implementation and at the end of their period of validity (ex-post). Currently, the SEA is applied in Italy in fields (ex. articles 6 and 7 of Legislative Decree no. 152/2006), such as: water management, telecommunications, tourism, town and country planning or land use. It also supports the planning process at a large scale (at city, regional, national level, etc.).

EIA is referred to the design and authorization of specific projects, even at territorial scale, and aims to assess their environmental impacts (i.e. the changes in status of the environmental components) normally linked to an authorization process.

These evaluations have a static character and do not consider the interaction of several variables and how they react in relation to changes caused by external factors. In particular, in the case of interventions for new developments (greenfield) or re-development (brownfield): construction of new districts, re-functionalization of old industrial buildings, processing of degraded parts of the city, which in the context of environmental assessments are considered interventions at a microscale, no practical tools are available supporting the analysis of the consequences due to design choices and the actions needed accordingly.

As the design activity is affected by the lack of an integrated view (Malatras, Asgari, Baugè, & Irons, 2008), in the current practice the interventions are conducted by specialists and are the sum of specific contributions and not the best result of a real and effective integration of skills. A real integration can only be realized if we can systematize the process of assessing, checking and evaluating the results of the different contributions.

Useful tools are more and more being adopting, especially at regional level, such as atmospheric models, that can now reliably reproduce the spatial distribution of concentrations of air pollutants and therefore are used extensively, for example in the preparation of the annual report on air quality or in the planning and evaluation of limitation measures.

The planning of measures for the remediation of ambient air quality is therefore up to now carried out at the regional scale (e.g. incentives for markets of cleaner technologies) or urban scale (e.g. pollution charge for accessing low emission zones), but little is worked out at the finer scale of a district, also because of the supposed lack of operating analytical tools; the perception, on the part of authorities, planners and citizens, that the climate may be somewhat influenced by urban planning at district scale, is still weak. This theme is generally regarded as pertaining to a global scale - obviously the most appropriate - and operations such as a SEA of a neighbourhood tend to contribute to the theme with the quantification of the contributes to global greenhouse gas budget (Roper, & Beard, 2006).

However, assessment tools are now available able to clarify the role of the planning of actions to the scale of neighbourhood on local climate. The nearest instrument to this approach currently existing is the LEED neighbourhood protocol (Leadership in Energy and Environmental Design), the American rating system for the design, construction and operation of high performance green buildings, homes and neighbourhoods. The main weakness of this protocol lies in the fact that it originates in a territorial context deeply different from the European and some of the indicators are not very applicable to the Italian and European context (for example, the protocol positively considers urban density, while in the Italian context, very densely built, high quality housing developments are characterized by low density spaces and greater distances between buildings).

Understanding how the design of the neighbourhood affects the local microclimate is not an end to itself, if we think, for example to the increase in hospitalizations and deaths that usually occurs at extreme temperature events because of summer heat waves, events more and more frequent, at European latitudes, because of the climate change: the introduction of urban design elements that reduce the air temperature even just a few degrees would improve not only the wellbeing state of the neighbourhood inhabitants but also their health (Lee, & Chan, 2009). In this sense, also the improvement of micro-climate must be considered within a protocol aimed at quantifying the design good practices and, as demonstrated by tests carried out by the working group on CityLife district of Milan, tools to quantify changes in the microclimate induced by different design solutions at the neighbourhood scale are now available.

FINDINGS AND METHODS

A new approach, which aims to simulate through specific modelling the constructions and development projects, could be able to give concrete suggestions to improve their quality and, above all, really contribute to an improvement in citizens' quality of life. This goal can be achieved only by identifying indicators that can be assessed in an integrated way and suggesting corrective actions/modification and/or integration of the projects.

The project team[1] received in 2012 a private commission, requiring to assess the environmental performances of a large urban project in the City of Milan: the CityLife district.

The work is focused on 3 main areas of research:

1. Three-dimensional rendering of the project area and its surroundings;
2. Measurement of air quality and wellbeing of the microclimate in the finished district, through specific modelling;
3. Measurement of acoustic comfort in the finished district, through specific modelling.

In particular, the parameters related to air quality, acoustic comfort and well-being of the microclimate were compared with an area similar to that of the project analysed.

The approach investigated aims to simulate, through specific modelling, the construction and development projects. It could be able to give concrete suggestions to improve their quality and, above all, really contribute to an improvement in citizens' quality of life. This goal can be achieved only identifying indicators that can be assessed in an integrated way and suggesting corrective actions/modification and/or integration of the projects.

Before the works were carried out, in collaboration with the client, with Milan Council and a series of satellite image display tools, information and data were collected in relation to the project based on a checklist prepared by the work group, with particular reference to:

- Technological solutions, plants and buildings that characterise the features of the urban district CityLife;
- The district's green spaces and parking system;
- The transport system and infrastructure within the urban district CityLife and the network to connect it with the city;
- The services in the area, etc.

The success of the work is represented by specific indicators (air quality, acoustic performance and microclimate) that relate to the district, considering it completely built, and that can be compared with average values in other context within the city. For this reason, it was necessary to model the neighbourhood (in 3D to see it completed) and dynamically simulate all external factors (vehicular traffic, general neighbourhood activity, pollutant emissions, etc.) corresponding to the different seasons.

This system may constitute a protocol able to bring benefits to local authorities; in fact, this type of assessment could be requested of developers/builders for complex projects, resulting in changes to the initial plan if the assessment identifies critical issues related to the design choices:

- Orientation of buildings;
- Quality and presence of green areas, and;
- Traffic emissions inside the urban district, etc.

with the ultimate goal of creating neighbourhoods with better environmental conditions.

THE CITYLIFE PROJECT

This approach of analysis has been tested on an area of about 255.000 square metres, located in the former historic district called "Fiera di Milano", where a series of typical urban functions (residential, commercial and trade) are inserted within a large public park.

CityLife is the company committed to redeveloping the ex-historical district of the Fair in Milan. This area, free by the move of the Fair plant in Rho-Pero zone, has been object of an international tender of urban qualification, that has involved companies, financiers and great names of the international architecture. The contest winner was CityLife, with a project signed by Zaha Hadid, Arata Isozaki and Daniel Libeskind. CityLife has assured the realization of the plan thanks to a shareholder, participated by 2 main insurance groups in the world: Generali Real Estate and Allianz. The Project started in 2009 and It will be completed in 2023.

The residential buildings designed by Hadid are characterised by:

- 5-13 floors;
- Energy Class A;
- 225 housing units;
- 700 living residents (700);
- 35.000 square metres of Gross Floor Area.

The residential buildings designed by Libeskind are characterised by:

- 4-13 floors;
- Energy Class A;

- 382 housing units;
- 1.000 living residents;
- 50.000 square metres of Gross Floor Area.

In the area of the project there are 3 towers, one of which is the tallest in Milan.

Hadid tower is characterised by: 170 metres high, 44 floors (39 floors for office spaces), 3.200 people ant it was pre-certification LEED with result gold.

Libeskind tower is characterised by: 150 metres high for office spaces and about 2.000 people.

Isozaki tower is characterised by: 202 metres high (infect it is the tallest tower in Milan), 50 floors (46 floors for office spaces), 3.800 people and it was pre-certification LEED with result gold.

Trading area, in the centre of the project, is characterised by:

- A subway stops called M5, it is the fifth subway line of Milan, the violet line;
- 20.000 square metres of Gross Leasable Area;
- About 100 stores;
- Three sectors:
    - The gallery in the middle of 3 Tower square, with the most important brands;
    - The gallery in the place of underlying square, with a mix of shops and public utility services;
    - The external square, which are surrounded by coffee shops and restaurants.

THE IMPLEMENTATION OF A SYSTEM, REGARDING THE ENVIRONMENTAL PERFORMANCE ASSESSMENT FOR URBAN DISTRICTS

Three-Dimensional Rendering of the Project Area and Its Surroundings

In order to carry out the work, the project area and the area for comparison were rendered in 3D using a shape file, a file containing a plan of building perimeters with their height as an associated attribute. Thanks to the collaboration with the owners and the use of satellite and map image display tools, all necessary data for the three-dimensional rendering were collected, mainly in reference to building perimeters and heights, road networks, green areas, etc.

Starting from the shape file previously developed, the three-dimensional topography was created at high resolution (to the order of one metre) of the land and buildings of the project area and of the area for comparison, nearby and not more than 1 km away.

The software used were:

- Micro Swift Spray for the measurement of the air quality;
- ENVI-Met 3.1 for the measurement of Wellbeing;
- CATT ACOUSTIC for the measurement of acoustic comfort.

The three-dimensional rendering of the project area and its surroundings is shown in Figure 1, Figure 2, and Figure 3.

Microscale Assessment of Microclimate Wellbeing and Air Quality in the Urban District CityLife for the Purpose of Comparison With Average Values

The objective of this phase is the development of specific micro-simulations to assess the benefits contributing to the wellbeing of the microclimate and air quality at a local level in the CityLife project simulating traffic issues and heating. The aim is to compare the results with another district in Milan, Garibaldi district. To do this was used the Model Micro-Swift-Spray (MSS), which allows the microscale reconstruction of pollutants spread into the atmosphere.

Figure 1. The three-dimensional rendering of the project area and its surroundings

Figure 2. The three-dimensional rendering of the project area and its surroundings

**For a more accurate representation see the electronic version.*

Figure 3. The three-dimensional rendering of the project area and its surroundings

**For a more accurate representation see the electronic version.*

A number of parameters have been identified to represent the district in terms of wellbeing, comfort and quality and compared with another district in Milan, which was urbanised in a more traditional way. These parameters refer to practices adopted by individuals and institutional sources for the representation of environmental/urban quality (i.e. "*Report on the State of the Environment in Lombardy*", ARPA Lombardy, "*Quality of the Urban Environment*" ISPRA; "*Urban Ecosystem*" Legambiente, etc.).

In particular, the calculations made allow us to:

- Quantify the benefits, from the viewpoint of air quality, of the solutions adopted in the urban district project compared to those estimated within a more traditional urban context;
- Highlight the major / minor environmental critical points of the district (relative to air quality and microclimate) transmitting, if necessary, the design choices that are still works in progress (ideal location of cycle and pedestrian paths, play areas, public transport stops, etc.).

The steps of this phase are:

- Selection of the weather-dispersive episodes and weather simulations;
- Definition of the pollution sources and calculation of emissions;
- Study of air quality in microscale;
- Study of the microclimate;
- Study of acoustic comfort, and;
- Comparison with the sampling area.

Selection of the Weather-Dispersive Episodes and Weather Simulations

Simulations were carried out considering two episodes lasting one-two days each, chosen from those typical and / or critical from a weather / dispersive viewpoint for the city of Milan. To select these episodes (winter and summer) one year of meteorological data obtained from a weather station of the regional network were analysed, significant with respect to the location of the project district.

Simulations were carried out on January 29th, 2010 for winter meteorological episodes. In this period:

- Mean wind speed is 2-3 metres per seconds, and;
- Wind direction is East and West.

The other simulations were carried out on 12th July 2010 for summer meteorological episodes. In this period:

- Mean wind speed is 3 metres per seconds, and;
- Wind direction is North-NorthWest.

Taking the precise meteorological time series as the starting point, simulations were performed to reconstruct wind fields and turbulence for the area in question, in relation to the two selected episodes.

Figure 4. Selection of simulation period: winter simulation period: 29 January 2010 (wind 0 speed and direction)

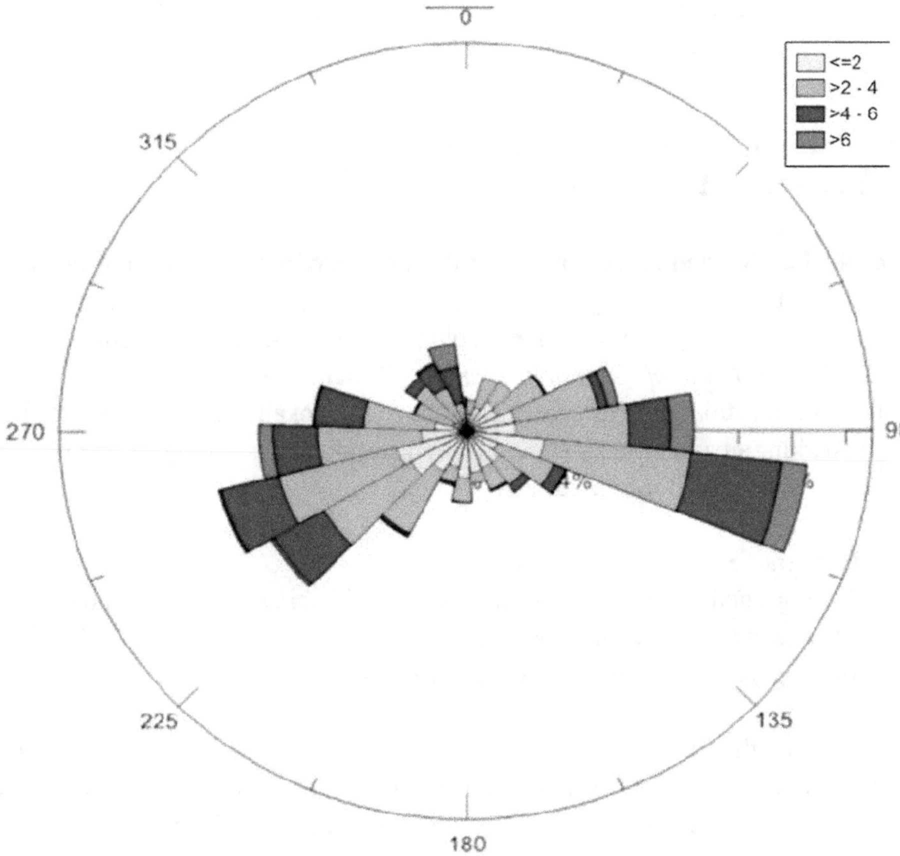

**For a more accurate representation see the electronic version.*

Figure 5. Selection of simulation period: winter simulation period: 29 January 2010 (wind 0 speed and direction)

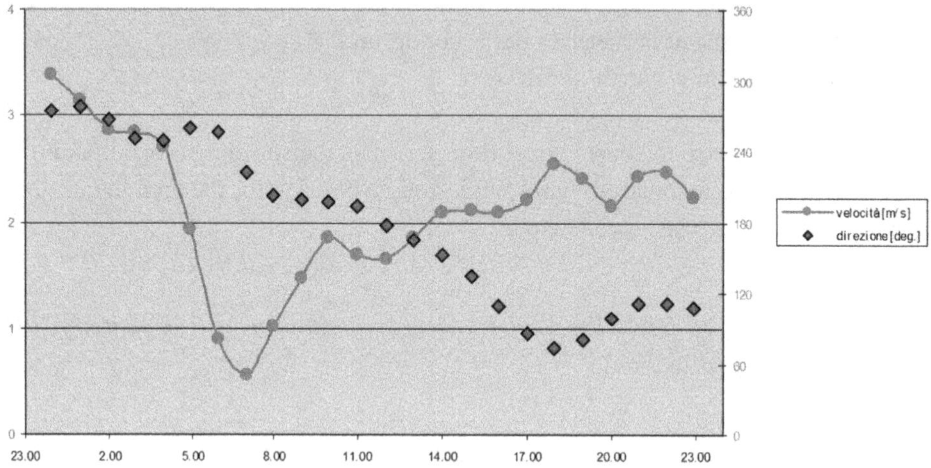

Definition of the Pollution Sources and Calculation of Emissions

Once the location and characteristics of the sources of air pollutants present were defined (roads, parking areas, power plants, etc.) the corresponding polluting emissions were calculated using standard methodologies at European level documented in *EPER/EEA Air Pollutant Emission Inventory Guidebook*.

Once the location and characteristics of the sources of air pollutants were defined (roads, parking areas, power plants, etc.) the corresponding polluting emissions were calculated. The pollution sources are:

- Road traffic;
- Underground road traffic: (only for CityLife area). The emissions were calculated through metal grids, and;
- Heating: calculated only for winter case.

In regard to the road traffic, emissions (CO, NOx, VOC, N_2O, NH_3, SO_2, CO_2, CH_4, Pb, PM, Pb, HM, NMVOC) were calculated using the program TREFIC, developed on the basis of the most up-to-date methodology COPERT (*Computer Program*

Figure 6. A traffic allocation model

**For a more accurate representation see the electronic version.*

to Calculate Emissions from Road Traffic). The work group has been involved in developing the distribution of vehicles in the COPERT categories on the basis of national data of registered vehicles and average distance travelled.

In regard to the road traffic, CityLife also provided data related to the traffic attracted/generated by the new functions. The emissions produced by the vehicles in the parking lot are linked to the heating metal grid emissions. In particular, there are 6 points of access of the traffic flows to the urban road. Finally, in regard to the heating emissions, all activities were:

- Calculation of the volumes;
- Individualization of the building function of (residential, not residential, tertiary, etc.);
- From System Informative Regional Energy Environment carved out the tipology of fuels in the city of Milan are used for heating and air conditioning (different for tertiary and residential sector);

- From site CESTEC (Center for the Technological Development, the Energy and the Competitiveness) was collected the middle values of primary energy requirement for the winter climatization, and;
- Calculation of the requirement in cubic metre (for built typology).

The simulations, representing a winter situation and a summer situation, take into account the contributions of both the city's local and background emissions.

Study of Air Quality in Microscale

Through the use of the Model MSS - Micro-Swift-Spray, which allows the microscale reconstruction of diffusion of pollutants into the atmosphere, up to four simulations have been performed, varying the two domains and the two weather-dispersive episodes.

A number of parameters have been identified to represent the district in terms of air comfort and quality. They are:

- **Nitrogen Dioxide (NO2):** Due to thermal power stations, domestic heating, gasoline and diesel cars.
- **Dust Particles (PM10):** Due to industrial activities (foundries, cement plants, constructions sites and mines), combustion processes relating to thermal power stations, traffic, tyre wear and wear of the brakes.
- Benzene.

In this way maps of the relevant concentration statistics were generated and comparisons made between domains and between episodes. The simulations, representative of a winter situation and a summer situation, take into account both the contributions of both the city's local and background emissions. In the following maps are represented the simulation of the pollutants in winter and summer in the project CityLife (Figure 7 and Figure 9) and in the area for comparison (Figure 8 and Figure 10).

In the following maps you can see the contributions of Nitrous Dioxide (NO2) and Dust Particles (PM10) concentrations with winter coming from the following emission sources: traffic (Figure 11), metal grids (Figure 12) and heating (Figure 13).

The simulations conducted on vertical section allows to:

- Monitor the development to the upper sections of concentrations, simulating what people can breathe in/or the air input flows into the apartments, and;
- Give an estimate of the vertical decadence of the concentrations.

Figure 7. Maximum hourly concentration of Nitrous Dioxide (NO2) – Summer case

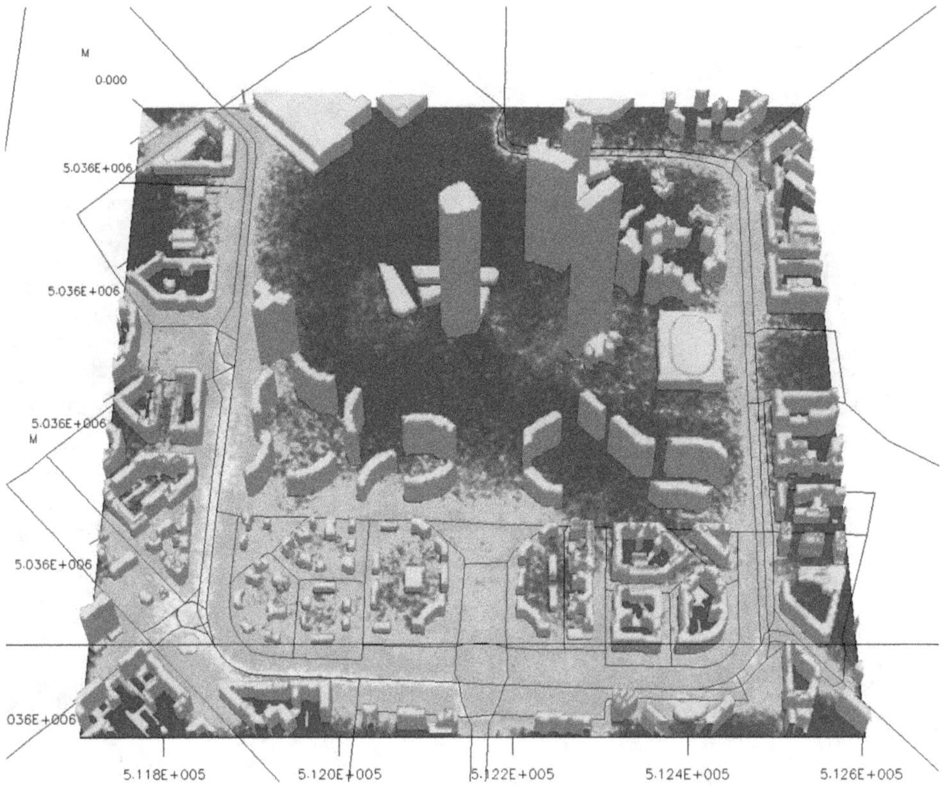

**For a more accurate representation see the electronic version.*

In CityLife district, near to the residential area, at 35-40 metres from the ground, the concentrations of Nitrous Oxide (NOx) are approximately one third of those at the ground. In Garibaldi significant environmental impact has been calculated in the canyon areas and also upstairs, at 35 metres from the ground. Here the concentrations of Nitrous Oxide (NOx) are one sixth of those at the ground.

In conclusion of this phase we can say that:

- The pollution of CityLife is better than in the area for comparison;
- Air pollution levels are affected by the fact that the district is situated inside the city;
- The concentrations of nitric oxide (NOx) in CityLife are better than in Garibaldi: the percentage of improvement is 15% in winter and 40-50% in summer.

Figure 8. Maximum hourly concentration of Nitrous Dioxide (NO2) – Summer case

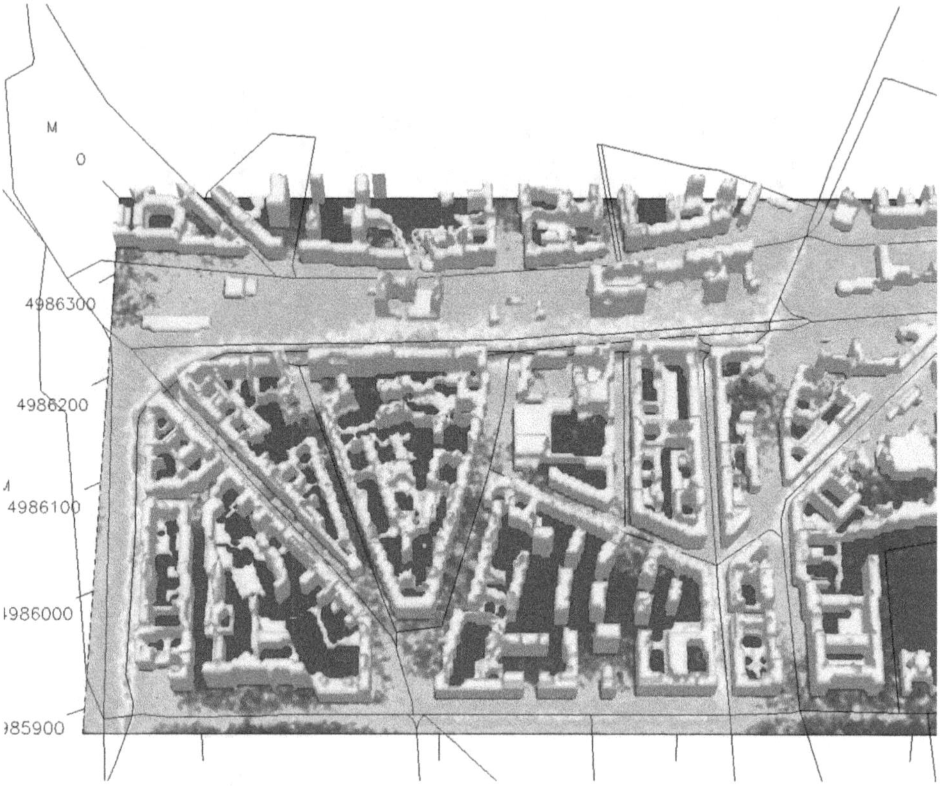

**For a more accurate representation see the electronic version.*

- The percentage of improvement for Dust Particles (PM10) is lower than Nitric Oxide (NOx). They are: 5-6% in winter and 25-30% in summer.
- The levels of pollution in CityLife are superior for the zones which look out the lot and smaller in the lot.

Study of the Microclimate

Through the use of ENVI-met, three-dimensional model of the microclimate designed to simulate the surface-plant-air interactions in an urban environment, modelling studies were conducted for the reconstruction of the microclimate in the presence of the architectural elements characteristic of the urban intervention in question and of that used for comparison. The simulations conducted for the microclimate study take into account some initial parameters, in particular:

Figure 9. Maximum hourly concentration of Nitrous Dioxide (NO2) – Winter case

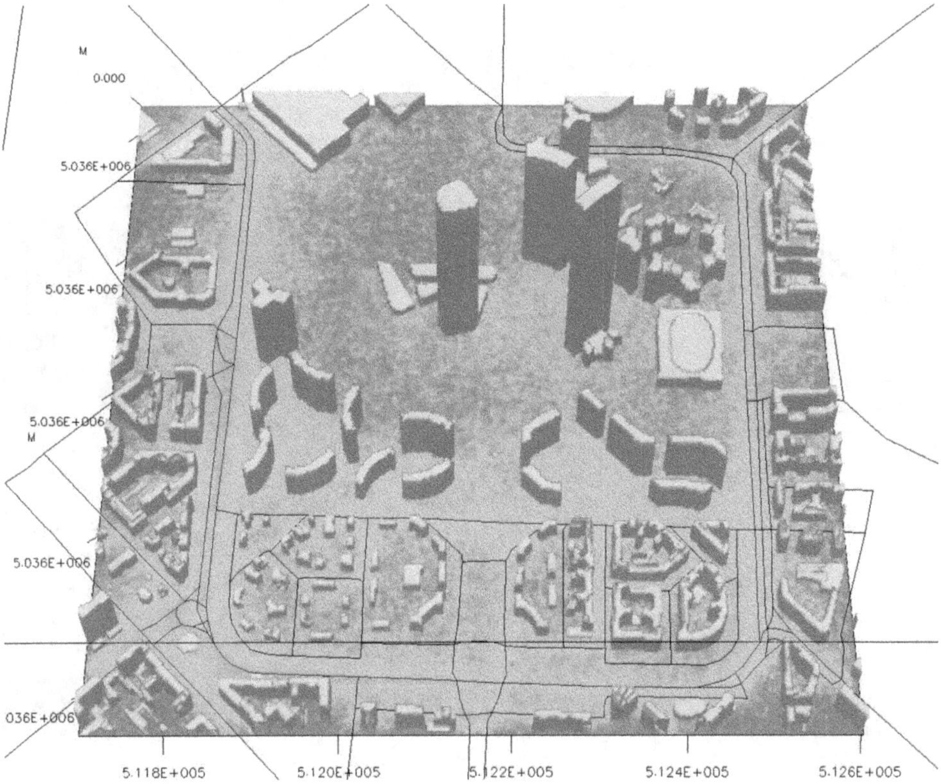

**For a more accurate representation see the electronic version.*

- In winter:
    - The simulation was carried out in10 January two thousand and ten (2010);
    - In this period the temperature is two hundred and seventy-five point five degrees' kelvin (275,5);
    - Relative humidity at 2 metres from the ground is 81%;
    - Cloud cover is none;
    - Wind speed is 3 metres per second;
    - Wind direction is three hundred and thirty-seven point five degrees (337.5).
- In summer:
    - The simulation was carried out in10 July two thousand and ten (2010);
    - In this period the temperature is two hundred and ninety-seven degrees' kelvin (297);

Figure 10. Maximum hourly concentration of Nitrous Dioxide (NO2) – Winter case

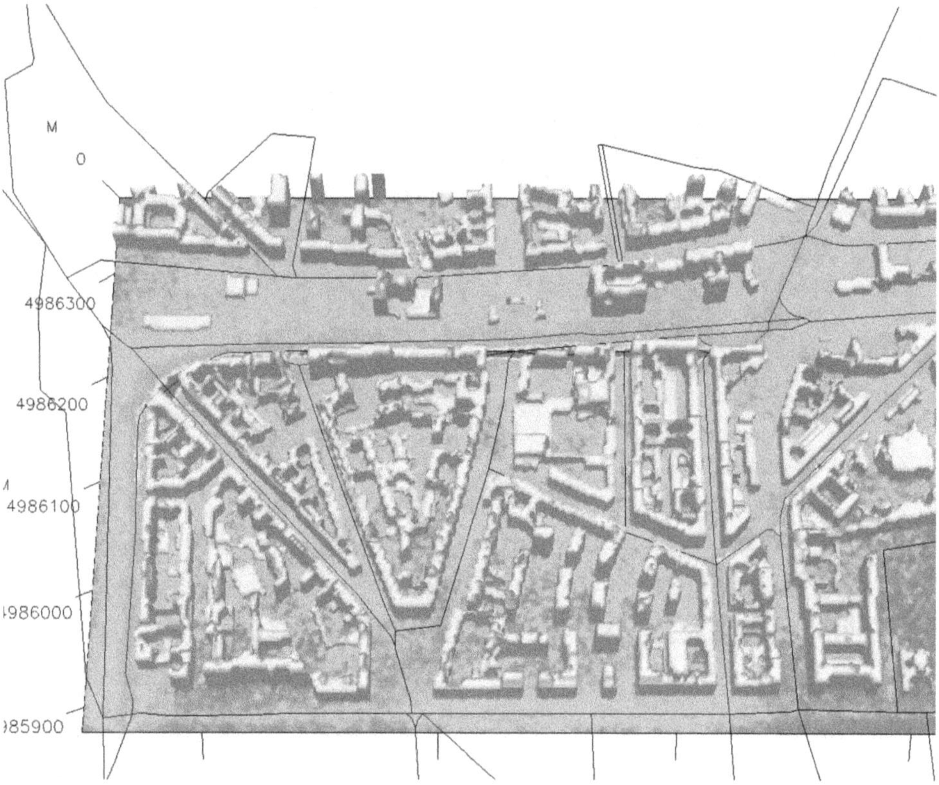

**For a more accurate representation see the electronic version.*

- ◦ Relative humidity at 2 metres from the ground is 66%;
- ◦ Cloud cover is none;
- ◦ Wind speed is 3 metres per second;
- ◦ Wind direction is three hundred and forty-seven point five degrees (347.5).

These were chosen based on the series of climatic conditions measured at Linate (Milan) weather station and, with regard to the wind, on simulations of the national project QualeAria.

The simulations take into account the following parameters.

For the green areas have been taken into account the following parameters:

- Plant species: with 2, 10, 15 e 20 metres height;
- Conifers;

Figure 11. Contributions from Nitric oxide concentrations (background 145 mg/m3) in winter coming from the following emission source: traffic

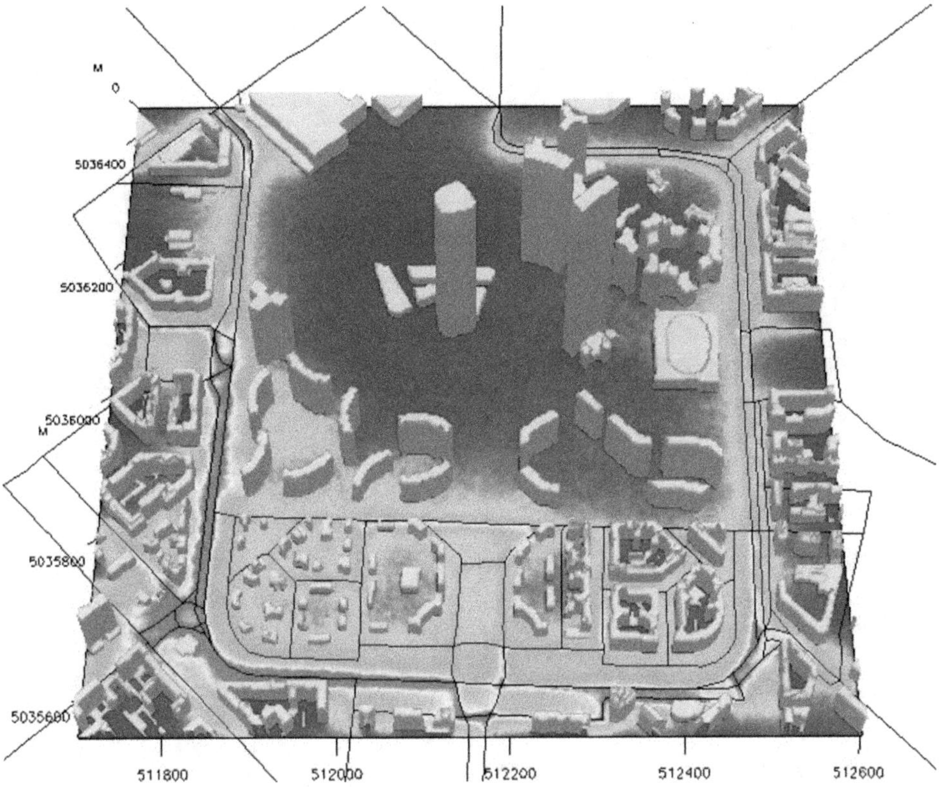

**For a more accurate representation see the electronic version.*

- Deciduous plants in summer, in particular:
    - Plants with dense crown;
    - Plants with very dense crown.
- Deciduous plants in winter with very light crown, and;
- Lawn.

For the land have been taken into account the used profiles, in particular:

- Concrete pavement;
- Cobblestones;
- Asphalt, and;
- Soil, etc.

Figure 12. Contributions from Nitric oxide concentrations (background 145 mg/m3) in winter coming from the following emission source: grates

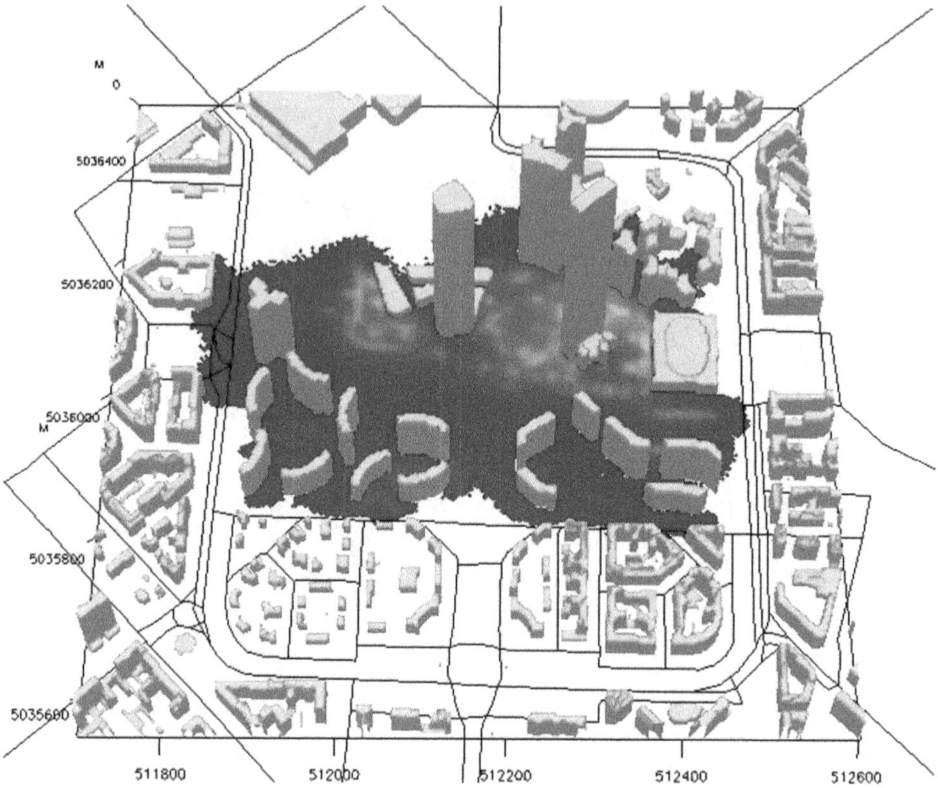

**For a more accurate representation see the electronic version.*

The simulations of the temperature have been developed at 1,6 metres from the ground to know what breathed by people.

The results were presented as three-dimensional maps (at different altitudes) of meteorological parameters calculated, highlighting the presence of points of particular criticality or, vice versa, particularly advantageous from a microclimate viewpoint.

In the following maps are represented the simulation of the temperature in summer and in winter in the project CityLife and in the area for comparison in different times of the day at 7 A.M. and at 1-3 and 7 P.M:

- Blue part indicates a low temperature;
- Green and white part indicate a medium temperature, and;
- Red and violet part indicate a high temperature.

Figure 13. Contributions from Nitric oxide concentrations (background 145 mg/m3) in winter coming from the following emission source: heating

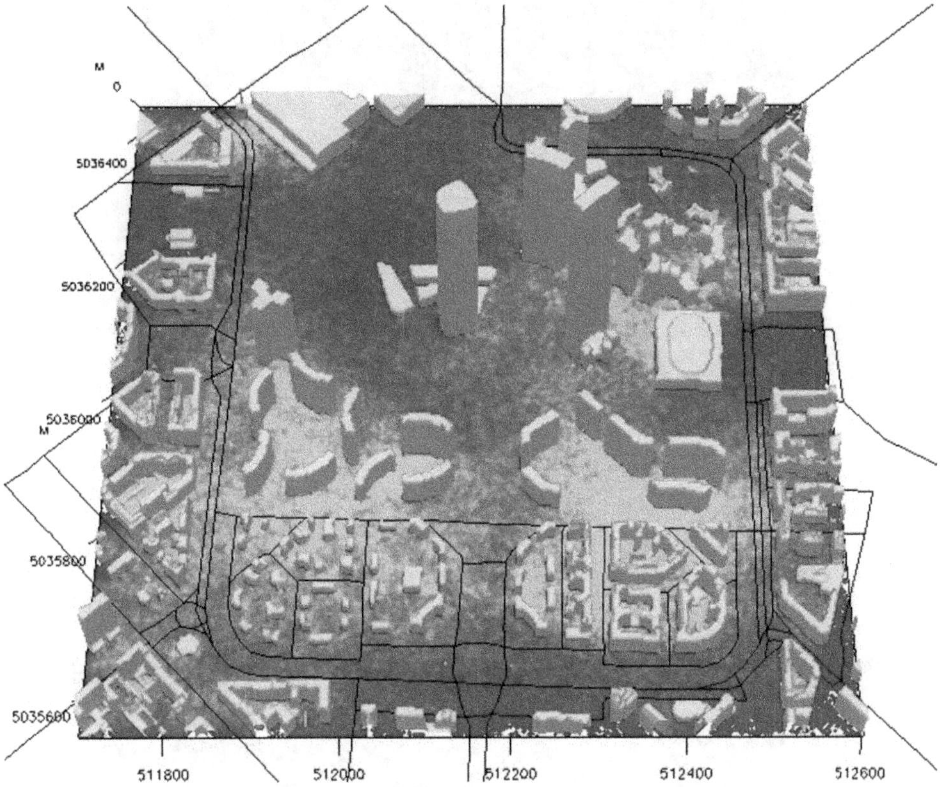

**For a more accurate representation see the electronic version.*

The work has continued with the simulation of Predicted Mean Vote. Global comfort is quantitatively valued through Predicted Mean Vote (PMV) index. It represents the value of the average vote expressed by a wide sample of people in regard to the environment in examination. It's based on the equilibrium between the heat produced inside the body and the heat dissipated by the body.

- It shall take account of:
- Air temperature;
- Average radiant temperature;
- Wind speed;
- Atmospheric pressure;
- Heat produced inside the body, and;
- Thermal insulation of clothing.

Figure 14. Decay of average Nitrous Oxide (NOx) concentrations in CityLife

**For a more accurate representation see the electronic version.*

Figure 15. Decay of average Nitrous Oxide (NOx) concentrations in Garibaldi

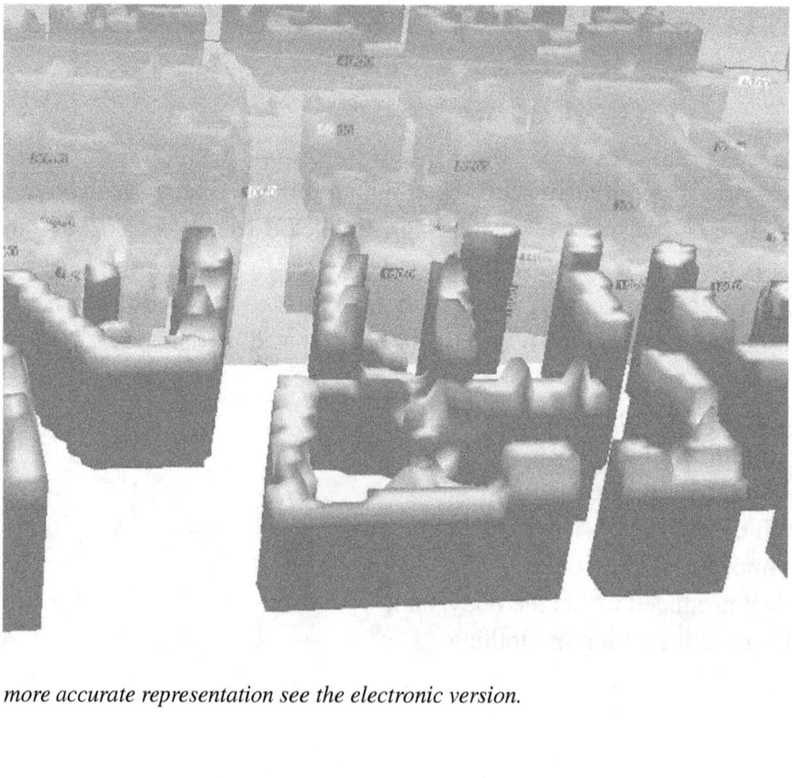

**For a more accurate representation see the electronic version.*

Figure 16. Summer temperature at 1:00 P.M.

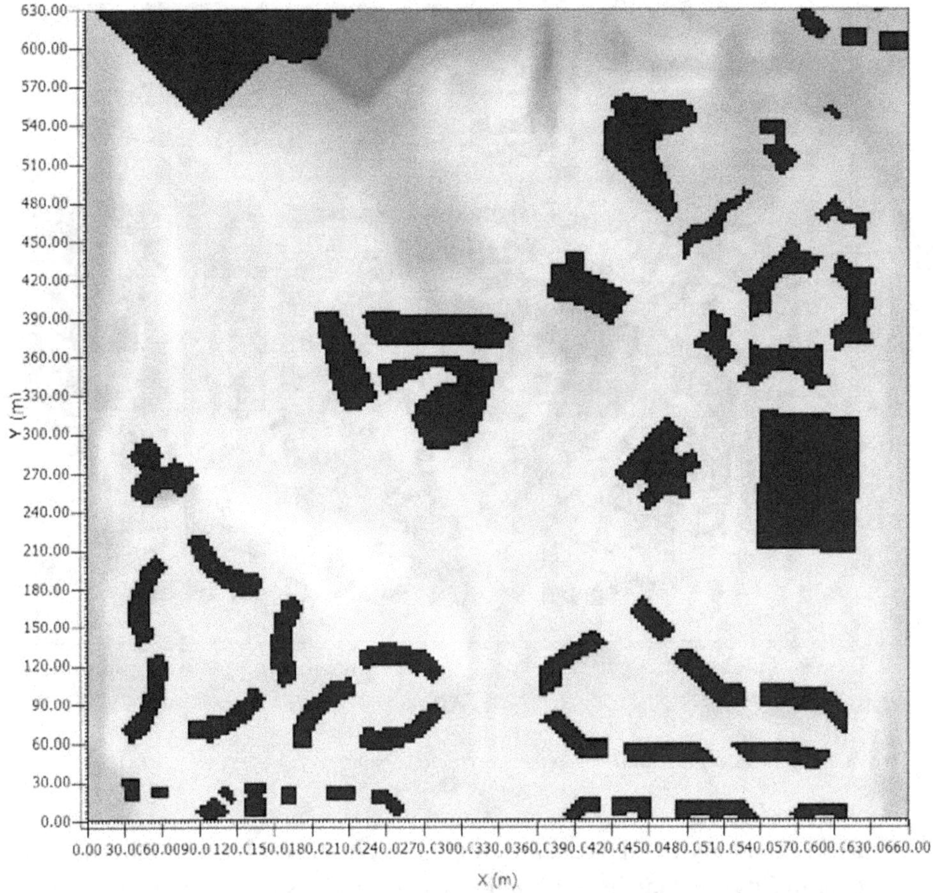

**For a more accurate representation see the electronic version.*

The scale of Predicted Mean Vote is from less 3 (very cold) to plus 3 (very hot).

In the following maps are represented the simulation of Predicted Mean Vote in summer and in winter in the project CityLife and in the area for comparison in different times of the day at 7-9 and 11 A.M. and at 1-3-5-7 and 9 P.M:

- Blue part indicates a low value of Predicted Mean Vote (very cold);
- White part indicates a medium value of Predicted Mean Vote (zero impact for the people);
- Orange part indicates a high value of Predicted Mean Vote (very hot).

Figure 17. Summer temperature at 1:00 P.M.

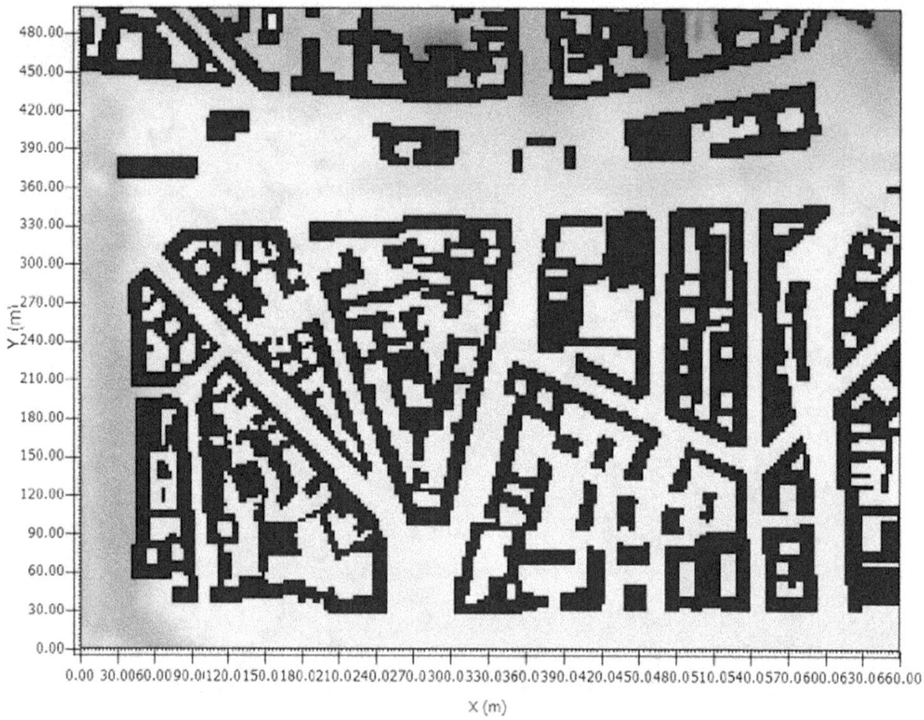

**For a more accurate representation see the electronic version.*

In all different times of the day, the simulation shows a value of Predicted Mean Vote better in CityLife than in Garibaldi. Infect the maps show more areas in white, which represents a good condition for the people.

In conclusion of this phase we can say that:

- The green areas of the project contribute to general improvement of the micro-climate in the CityLife district.
- The improvements are evident in CityLife in the following parameters:
  - Air temperature;
  - Predicted Mean Vote index.
- The wellbeing of the microclimate is better in CityLife thanks to the values of temperature and Predicted Mean Vote.
- In CityLife district there are some heat isles.

Figure 18. PMV at 9:00 A.M. – Summer case

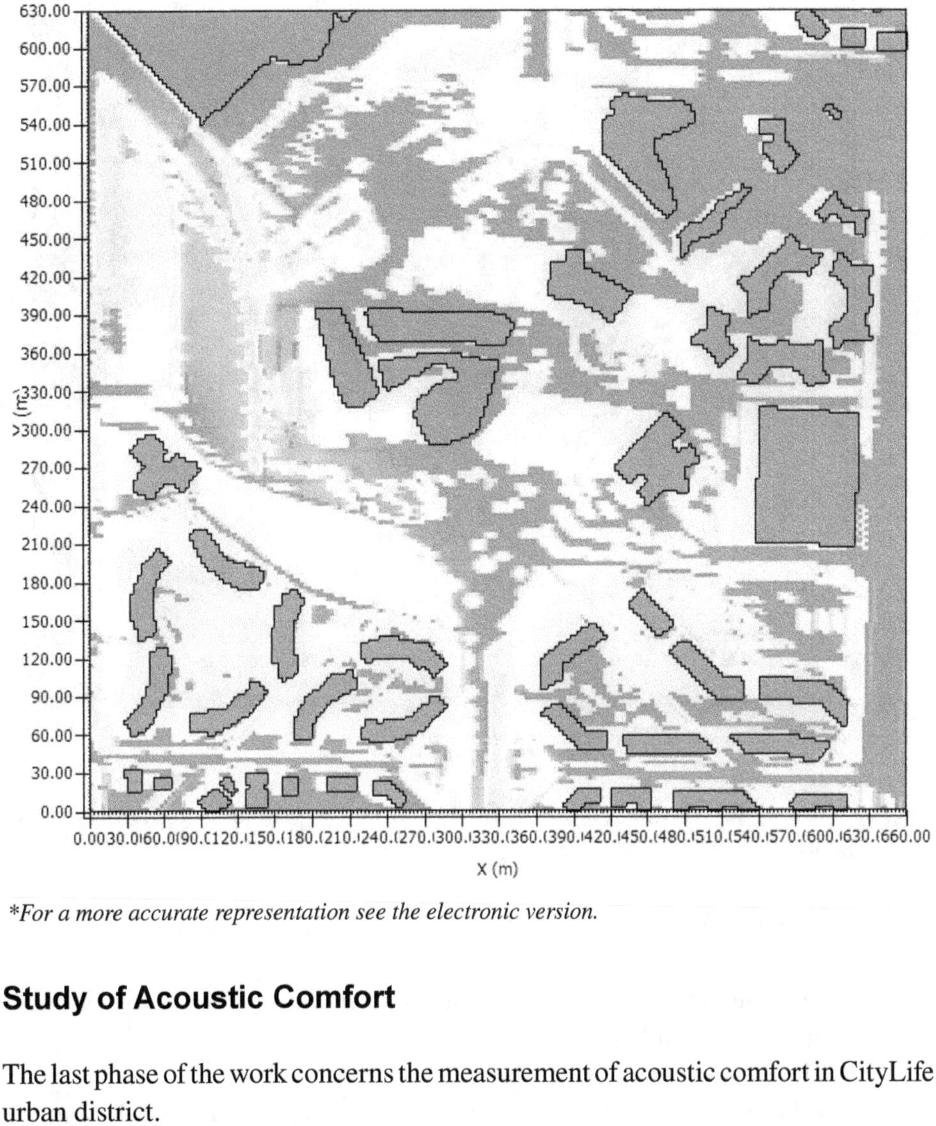

**For a more accurate representation see the electronic version.*

Study of Acoustic Comfort

The last phase of the work concerns the measurement of acoustic comfort in CityLife urban district.

The acoustic comfort evaluation aims to analyse potential acoustic dependence that could be generated in confined environments. Sounds and noises in confined environments must have low intelligibility, clarity and definition so they are not harmful and do not constitute a clear limitation of quality of life and a significant loss of privacy. These parameters can be determined at the planning stage from the position of buildings in the territorial context, the form of each building and the materials used on the exteriors.

Figure 19. PMV at 9:00 A.M. – Summer case

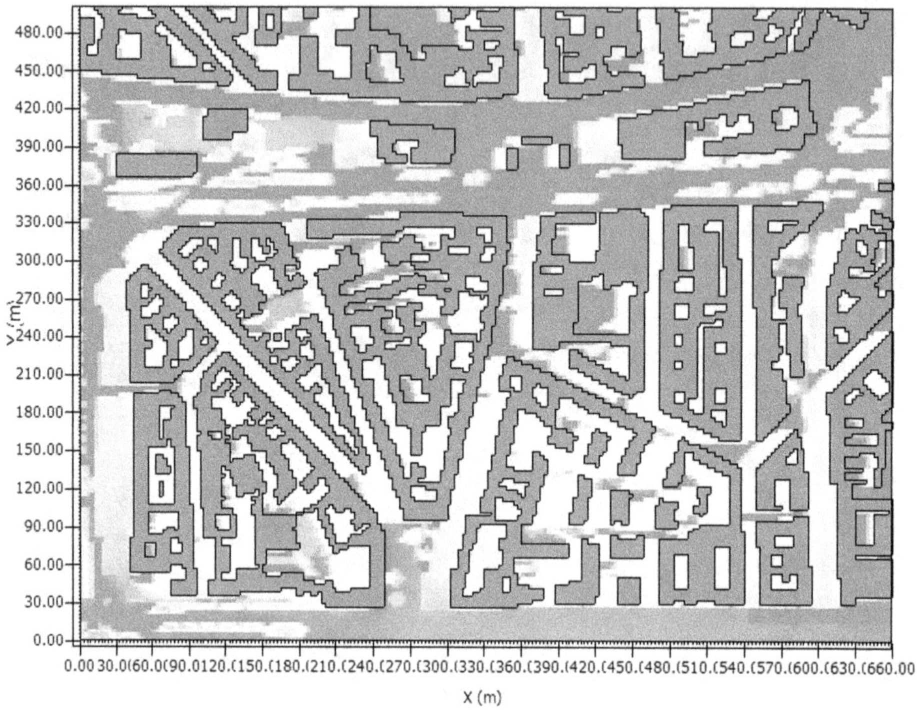

**For a more accurate representation see the electronic version.*

The current legislation on the acoustics in Italy provides for two types of assessments:

- **On Urban Scale:** The respect of the limits provided by acoustic classification plan;
- **On Building Scale:** The respect of the limits provided by the passive acoustic requirements of buildings.

In mixed used area the emission limit value is 55 decibels during the day and 45 decibels during the night. In areas of intense human activity, the emission limit value is 60 decibels during the day and 55 decibels during the night.

Quantifying acoustic descriptors for sound levels and intelligibility of speech has enabled the degree of privacy and comfort of the site to be established.

To achieve this objective a study will be carried out consisting of four separate sequential parts:

- Virtual modelling designed using the customized proprietary software CATT ACOUSTIC produced by CATT in Gothenburg (Sweden) and simulation of the whole representative built context. This phase of the simulation includes a section on the current scenario simulation, using a source of white noise with the same sound level across the entire frequency range;
- A series of measurements on site using the white noise source mentioned above. Calibration of the model calculated by comparing experimental data with simulated data. The calibration of the model was taken by specific measure carried out on the spot. First of all, it was placed a microphone into the courtyards of the residences and then a Sound level metre was placed to 2,5 – 5 and 9 metres high on the façade of residential buildings
- Future scenario simulation conducted using a source with human voice characteristics.

Computerized acoustic simulation of indoor spaces has been conducted using special software based on principles of geometric acoustics.

The first step has been to reproduce the correct geometry of the environment to be modelled in the software. In a second stage, according to the type of material chosen, each surface has been assigned a coefficient of acoustic absorption (α), corresponding to the assumed technical characteristics.

Starting with a specific position of the sound source, which can be positioned at will, the computer has simulated the contemporaneous emission of many thousands of sound rays, using a technique called raytracing:

- If the sound ray hits an absorbent material, the sound is reduced, and;
- If the sound ray hits a reflecting material, the sound is transmitted into the environment.

In these types of studies, unlike in energy studies, the wall construction is not important, just the layer that comes into direct contact with the sound wave. The precision and value of the results obtained depend to a large extent on:

- Surface modelling;
- Absorption characteristics, and;
- Transmission characteristics.

As Catt Acoustic does not have a graphic interface and autonomous modelling system the graphic representation of a confined environment is produced in the form of 3 files in.TXT format:

1. The .GEO file contains all the information needed to reconstruct the geometry of the model and the assignation of a material to every surface, identifying it by its absorption values (ABS) at different frequencies.;
2. The .SRC file provides information regarding the sound sources (positions and spectrum of sound pressure that the source produces for every frequency);
3. The .REC file contains information regarding the positioning of the receptors using Cartesian coordinates.

These files can be produced manually or using special CAD plugins (for example ArchiCAD).

This software can also be used to check the influence of sound transmission on objective parameters such as:

- T60 (reverberation time),
- C80 (clarity index),
- D50 (definition index): it is a percentage of the balance between direct and reflect waves and It controls the communication flow;
- STI (speech intelligibility);
- SPL (Sound Pressure Level): measures (in decibels) the deviation from the environmental pressure of the air caused by a sound wave;

which form the basis of architectural acoustics.

Using this software, it is possible to carry out mapping on confined environments in order to:

- Identify areas with different degrees of reverberation in various scenarios;
- Identify the intelligibility, clarity and definition of the sound from different positions in the room;
- Design a graphic scale that assigns a colour to each value in relation to the objective criterion under evaluation;
- Determine case by case the scale of values best suited to the criterion under evaluation;
- Adapt the geometry or materials to reach an optimal acoustic level.

With this kind of representation, the results of the simulation will also be easily understood by the layman.

The study carried out showed that:

1. Areas with Leq values of 45 decibels or less, the areas highlighted in yellow, mostly internal courtyards of the houses, are protected from road traffic noise, and;
2. Areas with Leq values between 50 and 55 decibels, which are almost all of the pedestrian areas inside the settlement, correspond to especially protected areas. These pressure levels allow sound to be perceived but this perception is equivalent to a single source of road noise in an open field, located 250 meters away. The settlement in question appears to be, instead, surrounded by noise sources road away from the receptor, in some cases, even less than 200 meters.

Figure 20. Leq levels in the CityLife urban district

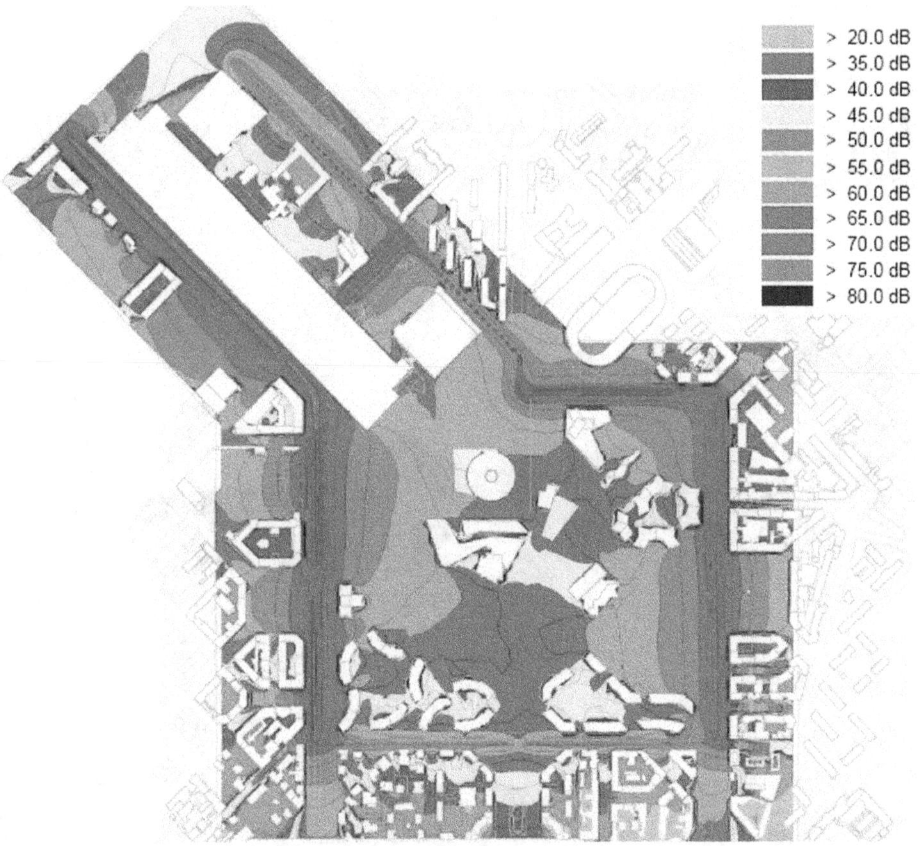

**For a more accurate representation see the electronic version.*

Comparison With the Sampling Area

The area for comparison, characterised by traditional urban solutions, has been considered as an area representative of Milan for comparison with the project area. The simulations were carried out in order to follow the temporal evolution of the main physical parameters over a span of 24 hours, and in order to compare the parameters that characterise the microclimatic comfort and local air quality of the two domains. They are, therefore, simulations of very specific episodes; although representative of typical winter and summer conditions, but have no statistical value. The average values were calculated on the basis of simulation results at 1 pm, for both the winter and summer scenarios scenery, so as to maximize the effect of solar radiation.

At the end of this analysis, it has been defined a comparison between different domains and similar episodes of hourly Dust Particles (PM10) concentrations in winter time.

- Events area of CityLife was compared to Gioia Monte Grappa, that it is a park area in Garibaldi with play possibilities for the children:

Figure 21. Simulation of Sound Pressure Level - SPL [dB] in CityLife

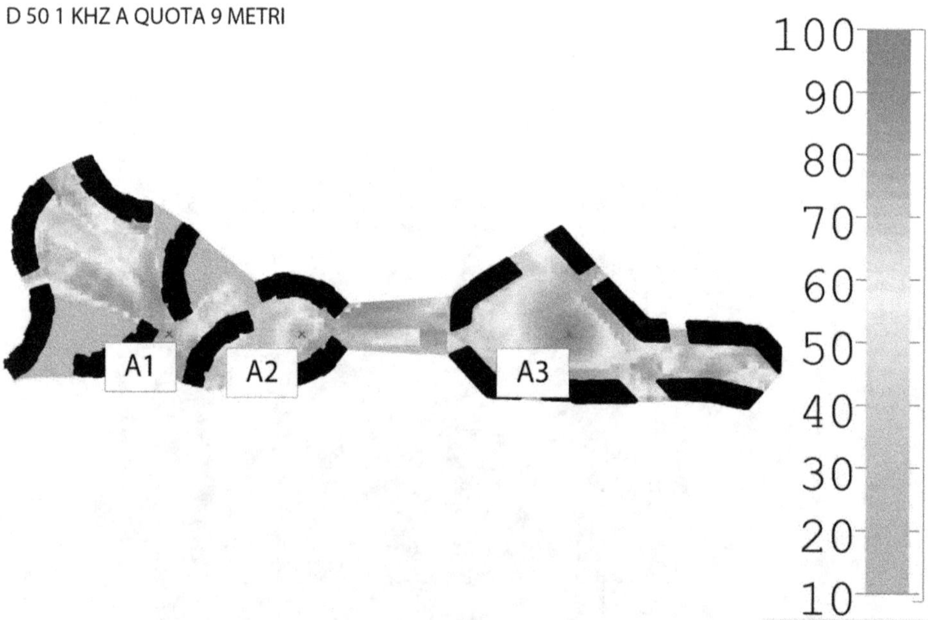

**For a more accurate representation see the electronic version.*

- Commercial gallery in CityLife was compared to Corso Garibaldi, that it is a zone with retail outlets;
- 3 Tower square in CityLife was compared to Corso Como, that it is a public space in Garibaldi.

CONCLUSION

This is a project aiming at promoting innovative application of technologies and systems in support of '"urban environment" in order to assess the environmental quality at a microscale level in a dynamic way.

The model represented by a set of indicators, refers to the three main areas of activity:

- The measurement of microclimate comfort;
- The measurement of air quality, and;
- The measurement of acoustic comfort.

A specific-oriented modelling will simulate the real conditions of the neighbourhood/district by considering all the internal and external factors.

The model is able to bring a benefit both to the Local Public Administration, which may require this type of assessment on the occasion of complex projects and to the promoters/developers, which could possibly make changes to the initial project should the evaluation identify critical issues related to the design choices (orientation of buildings, quality and presence of green, traffic emissions inside the neighbourhood, etc.).

The main result of the project is the delivery of a protocol for environmental good practice in urban planning and design at the district scale. The project will also monitor the socio-economic impact under several points of view.

During the project, will be quantified the ability:

- To attract and mix highly professional figures from several fields (architecture, city planning, engineering, environmental sciences, physics, etc.), in other words create a new consultancy market in the local territory of Milan metropolitan area where most of the partners are hosted;
- To play an active role in the planning and design process by guiding it towards virtuous interaction with the urban planning and design world, and;
- To increase social awareness.

The protocol that will be developed inside the project is aimed at pointing out and encouraging the best design practice, at the district scale, for improving the protection of the environment and the quality of life of the citizens.

Activities will be performed to monitor the extent of the protocol's capability to drive the district design towards positive environmental consequences.

This experience highlighted the lack of appropriated integrated instruments – applying a general approach – for this scope and, above all, the total lack of validated protocols aimed at guiding both local administration and real estate operators throughout the combined assessment and designing process.

REFERENCES

Ali-Toudert, F., & Mayer, H. (2007). Effects of asymmetry, galleries, overhanging facades and vegetation on thermal comfort in urban street canyons. *Solar Energy*, *81*(6), 742–754. doi:10.1016/j.solener.2006.10.007

Arpa (Agenzia Regionale per la Protezione dell'Ambiente) Lombardia. (2011). *Rapporto sullo stato dell'ambiente in Lombardia, 2010-11*. Author.

Bruse, M., & Fleer, H. (1998). Simulating surface-plant-air interactions inside urban environments with a three dimensional numerical model. *Environmental Modelling & Software*, *13*(3-4), 373–384. doi:10.1016/S1364-8152(98)00042-5

Caserini, S., Giugliano, M., & Pastorello, C. (2007). Scenari di emissioni da traffico in Lombardia nel periodo 1998-2015. *Ingegneria Ambientale*, *36*(3), 115–118.

Centro Nazionale di Meteorologia e Climatologia dell'Aeronautica Militare. (2009). (Vol. 3). Atlante Climatico d'Italia.

Chan, E. H. W., & Lee, G. K. L. (2008). Critical factors for improving social sustainability of urban renewal projects. *Social Indicators Research*, *85*(2), 243–257. doi:10.100711205-007-9089-3

TRM Engineering. (2012). *Studio viabilistico*. Atto integrativo all'accordo di programma 1994 per la qualificazione e lo sviluppo del sistema fieristico lombardo, Procedura di VAS, Rapporto Ambientale, allegato 3a.

ISPRA (Istituto Superiore per la Protezione e la Ricerca Ambientale). (2012). *Qualità dell'ambiente urbano - VIII Rapporto, Edizione 2012*. Author.

LEED 2009 for Neighborhood Development Rating System. (2011). Congress for the New Urbanism, Natural Resources Defense Council, and the U.S. Green Building Council.

Legambiente. (2011). *Ecosistema urbano* (19th ed.). Author.

Malatras, A., Asgari, A. H., Baugé, T., Irons, M., (2008). A service-oriented architecture for building services integration. *Journal of Facilities Management, 6*, 132-151.

MATTM. (n.d.). *Geoportale Nazionale*. Available at: http://www.pcn.minambiente.it/GN/

Moussafir, J., Oldrini, O., Tinarelli, G., Sontowski, J., & Dougherty, C. (2004). A new operational approach to deal with dispersion around obstacles: the MSS (Micro-Swift-Spray) software suite. *Proc. of 9th International Conference on Harmonisation within Atmospheric Dispersion Modelling for Regulatory Purposes*, 1-4.

Roper, O.K., & Beard, J. L. (2006). Justifying sustainable buildings- championing green operations. *Journal of Corporate Real Estate*, 91-103.

Rosheidat, A., Hoffman, D., & Bryan, H. (2008). *Visualizing pedestrian comfort using ENVI-met*. Third National Conference of IBPSA-USA.

Tinarelli, G., Anfossi, D., Bider, M., Ferrero, E., & Trini Castelli, S. (2000). A new high performance version of the Lagrangian particle dispersion model SPRAY, some case studies. In S. E. Gryning & E. Batchvarova (Eds.), *Air Pollution Modelling and its Applications XIII* (pp. 499–507). New York: Kluwer Academic/Plenum Press. doi:10.1007/978-1-4615-4153-0_51

Tinarelli, G., Brusasca, G., Oldrini, O., Anfossi, D., Trini Castelli, S., & Moussafir, J. (2004). Micro-Swift-Spray (MSS) a new modelling system for the simulation of dispersion at microscale, general description and validation. Proc. of the 27th CCMS-NATO meeting, 25-29.

Trini Castelli, S., & Piefel, M. (2010). *Evaluation of the Envi-met model simulating the flow around a single obstacle: the CEDVAL A1-1 test case*. Internal Report ISAC-TO/01-2010.

Wania, A., Bruse, M., Blond, N., & Weber, C. (2012). Analysing the influence of different street vegetation on traffic-induced particle dispersion using microscale simulations. *Journal of Environmental Management, 94*, 91-101.

ENDNOTE

1 This research work has been done by the Department of Architecture, Built Environment and Construction Engineering, Gesti.Tec Laboratory of Politecnico of Milan; ARIANET Srl: which is a Consultancy firm that operates in the environmental field. It studies the transport and the dispersion of pollutants in atmosphere; SIMULARIA Srl: which is a Services company of environmental modeling. It studies the transport and the dispersion of pollutants in atmosphere.

Compilation of References

Aanesen, K., Heck, S., & Pinner, D. (2017). *Solar power: Darkest before dawn.* McKinsey on Sustainability & Resource Productivity.

Abott, A. (2013, November 8). Novel process to manufacture starch-based recyclable medium density fiberboards. *Industrial Processing*.

Achal, V., Mukherjee, A., Basu, C., & Reddy, S. (2009). Lactose mother liquor as an alternative nutrient source for microbial concrete production by *Sporosarcina pasteurii. Journal of Industrial Microbiology & Biotechnology*, *36*(3), 433–438. doi:10.100710295-008-0514-7 PMID:19107535

Ahmad, Z. (2006). *Principles of Corrosion Engineering and Corrosion Control.* New York, NY: Elsevier, Ltd.

Albrecht-Saavedra, J., & Lippelt, J. (2016). Climate notes: Is sand running between our fingers? *CESifo Forum*, *17*(1), 92.

Ali-Toudert, F., & Mayer, H. (2007). Effects of asymmetry, galleries, overhanging facades and vegetation on thermal comfort in urban street canyons. *Solar Energy*, *81*(6), 742–754. doi:10.1016/j.solener.2006.10.007

Allen, E., Thallon, R., & Schreyer, A. (2017). *Fundamentals of residential construction*. Retrieved from https://books.google.com/books?id=CU4aDgAAQBAJ

Allwood, J., & Cullen, J. (2012). *Sustainable materials: With both eyes open.* Cambridge, UK: UIT Cambridge.

Aloisi, S. (2009). Italy earthquake exposes poor building standards. *Thomson Reuters*. Retrieved from https://uk.reuters.com/article/us-italyquake-buildings/italy-earthquake-exposes-poor-building-standardsidUKTRE5364T720090407

Alvarez, J. (2015). Are 'ghost homes' becoming a problem? *Embarcadero Media*. Retrieved from https://www.paloaltoonline.com/news/2015/04/24/are-ghost-homes-becoming-a-problem

Alvarez, L. (2013, August 24). Where sand is gold, the reserves are running dry. *The New York Times*.

Anatsui, E. (2014). *Ascension*. Retrieved from http://el-anatsui.com/artworks/

Andersen Windows and Doors. (2012). *A-Series Double-Hung Windows (2008-Present)*. Retrieved from https://www.andersenwindows.com/-/media/aw/files/technical-docs/parts-catalog/partscatalog-windowsincludingstormwatch-2008-present--a-series-hung.pdf

Ariyanti, D., & Handayani, N. (2011). An overview of biocement production from microalgae. *International Journal of Science and Engineering*, *2*(2), 30–33.

Arpa (Agenzia Regionale per la Protezione dell'Ambiente) Lombardia. (2011). *Rapporto sullo stato dell'ambiente in Lombardia, 2010-11*. Author.

Ashby, M. (2012). *Materials and the Environment: Eco-Informed Material Choice* (2nd ed.). Amsterdam: Elsevier Publishers.

Ashby, M., Shercliff, H., & Cebon, D. (2007). *Materials Engineering, Science, Processing and Design*. Burlington, MA: Elsevier Ltd.

Asociación de Fabricantes de Cemento Portland. (Ed.). (2010). *La industria del cemento y la sostenibilidad*. Author.

Athena Impact Estimator for Buildings 5.2.01, Version 5.2 Build 0119. (2017). Athena Sustainable Materials Institute. Retrieved from https://calculatelca.com/software/impact-estimator/

Atkins, P. (2015). *Chemistry: A Very Short Introduction*. Oxford, UK: Oxford University Press. doi:10.1093/actrade/9780199683970.001.0001

Baker-Brown, D. (2017). *The Re-Use Atlas*. London, UK: RIBA Publishing.

Ballard Bell, V., & Rand, P. (2006). *Materials for architectural design*. London: Laurence King.

Bambridge, J. (2017). Dubai concrete pour breaks Guinness world record. *Construction Week Online*. Retrieved from http://www.constructionweekonline.com/article-46299dubai-concrete-pour-breaks-guinness-world-record/

Bansal, S., & Singh, S. K. (2014). A Sustainable Approach towards the Construction and Demolition Waste. *IJIRSET, 3*(2), 9226–9235.

Barker, J., Turner, A., Napier-Moore, A., Clark, M., & Davison, E. (2009). CO_2 capture in the cement industry. *Energy Procedia, 1*(1), 87–94. doi:10.1016/j.egypro.2009.01.014

Bastioni, S., Galli., A., Niccolucci, V., & Pulselli, R. M. (2006). The ecological footprint of building construction. *WIT Transactions on Ecology and the Environment, 93*.

Beekmans, J., & de Boer, J. (2014). *Pop-up city: City making in a fluid world.* Amsterdam: BIS.

Beiser, V. (2017a). Sand mining: The global environmental crisis you've never heard of. *Guardian News and Media*. Retrieved from https://www.theguardian.com/cities/2017/feb/27/sand-mining-global-environmentalcrisis-never-heard

Beiser, V. (2017b). He who controls the sand: Killing each other to control cities. *The Guardian*. Retrieved from https://www.theguardian.com/cities/2017/feb/28/sand-mafias-killing-each-other-build-cities

Belie, N. (2010). Microorganisms versus stony materials: A love-hate relationship. *Materials and Structures, 43*(9), 1191–1202. doi:10.161711527-010-9654-0

Benjamin, W. (1936). The Work of Art in the Age of Mechanical Reproduction. In H. Arendt (Ed.), *Illuminations* (pp. 219–253). New York: Schocken Books.

Berman, D. (2017). Everything you need to know about Dubai's man made islands. *Time*. Retrieved from http://www.travelandleisure.com/tripideas/island-vacations/dubai-man-made-islands-facts

Bhattacharyya, S.K., Minocha, A.K., Garg, M., Singh, J., Jain, N., Maiti, S. & Singh, S.K. (n.d.). Demolition Wastes as Raw Materials for Sustainable Construction Products. *CSIR-CBRI News Letter, 33*(2), 1-2.

Biobaku, S. (2017). Building collapse in Nigeria: 6 early warning signs. *Private Property*. Retrieved from https://www.privateproperty.com.ng/news/buildingcollapse-in-nigeria-6-early-warning-signs/

Bird, E., & Lewis, N. (2015). Causes of beach erosion. In E. Bird & N. Lewis (Eds.), Beach Renourishment (pp. 7–28). Cham, Switzerland: Springer International Publishing. 2 doi:10.1007/978-3-319-09728-2_2

Bokern, A. (2008). Message in a Bottle. In A. Bahamon & M. C. Sanjines (Eds.), *Remateiral: from waste to architecture* (pp. 24–27). New York, NY: W.W. Norton & Company.

Bossink, B. A. G., & Brouwers, H. J. H. (1996). Construction waste: Quantification and source evaluation. *Journal of Construction Engineering and Management*, *122*(1), 55–60. doi:10.1061/(ASCE)0733-9364(1996)122:1(55)

Bouquet, E., Boronat, A., & Ramos-Cormenzana, A. (1973). Production of calcite (calcium carbonate) crystals by soil bacteria is a general phenomenon. *Nature*, *246*(5434), 527–529. doi:10.1038/246527a0

Bowley, W., & Mukhopadhaya, P. (2017). A sustainable design for an off-grid passive container house. *International Review of Applied Sciences and Engineering*, *8*(2), 145–152. doi:10.1556/1848.2017.8.2.7

Braham, W., & Hale, J. (2007). *Rethinking technology: A reader in architectural theory*. Abingdon, UK: Routledge.

Brand, S. (1994). *How Buildings Learn: What Happens After They're Built*. London: Penguin Books.

Bravo, M., De Brito, J., Pontes, J., & Evangelista, L. (2015). Durability performance of concrete with recycled aggregates from construction and demolition waste plants. *Construction & Building Materials*, *77*, 357–369. doi:10.1016/j.conbuildmat.2014.12.103

Brito, D. J., & Saikia, N. (2013). *Recycled concrete in aggregate*. Springer. doi:10.1007/978-1-4471-4540-0

Brownell, B. (2010). Material Ecologies in Architecture. In L. Tilder & B. Blostein (Eds.), *Design Ecologies: Essays on the Nature of Design*. New York, NY: Princeton Architectural Press.

Brunatti, C. A. (2014). *La industria del cemento Portland y la sostenibilidad. Ciudad Autónoma de Buenos Aires*. Asociación de Fabricantes del Cemento Portland.

Bruse, M., & Fleer, H. (1998). Simulating surface-plant-air interactions inside urban environments with a three dimensional numerical model. *Environmental Modelling & Software*, *13*(3-4), 373–384. doi:10.1016/S1364-8152(98)00042-5

Buitenkamp, M., Venner, H., & Wams, T. (1993). *Action Plan Sustainable Netherlands*. Amsterdam: Dutch Friends of the Earth.

Buratti, C., & Moretti, E. (2012). Glazing systems with silica aerogel for energy savings in buildings. *Applied Energy*, *98*, 396–403. doi:10.1016/j.apenergy.2012.03.062

Bureau of Planning and Sustainability. (2016). *City of Portland 2016 Recycling Program Summary*. Portland, OR: Bureau of Planning and Sustainability.

Burry, J., & Burry, M. (2010). *The new mathematics of architecture*. London: Thames & Hudson.

Cabeza, L. F., Castellon, C., Nogues, M., Medrano, M., Leppers, R., & Zubillaga, O. (2007). Use of microencapsulated PCM in concrete walls for energy savings. *Energy and Building*, *39*(2), 113–119. doi:10.1016/j.enbuild.2006.03.030

Calas, G. (2017). Mineral resources and sustainable development. *Elements: A Magazine of Mineralogy. Geochemistry, and Petrology*, *13*(5), 301–306.

Cantat, I. (2013). Uses of Foam. In *Foams: Structure and Dynamics*. Oxford, UK: Oxford University Press. doi:10.1093/acprof:oso/9780199662890.003.0001

Caserini, S., Giugliano, M., & Pastorello, C. (2007). Scenari di emissioni da traffico in Lombardia nel periodo 1998-2015. *Ingegneria Ambientale*, *36*(3), 115–118.

Centre for Sustainable Tourism and Transport, Breda University of Applied Sciences. (2015). *FACTSHEET CARMATOP: Carbon Management for Tour Operators*. Retrieved from https://www.cstt.nl/userdata/file/factsheetprojectcarmatop2014.pdf

Centre for Sustainable Tourism and Transport, Breda University of Applied Sciences. (2016). *Travelling large in 2016: The carbon footprint of Dutch holidaymakers in 2016 and the development since 2002*. Retrieved from https://www.cstt.nl/userdata/documents/travellinglarge2016-web.pdf

Centro Nazionale di Meteorologia e Climatologia dell'Aeronautica Militare. (2009). (Vol. 3). Atlante Climatico d'Italia.

Chan, K. (2012 May 8). *Pop-Up Populism: How the Temporary Architecture Craze is Changing Our Relationship to the Built Environment*. Retrieved from http://www.blouinartinfo.com/news/story/802841/pop-up-populism-how-the-temporary-architecture-craze-is

Chanchaichujit, J., & Saavedra-Rosas, J. F. (2018). *Using Simulation Tools to Model Renewable Resources: The Case of the Thai Rubber Industry*. Palgrave Macmillan, Springer Nature. doi:10.1007/978-3-319-55816-5

Chan, E. H. W., & Lee, G. K. L. (2008). Critical factors for improving social sustainability of urban renewal projects. *Social Indicators Research*, *85*(2), 243–257. doi:10.100711205-007-9089-3

Chatterjee, K. (2009). Sustainable construction and green buildings on the foundation of building ecology. *Indian Concrete Journal*, *83*(5), 27–30.

Chele, E.-S., Ricardo, M.-C., Ana, P.-M., & Teresa, M.-R. (2012). Bamboo, from Traditional Crafts to Contemporary Design and Architecture. *Procedia: Social and Behavioral Sciences*, *51*, 777–781. doi:10.1016/j.sbspro.2012.08.239

Chervinskaya, A. V., & Zilber, N. A. (1995). Halotherapy for treatment of respiratory diseases. *Aerosol Medicine*, *8*(3), 221–232. doi:10.1089/jam.1995.8.221 PMID:10161255

Chou, S. K., Chua, K. J., & Ho, J. C. (2009). A study on the effects of double skin facades on the energy management in buildings. *Energy Conversion and Management*, *50*(9), 2275–2281. doi:10.1016/j.enconman.2009.05.003

Chun, Y.-M., Claisse, P., Naik, T. R., & Ganjian, E. (2007). *Sustainable Construction Materials and Technologies*. London: Taylor & Francis.

Ciancio, D. (2015). *Rammed Earth Construction: Cutting-Edge Research on Traditional and Modern Rammed Earth*. London: Taylor & Francis, Ltd. doi:10.1201/b18046

Citimetric Staff. (2015 April 15). *9 building materials made entirely from waste products. CitiMetric*. Retrieved from https://www.citymetric.com/skylines/9-building-materials-made-entirely-waste-products-932

Claisse, A., Elsayad, A., & Shaaban, G. (1997). Absoprtion and sorptivity of cover concrete. *Journal of Materials in Civil Engineering*, *9*(3), 105–110. doi:10.1061/(ASCE)0899-1561(1997)9:3(105)

Clariá, M., Irassar, E. F., López, R., & Bonavetti, V. (2012). Cements. In N. Maldonado & M. F. Carrasco (Eds.), *Ese material llamado hormigón* (pp. 19–78). Buenos Aires: Asociación Argentina de Tecnología del Hormigón.

Cobirzan, N., Balog, A. A., Belean, B., Borodi, G., Dadarlat, D., & Streza, M. (2016). Thermophysical properties of masonry units: Accurate characterization by means of photothermal techniques and relationship to porosity and mineral composition. *Construction & Building Materials*, *105*, 297–306. doi:10.1016/j.conbuildmat.2015.12.056

Coelho, A., & De-Brito, J. (2013). *Preparation of concrete aggregate from construction and demolition waste (CDW). In Handbook of recycled concrete and demolition waste*. Woodhead Publishing Limited.

Comaroff, J. (2014). Built on sand: Singapore and the new state of risk. *Harvard Design Magazine: Architecture, landscape architecture, urban design and planning*, (39), 138.

Constable, H. (2017). *Kenya's sand wars: Communities are pitted against sand harvesters, powerful cartels and one another as demand for sand in Kenya grows*. Retrieved from https://interactive.aljazeera.com/aje/2017/kenya-sand-wars/index.html

Cook, P. (2008). *Drawing: the motive force of architecture*. Chichester, UK: Architectural Design Primers, John Wiley & Sons.

Costanzo, V., Evola, G., & Marletta, L. (2016). Thermal and visual performance of real and theoretical thermochromic glazing solutions for office buildings. *Solar Energy Materials and Solar Cells*, *149*, 110–120. doi:10.1016/j.solmat.2016.01.008

Cotana, F., Pisello, A.L., Moretti, E., & Buratti, C. (2014). *Multipurpose characterization of glazing systems with silica aerogel: In-field experimental analysis of thermal-energy, lighting and acoustic performance*. Academic Press.

Cragg, T. (2011). *Dice Sculptures*. FAIC 2011. Retrieved from http://www.thisiscolossal.com/2011/10/dice-sculptures-by-tony-cragg/

Croaker, T. (2013 March 30). Build It and They Will Go. *Sydney Morning Herald*. Retrieved from https://www.smh.com.au/entertainment/art-and-design/build-it-and-they-will-go-20130327-2gsvc.html

Danger, T. (2016). The history of Fort Bragg glass beach. *Roadtrippers, Inc.* Retrieved from https://roadtrippers.com/trips/14332753

Davies, M. G. (2004). *Building Heat Transfer*. West Sussex, UK: John Wiley & Sons, Ltd. doi:10.1002/0470020555

Day, J. W., Boesch, D. F., Clairain, E. J., Kemp, G. P., Laska, S. B., Mitsch, W. J., ... Shabman, L. (2007). Restoration of the Mississippi delta: Lessons from hurricanes Katrina and Rita. *Science*, *315*(5819), 1679–1684. doi:10.1126cience.1137030 PMID:17379799

DeSilver, D. (2015). As American homes get bigger, energy efficiency gains are wiped out. *Pew Research Center.* Retrieved from http://www.pewresearch.org/fact-tank/2015/11/09/as-american-homes-get-bigger-energy-efficiency-gains-are-wiped-out/

Dickson, M., & Parker, D. (2014). *Sustainable timber design.* New York, NY: Routledge.

Ding, T., & Xiao, J. (2014). Estimation of building related construction and demolition waste in Shanghai. *Waste Management (New York, N.Y.), 34*(11), 2327–2334. doi:10.1016/j.wasman.2014.07.029 PMID:25164857

Dolley, T. P. (2012). Silica. *U.S. Department of the Interior, U.S. Geological Survey.* Retrieved from https://minerals.usgs.gov/minerals/pubs/ commodity/silica/myb1-2010-silic.pdf

Donovan, T. (2018). *Hyperobjects.* Retrieved from https://www.pacegallery.com/news/3019/tara-donovan-in-hyperobjects-at-ballroom-marfa

Dreyer, J. (2015). *China's political system.* Retrieved from https://books.google.se/books?id=j%5CooCgAAQBAJ

Duffy, F., & Henney, A. (1989). *The Changing City.* London: Bulstrode Press.

Dunn, N. (2012). *Digital fabrication in architecture.* London: Laurence King.

Dworkin, M., Falkow, S., Rosenberg, E., Schleifer, K., & Stackebrandt, E. (2006). The prokaryotes: A handbook on the biology of bacteria. In M. Dworkin, S. Falkow, E. Rosenberg, K. Schleifer, & E. Stackebrandt (Eds.), *Archaea Bacteria: Firmicutes, Actinomycetes* (3rd ed.; Vol. 3). New York: Springer-Verlag.

Dyer, J. (2016). Sand: It's as precious as oil - and its extraction is destroying the world's coasts and rivers. *Vice Media LLC.* Retrieved from https:// news.vice.com/article/sand-its-as-precious-as-oil-and-its-extractionis-destroying-the-worlds-coasts-and-rivers

Ehrlich, H. (1999). Microbes as geologic agents: Their role in mineral formation. *Geomicrobiology Journal, 16*(2), 135–153. doi:10.1080/014904599270659

Eisenman, P. (2009). RE: RE: Architecture or Design: Wither the Discipline? In C. O'Donnell (Ed.), *Journal of Architecture, Issue 8* (pp. 178–181). Ithaca, NY: AAP Publications/Actar.

Elipe, M. G., & Lo'pez-Querol, S. (2014). Aeolian sands: Characterization, options of improvement and possible employment in construction – the state-of-the-art. *Construction & Building Materials, 73*(Supplement C), 728–739. doi:10.1016/j.conbuildmat.2014.10.008

Environmental Justice Atlas. (2017). *Offshore sea sand harvest in Kwale district, south of Mombasa, Kenya*. Retrieved from https://ejatlas.org/conflict/offshore-sea-sand-harvest-from-in-south-coast-of-the-indian-ocean-kenya, December 28.

Estokova, A., & Porhincak, M. (2015). Environmental analysis of two building material alternatives in structures with the aim of sustainable construction. *Clean Technologies and Environmental Policy, 17*(1), 75–83. doi:10.100710098-014-0758-z

Evans, R. (1995). *The projective cast: Architecture and its three geometries*. Cambridge, MA: MIT Press.

Ewell, C., Cullis-Suzuki, S., Ediger, M., Hocevar, J., Miller, D., & Jacquet, J. (2017). Potential ecological and social benefits of a moratorium on transshipment on the high seas. *Marine Policy, 81*, 293–300. doi:10.1016/j.marpol.2017.04.004

Farzadkia, M., Gholami, M., Abouee, E., Asadgol, Z., Sadeghi, S., Arfaeinia, H., & Noradini, M. (2016). The impact of excited pollutants of cement plant on the soil and leaves of tree species: A case study in Golestan Province. *Open Journal of Ecology, 6*(7), 404–411. doi:10.4236/oje.2016.67038

Filimonau, V., Dickinson, J., Robbins, D., & Huijbregts, M. (2011). Reviewing the carbon footprint analysis of hotels: Life Cycle Energy Analysis (LCEA) as a holistic method for carbon impact appraisal of tourist accommodation. *Journal of Cleaner Production, 19*(17-18), 1917–1930. doi:10.1016/j.jclepro.2011.07.002

Fischetti, M. (2001). Drowning New Orleans. *Scientific American, 285*(4), 76–85. doi:10.1038cientificamerican1001-76 PMID:11570045

Formentini, M., & Lenci, S. (2018). An innovative building envelope (kinetic façade) with Shape Memory Alloys used as actuators and sensors. *Automation in Construction, 85*, 220–231. doi:10.1016/j.autcon.2017.10.006

Freer-Hewish, R., Ghataora, G., & Niazi, Y. (1999). Stabilization of desert sand with cement kiln dust plus chemical additives in desert road construction. *Proceedings of the Institution of Civil Engineers – Transport.* 10.1680/itran.1999.31285

Fretwell, S. (2016). Marine life dwindles after beach renourishment at folly, report says. *The State*. Retrieved from http://www.thestate.com/ news/local/article96794212.html

Frondel, M., Grosche, P., Huchtemann, D., Oberheitmann, A., & Peters, J. (2007). *Trends der Angebots- und Nachfragesituation bei mineralischen Rohstoffen: Endbericht. Forschungsprojekt Nr. 09/05 des Bundesministeriums für Wirtschaft und Technologie (BMWi)*. Retrieved from https://www.econstor.eu/bitstream/10419/70880/1/737674393.pdf

Gadano, P. (2016). Architecture Activating Awareness. In C. O'Donnell & S. Chodoriwsky (Eds.), *This Is Not A Wall* (p. 256). Ithaca, NY: Cornell AAP Publications.

Gaggino, R. (2012). Water-resistant panels made from recycled plastics and resin. *Construction & Building Materials*, *35*, 468–482. doi:10.1016/j.conbuildmat.2012.04.125

Galli, A., Wiedmann, T., Ercin, E., Knoblauch, D., Ewing, B., & Giljim, S. (2012). Integrating Ecological, Carbon and Water footprint into a "Footprint Family" of indicators: Definition and role in tracking human pressure on the planet. *Ecological Indicators*, *16*, 100–112. doi:10.1016/j.ecolind.2011.06.017

Gebler, M., Uiterkamp, A. J. M. S., & Visser, C. (2014). A global sustainability perspective on 3D printing technologies. *Energy Policy*, *74*, 158–167. doi:10.1016/j.enpol.2014.08.033

Ghaffarianhoseini, A., Ghaffarianhoseini, A., Berardi, U., Tookey, J., Li, D. H. W., & Kariminia, S. (2016). Exploring the advantages and challenges of double-skin facades (DSFs). *Renewable & Sustainable Energy Reviews*, *60*, 1052–1065. doi:10.1016/j.rser.2016.01.130

Ghaffari, K., Habibzadeh, T., Asfad, M. N., & Mousazadeh, R. (2017). Construction of artificial island in southern coast of the Persian Gulf from the viewpoint of international environmental law. *Journal of Politics and Law*, *10*(2), 264. doi:10.5539/jpl.v10n2p264

Ghosh, P., Biswas, M., Chattopadhyay, B., & Mandal, S. (2009). Microbial Activity on the microstructure of bacteria modified mortar. *Cement and Concrete Composites*, *31*(2), 93–98. doi:10.1016/j.cemconcomp.2009.01.001

Gibson, I., Rosen, D., & Stucker, B. (2014). *Additive manufacturing technologies: Rapid prototyping to direct digital manufacturing* (2nd ed.). New York: Springer.

Gibson, J. J. (1986). *The Ecological Approach to Visual Perception*. Lawrence Erlbaum Assoc. Inc. Publishers.

Gibson, L. J. (2012). The hierarchical structure and mechanics of plant materials. *Journal of the Royal Society, Interface*, *9*(76), 2749–2766. doi:10.1098/rsif.2012.0341 PMID:22874093

Ginley, D. S., & Cahen, D. (Eds.). (2012). *Fundamentals of Materials for Energy and Environmental Sustainability*. New York, NY: Cambridge University Press.

Giordano, R., Serra, V., Totalla, E., Valentini, V., & Aghemo, C. (2015). Embodied Energy and Operational Energy assessment in the framework of Nearly Zero Energy Building and Building Energy Rating. *Energy Procedia*, *78*, 3204–3209. doi:10.1016/j.egypro.2015.11.781

Glavind, M. (2009). Sustainability of cement, concrete and cement replacement materials in construction. In Sustainability of construction materials. Woodhead Publishing Limited.

González-Vallejo, P., Marrero, M., & Solís-Guzmán, J. (2014). The ecological footprint of dwelling construction in Spain. *Ecological Indicators*, *52*, 75–84. doi:10.1016/j.ecolind.2014.11.016

Gopalakrishnan, S., Smith, M. D., Slott, J. M., & Murray, A. B. (2011). The value of disappearing beaches: A hedonic pricing model with endogenous beach width. *Journal of Environmental Economics and Management*, *61*(3), 297–310. doi:10.1016/j.jeem.2010.09.003

Gore, A. (2013). *The future*. New York: WH Allen.

Gramazio, F., Kohler, M., & Willmann, J. (2014). *The robotic touch: How robots change architecture*. Zurich, Switzerland: Park Books.

Griggs, G. (2017). *Coasts in crisis: A global challenge*. Berkeley, CA: University of California Press.

Grossman, J., & McCourt, T. (Producers), & Grossman, J. (Director). (2012). Drop City [Documentary]. United States: Pinball Films.

Guettala, S., & Mezghiche, B. (2011). Compressive strength and hydration with age of cement pastes containing dune sand powder. *Construction & Building Materials*, *25*(3), 1263–1269. doi:10.1016/j.conbuildmat.2010.09.026

Guggenheim Bilbao. (2015). *Press release: 1,103,211 people visited the Guggenheim Museum Bilbao in 2015, the second-best figure in its history*. Retrieved from https://prensa.guggenheim-bilbao.eus/src/uploads/2016/01/NP_Balance-2015_EN.pdf

Guo, H., Liu, Y., Wen-Shao, C., Shao, Y., & Sun, C. (2017). Energy saving and carbon reduction in the operation stage of cross laminated timber residential buildings in china. *Sustainability*, *9*(2), 292. doi:10.3390u9020292

Gupta, P. (2015). Futures, fakes and discourses of the gigantic and miniature in 'the world' islands, Dubai. *Island Studies Journal*, *10*(2).

Habibi, Y., & Lucian, A. L. (Eds.). (2012). *Polysaccharide Building Blocks: A Sustainable Approach to the Development of Renewable Biomaterials*. Hoboken, NJ: Wiley. doi:10.1002/9781118229484

Hajkowicz, S. (2015). *Global megatrends: Seven patterns of change shaping our future*. Melbourne: CSIRO.

Hale, R., Godbold, J. A., Sciberras, M., Dwight, J., Wood, C., Hiddink, J. G., & Solan, M. (2017). Mediation of macronutrients and carbon by post disturbance shelf sea sediment communities. *Biogeochemistry*, *135*(1), 121–133. doi:10.100710533-017-0350-9

Hardin, G. (1968). The Tragedy of the Commons. *Science*, *162*(3859), 1243–1248. doi:10.1126cience.162.3859.1243 PMID:5699198

Hawken, P. (2017). *Drawdown: The most comprehensive plan ever proposed to reverse global warming*. London: Penguin Books.

Haynes, R. (2010). *Embodied Energy Calculations within Life Cycle Analysis of Residential Buildings*. Perth: ETool Global.

Hebel, D. E., Heisel, F., & Wisniewska, M. H. (2014). *Building from Waste: Recovered Materials in Architecture and Construction*. Basel, Switzerland: Birkhauser. doi:10.1515/9783038213758

Hedman, J. T., Hugg, J., Sandell, T. H., & Haahtela, T. (2006). The effect of salt chamber treatment on bronchial hyperresponsiveness in asthmatics. *Allergy*, *61*(5), 605–610. doi:10.1111/j.1398-9995.2006.01073.x PMID:16629791

Hee, W. J., Alghoul, M. A., Bakhtyar, B., Elayeb, O., Shameri, M. A., Alrubaih, M. S., & Sopian, K. (2015). The role of window glazing on daylighting and energy saving in buildings. *Renewable & Sustainable Energy Reviews*, *42*, 323–343. doi:10.1016/j.rser.2014.09.020

Heisel, F. (2018). Economics for a Circular Anthropogenic Environment. In C. O'Donnell & D. Pranger (Eds.), Cyclo: Architecture of Waste. Academic Press. (forthcoming)

Hendrickson, C., Lave, L., & Matthews, H. (2006). *Environmental life cycle assessment of goods and services: An input-output approach.* Washington, DC: Washington State, Resources of the Future.

HGTV. (2017). *The Benefits of Recycled Demin Insulation.* Retrieved from https://www.hgtv.com/remodel/mechanical-systems/the-benefits-of-recycled-denim-insulation

Homer-Dixon, T. (1991). On the threshold: Environmental changes as causes of acute conflict. *International Security, 6*(2), 76–116. doi:10.2307/2539061

Horowitz, S. (2010). Salt cave therapy: Rediscovering the benefits of an old preservative. *Alternative and Complementary Therapies, 16*(3), 158–162. doi:10.1089/act.2010.16302

Hot, J., Topalov, J., Ringot, E., & Bertron, A. (2017). Investigation on parameters affecting the effectiveness of photocatalytic functional coatings to degrade NO: TiO_2 amount on surface, illumination, and substrate roughness. *International Journal of Photoenergy*, 14.

Houle, G., Chhem, C., & Bougie, R. (2002). *Inventaire qu´eb´ecois des gaz `a effet de serre 1990-2000.* Quebec, Canada: Minist`ere de l'environnement du Qu´ebec.

Humes, E. (2012 April 13). Garbage: A Costly American Addiction. *Forbes.* Retrieved from https://www.forbes.com/sites/kerryadolan/2012/04/13/garbage-a-costly-american-addiction/#3ae122a65b1c

Hyndman, D., & Hyndman, D. (2016). *Natural hazards and disasters.* Retrieved from https://books.google.com/books?id=mChTCwAAQBAJ

INTI. (2005). CIRSOC 201 Reglamento Argentino de Estructuras de Hormigón. Buenos Aires, Argentina: INTI.

IPCC. (2007). Cambio climático 2007: Informe de síntesis. Contribución de los Grupos de trabajo I, II y III al Cuarto Informe de evaluación del Grupo Intergubernamental de Expertos sobre el Cambio Climático. Ginebra, Suiza.

IRAM. (2017). *Argentine Standards Catalogue.* IRAM.

ISPRA (Istituto Superiore per la Protezione e la Ricerca Ambientale). (2012). *Qualità dell'ambiente urbano - VIII Rapporto, Edizione 2012.* Author.

Ivanov, V., & Chu, J. (2008). Applications of microorganisms to geotechnical engineering for bioclogging and biocementation of soil in situ. *Reviews in Environmental Science and Biotechnology*, *7*(2), 139–153. doi:10.100711157-007-9126-3

Jia, R. (2013). *Essays on the political economics of China's development* (Unpublished doctoral dissertation). Department of Economics, Stockholm University, Stockholm, Sweden.

Jiménez-Saelices, C., Seantier, B., Cathala, B., & Grohens, Y. (2017). Effect of freeze-drying parameters on the microstructure and thermal insulating properties of nanofibrillated cellulose aerogels. *Journal of Sol-Gel Science and Technology*, *84*(3), 475–485. doi:10.100710971-017-4451-7

Jimeniez, J. R. (2013). *Recycled aggregate for roads. In Handbook of recycled concrete and demolition waste*. Woodhead Publishing Limited.

Kalay, Y. (2006). The impact of information technology on design methods, products and practices. *Design Studies*, *27*(3), 357–380. doi:10.1016/j.destud.2005.11.001

Kantro, D. L. (1980). Influence of water reducing admixture on properties of cement-paste –a miniature test. *Cement, Concrete and Aggregates*, *2*(2), 95–108. doi:10.1520/CCA10190J

Kara, H., & Bosia, D. (2016). *Design engineering refocused*. Chichester, UK: John Wiley & Son. doi:10.1002/9781119164838

Kathib, J. (2009). Sustainability of construction materials. Woodhead Publishing Limited.

Kenyon, W., Esch, S., & Buller, C. (2005). The curdlan-type exopolysaccharide produced by *Cellulomonas flavigena* KU forms part of an extracellular glycocalyx involved in cellulose degradation. *Antonie van Leeuwenhoek*, *87*(2), 143–148. doi:10.100710482-004-2346-4 PMID:15723175

Kerwin, L. (1997). *Potential applications for recycled glass in beach management: Emergency stabilization of erosional "hot spots" in Broward County, Florida* (Unpublished doctoral dissertation). Miami, FL: Florida Atlantic University.

Kevern, J. T. (2010, Spring). *How Do We Teach the Next Generation of Engineers Green Building?* Paper presented at the ACI Convention Spring 2010, Chicago, IL.

Khanafari, A., Khams, N., & Sepahy, A. (2011). An investigation of biocement production from hard water. *Middle East Journal of Scientific Research*, *7*(6), 964–971.

Kibert, C. (2005). *Sustainable construction: Green building delivery and design*. Hoboken, NJ: John Wiley & Son.

Kibert, C. (2013). *Sustainable Construction: Green Building Design and Delivery* (3rd ed.). Hoboken, NJ: John Wiley & Sons.

King, B. (2017). *New Carbon Architecture: Building to Cool the Planet*. Gabriola Island, BC: New Society Publishers.

Kiogi, B. (2017). Deadly and destructive: Kenya cartels fan sand wars. *Fair Planet*. Retrieved from https://www.fairplanet.org/story/deadly-and-destructive-kenya-cartels-fan-sand-wars/

Klammt, S., Neyer, A., & Muller, H. F. C. (2012). Redirection of sunlight by microstructured components – Simulation, fabrication and experimental results. *Solar Energy*, *86*(5), 1660–1666. doi:10.1016/j.solener.2012.02.034

Koppen, W., & Geiger, R. (2018). *World Maps of Koppen-Geiger Climate Classification*. Retrieved January 18, 2018 from http://koeppen-geiger.vu-wien.ac.at/

Kucharski, E., Winchester, W., Leeming, W., Cord-Ruwisch, R., Muir, C., Banjup, W., … Mutlaq, J. (2005). *Microbial biocementation*. Patent Application WO/2006/066326; International Application No. PCT/ AU2005/001927.

Kuniavsky, M. (2010). *Smart things: Ubiquitous computing user experience design*. Burlington, VT: Elsevier.

Kvan, T. (2004). *Modes of Representation. In Architectural Review Australia* (Vol. 90, pp. 82–83). Digital Architecture.

Langer, W. H., Giusti, C., & Barelli, G. (2003). Sustainable development of natural aggregate, with examples from Modena Province, Italy. *SME Transactions*, *314*, 138–144.

Laville, S., & Taylor, M. (2017 June 28). A million bottles a minute: world's plastic binge 'as dangerous as climate change'. *The Guardian*. Retrieved from https://www.theguardian.com/environment/2017/jun/28/a-million-a-minute-worlds-plastic-bottle-binge-as-dangerous-as-climate-change

Ledesma, E. F., Jimenez, J. R., Ayuso, J., Fernandez, J. M., & De-Brito, J. (2015). Maximum feasible use of recycled sand from construction and demolition waste for eco-mortar production e Part-I: Ceramic masonry waste. *Journal of Cleaner Production*, *87*, 692–706. doi:10.1016/j.jclepro.2014.10.084

Lee, J. (2014). Digging up the seafloor makes coral reefs sick. *National Geographic Society*. Retrieved from https://news.nationalgeographic.com/ news/2014/07/140716-australia-coral-reef-dredging-sediment-diseaseenvironment-ocean/

LEED 2009 for Neighborhood Development Rating System. (2011). Congress for the New Urbanism, Natural Resources Defense Council, and the U.S. Green Building Council.

Legambiente. (2011). *Ecosistema urbano* (19th ed.). Author.

Legge, K. (2012). *Doing it differently*. Sydney: Place Partners.

Leiva, C., Guzmán, J. S., Marrero, M., & Arenas, C. G. (2013). Recycled blocks with improved sound and fire insulation containing construction and demolition waste. *Waste Management (New York, N.Y.)*, *33*(3), 663–671. doi:10.1016/j.wasman.2012.06.011 PMID:22784475

Levy, S., & Helene, P. (2004). Durability of recycled aggregates concrete: A safe way to sustainable development. *Cement and Concrete Research*, *34*(11), 1975–1980. doi:10.1016/j.cemconres.2004.02.009

Long, L., & Ye, H. (2014). Discussion of the performance improvement of thermochromic glazing applied in passive buildings. *Solar Energy*, *107*, 236–244. doi:10.1016/j.solener.2014.05.014

Lucas, R., Wheeler, D., & Hettige, H. (1992). *Economic development, environmental regulation, and the international migration of toxic industrial pollution, 1960-88 (WPS1062)*. Washington, DC: World Bank. Retrieved from https://books.google.sc/books?id=3mJ4UHd0CjgC

Luo, F. J., He, L., Pan, Z., Duan, W. H., Zhao, X. L., & Collins, F. (2013). Effect of very fine particles on workability and strength of concrete made with dune sand. *Construction & Building Materials*, *47*, 131–137. doi:10.1016/j.conbuildmat.2013.05.005

Mairs, J. (2016 June 6). Alejandro Aravena uses over 90 tonnes of recycled waste for entrance rooms of Venice Biennale 2016. *Dezeen*. Retrieved from https://www.dezeen.com/2016/06/02/venice-architecture-biennale-2016-recycled-waste-exhibition-entrances-alejandro-aravena

Majumdar, N.B. (2014). *Experience of the first commercial scale pilot project for C &D Waste Management in India*. IL&FS Environmental Infrastructure & Services Ltd.

Makowski, C. & Rusenko, K. (2007). Recycled glass cullet as an alternative beach fill material: Results of biological and chemical analyses. *Journal of Coastal Research*, 545–552. doi:10.2112/06-0715.1

Makowski, C., Rusenko, K., & Kruempel, C. J. (2008). Abiotic suitability of recycled glass cullet as an alternative sea turtle nesting substrate. *Journal of Coastal Research*, *243*, 771–779. doi:10.2112/07-0864.1

Makowski, C., Thomson, G., Foye, P., & Higgins, S. (2007). Broward county beach demonstration project: From beers to beaches. In N. C. Kraus & J. D. Rosati (Eds.), *Coastal sediments' 07* (pp. 2265–2278). American Society of Civil Engineers. doi:10.1061/40926(239)178

Malatras, A., Asgari, A. H., Baugé, T., Irons, M., (2008). A service-oriented architecture for building services integration. *Journal of Facilities Management*, *6*, 132-151.

Maldonado, N. G., & Helene, P. (2005). ¿Hay diferencias entre métodos de dosificar hormigones de alta performance para evaluar durabilidad. In *Proceedings of Structural Concrete and Time, fib Symposium* (*vol. 1,* pp.163-170). La Plata, Argentina: AATH, AAHE & LEMIT.

Maldonado, N. G., & Helene, P. (2007). The importance of mixture proportioning in sustainable construction. In Y-M. Chun, P. Claisse, T.R. Naik, & E. Ganjian (Eds.), Sustainable Construction Materials and Technologies (pp. 223-230). Taylor & Francis.

Malkin, B. (2013). 'Poor quality construction materials to blame' for deadly Bangladesh factory collapse. *Telegraph Media Group*. Retrieved from http://www.telegraph.co.uk/news/worldnews/asia/bangladesh/10075098/Poor-quality-construction-materials-to-blame-for-deadlyBangladesh-factory-collapse.html

Mann, S. (1993). Molecular tectonics in biomineralization and biomimetic materials chemistry. *Nature*, *365*(6446), 499–505. doi:10.1038/365499a0

Marshall, G. (2016). From sandy beach to Kaby Lake: How sand becomes silicon. *TechRadar pro IT insights for business*. Retrieved from http://www.techradar.com/news/computing-components/processors/howsand-is-transformed-into-silicon-chips-599785

Marzouk, Y. O., Dheilly, R.M., Queneudec, M. (2007). Valorization of postconsumer waste plastic in cementitious concrete composites. *Waste Management*, (27), 310–318.

Masciola, C. (1992, November 30). He dreams of a beach–all covered with glass. *Los Angeles Times*. Retrieved from http://articles.latimes.com/1992-11-30/news/mn-9881crushed-glass

Matos, G. (2012). *Use of Raw Materials in the United States from 1900 Through 2010 (USGS)*. Washington, DC: United States Department of the Interior.

MATTM. (n.d.). *Geoportale Nazionale*. Available at: http://www.pcn.minambiente.it/GN/

McCarney-Castle, K., Voulgaris, G., & Kettner, A. J. (2010). Analysis of fluvial suspended sediment load contribution through anthropocene history to the south Atlantic bight coastal zone, U.S.A. *The Journal of Geology, 118*(4), 399–416. doi:10.1086/652658

McDonough, W., & Braungart, M. (2002). *Cradle to Cradle*. New York, NY: North Point Press.

McDonough, W., & Braungart, M. (2002). *Cradle to Cradle: Remaking the Way We Make Things*. New York: North Point Press.

McDonough, W., & Braungart, M. (2013). *The Upcycle: Beyond Sustainability – Designing for Abundance*. New York, NY: North Point Press.

McDonough, W., & Braungart, M. (2013). *The Upcycle: Beyond Sustainability—Designing for Abundance*. New York: North Point Press.

McDowell, R. (2015). AP tracks slave boats to Papua New Guinea. *Associated Press*. Retrieved from https://www.ap.org/explore/seafood-fromslaves/ap-tracks-slave-boats-to-papua-new-guinea.html

McGee, W., & Ponce de Leon, M. (Eds.). (2014). *Robotic Fabrication in Architecture, Art and Design 2014*. Springer International Publishing. doi:10.1007/978-3-319-04663-1

Mclean, J., & McDonald, B. (2012). Extravagant Dubai island project sinks under weight of the credit crunch. *Independent.ie*. Retrieved from https: //www.independent.ie/world-news/middle-east/extravagant-dubaiisland-project-sinks-under-weight-of-the-credit-crunch-26565663.html

McWhan, D. (2012). *Sand and silicon: Science that changed the world*. Oxford, UK: Oxford University Press. Retrieved from https://books.google.com/books?id=UZl4LVarWycC

Medina, C., Zhu, W., Howind, T., Frías, M., & Sánchez de Rojas, M. I. (2015). Effect of the constituents (asphalt, clay materials, floating particles and fines) of construction and demolition waste on the properties of recycled concretes. *Construction & Building Materials, 79*, 22–33. doi:10.1016/j.conbuildmat.2014.12.070

Mehta, P. K., & Monteiro, P. J. M. (2006). *Concrete, Microstructure, Properties and materials* (3rd ed.). McGraw-Hill. doi:10.1036/0071462899

Meijide, B. G., & Pérez, I. (2014). Effects of the use of construction and demolition waste aggregates in cold asphalt mixtures. *Construction & Building Materials, 51*, 267–277. doi:10.1016/j.conbuildmat.2013.10.096

Menges, A., Schwinn, T., & Krieg, O. D. (Eds.). (2017). *Advancing wood architecture: A computational approach*. London: Routledge.

Mibei, G. (2014). Introduction to types and classification of rocks. *Short Course IX on Exploration for Geothermal Resources by UNU-GTP, GDC and KenGen.*

Milanesi, C., Pappalardo, M., & Violini, D. (2008). Consideraciones sobre la aplicación del criterio canadiense para inhibir la reacción álcali-sílice. In *Memoria Los áridos como factor de desarrollo* (Vol. 1, pp. 475–481). Cámara Argentina de la Piedra.

Milicevic, I., Bjegovic, D., & Stirmer, N. (2015). Optimisation of concrete mixtures made with crushed clay bricks and roof tiles. *Magazine of Concrete Research, 67*(3), 109–120. doi:10.1680/macr.14.00175

Miller, H. (2007). *Tesis de maestría*. Facultad de Ingeniería y Medio Ambiente Construido, Universidad de Wolverhampton, Reino Unido.

Mimbacas, A. (2013). *Construction Sustentability. Boletín técnico Nº 7*. ALCONPAT Internacional.

Miodwonik, M. (2015). *Stuff matters: Exploring marvellous materials that shape our man-made world.* Seattle, WA: Mariner Books.

Mitchell, W. (2001). Roll over Euclid: How Frank Gehry designs and builds. In J. F. Ragheb (Ed.), Frank Gehry, Architect (pp. 352- 64). New York: Solomon R. Guggenheim Foundation.

Mitchell, R. (2014). *Tiny house living: Ideas for building and living well in less than 400 square feet*. Sacramento, CA: Betterway Home.

Mitchell, W. (1999). A Tale of Two Cities: Architecture and the Digital Revolution. *Science, 285*(5429), 839–841. doi:10.1126cience.285.5429.839

Moe, K., & Srinivasan, R. (2015). *The Hierarchy of Energy in Architecture: Emergy Analysis*. New York, NY: Routledge.

Morawski, C. (2006). Embodied energy. *Alternatives Journal*, *32*(1), 19.

Morel, J. C., Mesbah, A., Oggero, M., & Walker, P. (2001). Building houses with local materials: Means to drastically reduce the environmental impact of construction. *Building and Environment*, *36*(10), 1119–1126. doi:10.1016/S0360-1323(00)00054-8

Moreno, E. L., & Blanco, Z. G. (2014). Ghost cities and empty houses: Wasted prosperity. *American International Journal of Social Science*, *3*(2), 207–216.

Morris, R. (2017). What happens when your country drowns? *Mother Jones*. Retrieved from http://www.motherjones.com/environment/2009/11/tuvalu-climaterefugees/

Moussafir, J., Oldrini, O., Tinarelli, G., Sontowski, J., & Dougherty, C. (2004). A new operational approach to deal with dispersion around obstacles: the MSS (Micro-Swift-Spray) software suite. *Proc. of 9th International Conference on Harmonisation within Atmospheric Dispersion Modelling for Regulatory Purposes*, 1-4.

Mujah, D., Shahin, M., & Cheng, L. (2016). State-of-the-art review of biocementation by microbially induced calcite precipitation (MICP) for soil stabilization. *Geomicrobiology Journal*, *34*(6), 1–14.

Mumford, L. (1934). The Eotechnic Phase. In *Technics and Civilization*. San Diego, CA: Harcourt, Brace and Co.

Muthomi, S., Okoth, P., Were, E., & Vundi, S. (2015). An examination of the nature of sand harvesting conflicts and their influence on poverty alleviation initiatives in Makueni county, Kenya. *Journal of Education and Practice*, *6*(27), 28–36.

Muynck, W., Belie, N., & Verstraete, W. (2010). Microbial carbonate precipitation in construction materials: A Review. *Ecological Engineering*, *36*(2), 118–136. doi:10.1016/j.ecoleng.2009.02.006

Nair, A. S., & Bhattacharjee, N. (2017). Pullback in U.S. fracking sand use pressures producers. *Thomson Reuters*. Retrieved from https://www. reuters.com/article/us-usa-oil-sand/pullback-in-u-s-fracking-sanduse-pressures-producers-idUSKCN1AW07F

National League of Cities and United States Conference of Mayors, Solid Waste Management Task Force. (1973). *Cities and the Nations Disposal Crisis*. Washington, DC: U.S. Government Printing Office.

Nicholson, P. T. (2012). Stone…That Flows: Faience and Glass as Man-Made Stones in Egypt. *Journal of Glass Studies*, *54*, 11–23.

Nike. (2018). *Nike Grind*. Retrieved from https://www.nikegrind.com/about/

Nthambi, M. V., & Orodho, J. A. (2015). Effects of sand harvesting on environment and educational outcomes in public primary schools in Kathiani Sub-County, Machakos County, Kenya. *Journal of Education and Practice*, *6*(24), 88–97. Retrieved from https://files.eric.ed.gov/fulltext/EJ1078871.pdf

O'Connor, J. (2004). *Survey on actual service lives for North American buildings. Woodframe Housing Durability and Disaster Issues conference, Las Vegas, Nevada.* Vancouver, BC: Athena Sustainable Materials Institute / Forintek Canada Corporation.

Odum, H. T. (1983). *Systems Ecology: An Introduction*. New York, NY: John Wiley.

Ossa, A., García, J. L., & Botero, E. (2016). Use of recycled construction and demolition waste (CDW) aggregates: A sustainable alternative for the pavement construction industry. *Journal of Cleaner Production*, *135*, 379–386. doi:10.1016/j.jclepro.2016.06.088

Otlewska, A., & Gutarowska, B. (2016). Environmental parameters conditioning microbially induced mineralization under the experimental model conditions. *Acta Biochimica Polonica*, *63*(2), 343–351. doi:10.18388/abp.2015_1172 PMID:26894236

Ozalp̈, F., Yılmaz, H. D., Kara, M., Kaya, O., & Şahin, A. (2016). Effects of recycled aggregates from construction and demolition wastes on mechanical and permeability properties of paving stone, kerb and concrete pipes. *Construction & Building Materials*, *110*, 17–23. doi:10.1016/j.conbuildmat.2016.01.030

Padmalal, D., & Maya, K. (2014). Sand mining: The world scenario. In D. Padmalal & K. Maya (Eds.), Sand mining: Environmental impacts and selected case studies (pp. 57–80). Dordrecht: Springer Netherlands. doi:5 doi:10.1007/978-94-017-9144-1

Pandey, D., & Simba, K. (1990). Effect of cement kiln dust on chlorophyll in wheat leaf. *Environment and Ecology*, *8*(1B), 461–463.

Peduzzi, P. (2014). Sand, rarer than one thinks. *Environmental Development*, *11*, 208–218. doi:10.1016/j.envdev.2014.04.001

Pérez Gómez, A., & Pelletier, L. (1997). *Architectural Representation and the Perspective Hinge*. Cambridge, MA: The MIT Press.

Philips. (2018). *Circular Lighting at Schiphol airport*. Retrieved from http://www.lighting.philips.com/main/cases/cases/airports/schiphol-airport

Piesse, M. (2016). *Livelihoods and food security on the Mekong River*. Future Direction International, Strategic Analysis Paper.

Pilkey, O. H., & Cooper, J. A. G. (2014). Are natural beaches facing extinction? *Journal of Coastal Research*, *70*(sp1), 431–436. doi:10.2112/SI70-073.1

Pilkey, O. H., & Wright, H. L. III. (1988). Seawalls versus beaches. *Journal of Coastal Research*, 41–64.

Portilho, M., Matioli, G., Zanin, G., de Moraes, F., & Scamparini, A. (2006). Production of insoluble exopolysaccharide *Agrobacterium* sp. (ATCC 31749 and IFO 13140). *Applied Biochemistry and Biotechnology*, *131*(1-3), 864–869. doi:10.1385/ABAB:131:1:864 PMID:18563660

Rahim, L., & Rahim, L. (2010). *Singapore in the Malay world: Building and breaching regional bridges*. Routledge. Retrieved from https:// books.google.com/books?id=1utb8ZYyUeQC

Ramage, M. H., Burridge, H., Busse-Wicher, M., Fereday, G., Reynolds, T., Shah, D. U., ... Scherman, O. (2017). The wood from the trees: The use of timber in construction. *Renewable & Sustainable Energy Reviews*, *68*, 333–359. doi:10.1016/j.rser.2016.09.107

Rao, A., Jha, N. K., & Misra, S. (2006). *Use of Aggregates from recycled Construction and Demolition waste in Concrete. In Resources Conservation & Recycling*. Elsevier.

Rapoza, K. (2015). What will become of China's ghost cities? *Forbes Magazine*. Retrieved from https://www.forbes.com/sites/kenrapoza/2015/ 07/20/what-will-become-of-chinas-ghost-cities/#43d551cb2e7b

Rappeneau, G., & Mini, L. (Producers), & Delestrac, D. (Director). (2012). *Sand Wars* [Motion Picture]. France: PBS International.

Reddy, B. V. V., & Jagadish, K. S. (2003). Embodied energy of common and alternative building materials and technologies. *Energy and Building*, *35*(2), 129–137. doi:10.1016/S0378-7788(01)00141-4

Reynolds, M. (1990). *Earthship: How to Build Your Own* (Vol. 1). Solar Survival Architecture.

Rodríguez, C., Parra, C., Casado, G., Minano, I., Albaladejo, F., Benito, F., & Sanchez, I. (2016). The incorporation of construction and demolition wastes as recycled mixed aggregates in non-structural concrete precast pieces. *Journal of Cleaner Production*, *127*, 152–161. doi:10.1016/j.jclepro.2016.03.137

Rodriguez-Navarroa, C., Jroundib, F., Schiroa, M., Ruiz-Agudoc, E., & González-Munozb, M. (2012). Influence of substrate mineralogy on bacterial mineralization of calcium carbonate: Implications for stone conservation. *Applied and Environmental Microbiology*, *78*(11), 4017–4029. doi:10.1128/AEM.07044-11 PMID:22447589

Roper, O.K., & Beard, J. L. (2006). Justifying sustainable buildings- championing green operations. *Journal of Corporate Real Estate*, 91-103.

Rosheidat, A., Hoffman, D., & Bryan, H. (2008). *Visualizing pedestrian comfort using ENVI-met.* Third National Conference of IBPSA-USA.

Ross, A. (2016). *The Industries of the Future*. New York: Simon & Schuster.

Rostam, S. (1999). Durability. In Structural concrete, textbook on Behaviour, design and performance update knowledge of the CEB/FIP Model Code 1990. fib Bulletin 3.

Sadineni, S. B., Madala, S., & Boehm, R. F. (2011). Passive building energy savings: A review of building envelope components. *Renewable & Sustainable Energy Reviews*, *15*(8), 3617–3631. doi:10.1016/j.rser.2011.07.014

Saeed, H. Z., Khan, Q. Z., Khan, H. A., & Farooq, R. (2016). Experimental investigation of stress-strain behavior of CFRP confined Low Strength Concrete (LSC) cylinders. *Construction & Building Materials*, *104*, 208–215. doi:10.1016/j.conbuildmat.2015.12.061

Saeli, M., Piccirillo, C., Parkin, I. P., Binions, R., & Ridley, I. (2010). Energy modelling studies of thermochromic glazing. *Energy and Building*, *42*(10), 1666–1673. doi:10.1016/j.enbuild.2010.04.010

Sánchez, S., Oshiro, A., & Positieri, M. (2012). *Cálculo de la huella ecológica del hormigón. V Congreso Internacional.* 19ª Reunión Técnica de la Asociación Argentina de Tecnología del Hormigón, Bahía Blanca, Argentina.

Santhosh, K., & Ramachandran, V., & Ramakrishnan, S. (2001). Remediation of concrete using microorganisms. *ACI Materials Journal*, *98*(1), 3–9.

Schachermayer, E., Lahner, T., & Brunner, P. H. (2000). Assessment of Two different separate techniques for Building wastes. *Waste Management & Research*, *18*(1), 16–24. doi:10.1177/0734242X0001800103

Shameri, M. A., Alghoul, M. A., Sopian, K., Zain, M. F. M., & Elayeb, O. (2011). Perspectives of double skin façade systems in building and energy saving. *Renewable & Sustainable Energy Reviews*, *15*(3), 1468–1475. doi:10.1016/j.rser.2010.10.016

Shepard, W. (2015). *Ghost cities of China: The story of cities without people in the world's most populated country*. London: Zed Books Ltd.

Shutters, C., & Tufts, R. (2016, May 27). *LEED by the numbers: 16 years of steady growth*. Retrieved from https://www.usgbc.org/articles/leed-numbers-16-years-steady-growth

Silva, R. V., De Brito, J., & Dhir, R. K. (2015). Prediction of the shrinkage behaviour of recycled aggregate concrete: A review. *Construction & Building Materials*, *77*, 327–339. doi:10.1016/j.conbuildmat.2014.12.102

Simmons, A. (2016, April 22). The world's trash crisis, and why many Americans are oblivious. *Los Angeles Times*. Retrieved from http://www.latimes.com/world/global-development/ la-fg-global-trash-20160422-20160421-snap-htmlstory.html

Simonen, K. (2014). *Life Cycle Assessment*. Oxon, UK: Routledge Press.

Sinha, R., Lennartsson, M., & Frostell, B. (2016). Environmental footprint assessment of building structures: A comparative study. *Building and Environment*, *104*, 162–171. doi:10.1016/j.buildenv.2016.05.012

Smith, S. I. (2017). *Superporous Intelligent Hydrogels for Environmentally Adaptive Building Skins*. MRS Advances. doi:10.1557/adv.2017.429

Soares, D., De Brito, J., Ferreira, J., & Pacheco, J. (2014). *Use of coarse recycled aggregates from precast concrete rejects: mechanical and durability*. Academic Press.

Soleimani, M., & Shahandashti, M. (2017). Comparative process-based life-cycle assessment of bioconcrete and conventional concrete. *Journal of Engineering, Design and Technology*, *15*(5), 667–688. doi:10.1108/JEDT-04-2017-0033

Solid Waste Disposal Act of 1965, 42 U.S.C. §§ 6901-6908a (2012).

Solid waste: diversion, AB-341, 112th Cong. (2011).

Solis-Guzman, J., & Marrerro, M. (2015). Ecological Footprint Assessment of Building Construction. Sharjah, UAE: Bentham Science Publishers. doi:10.2174/97816810809871150101

Spencer, R. (2011). The world is sinking: Dubai islands 'falling into the sea'. *Telegraph Media Group*. Retrieved from http://www.telegraph.co.uk/ news/worldnews/middleeast/dubai/8271643/The-World-is-sinkingDubai-islands-falling-into-the-sea.html

Spengler, J. D., Samet, J. M., & McCarthy, J. F. (Eds.). (2001). *Indoor Air Quality Handbook*. New York, NY: McGraw-Hill.

Stahel, W. (2016). *The Circular Economy Package*. Retrieved from https://www.eesc.europa.eu/resources/docs/2016_01_26_presentation-stahel.pdf

Stahel, W., & Reday-Mulvey, G. (1981). *JOBS FOR TOMORROW: The Potential For Substituting Manpower For Energy*. New York, NY: Vantage Press, Inc.

Statista. (n.d.). *Annual hotel occupancy rate in London (UK) from 2008 to 2018*. Retrieved from https://www.statista.com/statistics/323719/annual-hotel-occupancy-rate-in-london-uk/

Sugiyama, T., Takahashi, S., Honda, M., & Sakai, E. (2007). Current State of the JSCE Standard on Test Method for Leaching of Trace Elements from Hardened Concrete. *Proceedings of Special Sessions Sustainable Construction Materials and Technologies*, *1*, 197–203.

T.E.A.M. (2017). *Clastic Order*. Retrieved from http://tpluseplusaplusm.us/clasticorder.html

Taborda, J. (2017). United states housing starts 1959-2018. *Trading Economics*. Retrieved from https://tradingeconomics. com/united-states/housing-starts

Taha, B. & Nounu, G. (2008). Properties of concrete contains mixed colour waste recycled glass as sand and cement replacement. *Construction and Building Materials*, *22*(5), 713–720. doi:.conbuildmat.2007.01.019 doi:10.1016/j

Téllez Martínez, L.A., Villarreal, U., Armenta Menchaca, C., Porsen Oveergard, R., & Bremer Bremer, M.H. (2014). *State of Play of Sustainable Building in Latin America*. UNEP – United Nations Enviroment Programme. Sustainable Buildings and Climate Initiative.

The 7th Environment Action Programme (EAP), 1383/2013/EU. (2013).

The Ragdale Foundation. (2018, March 8) *Call for proposals: The Ragdale Ring*. Retrieved from http://ragdale.org/wp/wp-content/uploads/2017/03/2018-Ragdale-Ring-RFP.pdf

Thomas, J., & Wilson, P. M. (2013). Construction waste management in India. *American Journal of Engineering Research*, *2*, 6-9.

Timberlake, K. (2008). *Cellophane House | Prefabricated Architecture & Design for Disassembly*. Retrieved May 6, 2018, from https://kierantimberlake.com/pages/view/14/cellophane-house/parent:3

Timothy, G. (1998). *The management and Environmental Impacts of C&D waste in Florida*. University of Florida.

Tinarelli, G., Brusasca, G., Oldrini, O., Anfossi, D., Trini Castelli, S., & Moussafir, J. (2004). Micro-Swift-Spray (MSS) a new modelling system for the simulation of dispersion at microscale, general description and validation. Proc. of the 27th CCMS-NATO meeting, 25-29.

Tinarelli, G., Anfossi, D., Bider, M., Ferrero, E., & Trini Castelli, S. (2000). A new high performance version of the Lagrangian particle dispersion model SPRAY, some case studies. In S. E. Gryning & E. Batchvarova (Eds.), *Air Pollution Modelling and its Applications XIII* (pp. 499–507). New York: Kluwer Academic/Plenum Press. doi:10.1007/978-1-4615-4153-0_51

Torgal, F. P., Tam, V. W. Y., Labrincha, J. A., Ding, Y., & De Brito, J. (2013). *Handbook of recycled concrete and demolition waste*. Woodhead Publishing Limited. doi:10.1533/9780857096906

Torres, A., Brandt, J., Lear, K., & Liu, J. (2017). A looming tragedy of the sand commons. *Science*, *357*(6355), 970–971. doi:10.1126cience.aao0503 PMID:28883058

Treating Construction and Demolition. (2004). *Waste Israel Environmental Bulletin*, 27.

Trini Castelli, S., & Piefel, M. (2010). *Evaluation of the Envi-met model simulating the flow around a single obstacle: the CEDVAL A1-1 test case*. Internal Report ISAC-TO/01-2010.

TRM Engineering. (2012). *Studio viabilistico*. Atto integrativo all'accordo di programma 1994 per la qualificazione e lo sviluppo del sistema fieristico lombardo, Procedura di VAS, Rapporto Ambientale, allegato 3a.

Tsang, P., Li, G., Brun, Y., Freund, L., & Tang, J. (2006). Adhesion of single bacterial cells in the micronewton range. *Proceedings of the National Academy of Sciences of the United States of America*, *103*(15), 5764–5768. doi:10.1073/pnas.0601705103 PMID:16585522

Tucker, M., Böhning-Gaese, K., Fagan, W., Fryxell, J., Van Moorter, B., Alberts, S., ... Mueller, T. (2018). Moving in the Anthropocene: Global reductions in terrestrial mammalian movements. *Science*, *359*(6374), 466–469. doi:10.1126cience.aam9712 PMID:29371471

U.S. Department of the Interior. (2017). *Mineral Commodity Summaries 2017*. Reston, VA: U.S. Geological Survey.

U.S. Energy Information Administration. (2016). *2012 Commercial Building Energy Consumption Survey: Energy Usage Summary*. Retrieved from https://www.eia.gov/consumption/commercial/reports/2012/energyusage/

U.S. Environmental Protection Agency. (2012). *Portfolio Manager Technical Reference: Energy Use Benchmarking*. Retrieved from https://www.energystar.gov/sites/default/files/buildings/tools/DataTrends_Energy_20121002.pdf

U.S. Environmental Protection Agency. (2016). *Portfolio Manager Technical Reference: U.S. Energy Use Intensity by Property Type.* Retrieved from https://portfoliomanager.energystar.gov/pdf/reference/US%20National%20Median%20Table.pdf

Umesh, M., & Murthy, A. (2014). Sand mining: Curbing the evil to the environment through sustainable substitution and legislative action. *Ontario International Development Agency International Journal of Sustainable Development*, *7*(3), 17–26.

United States Department of Labor, Bureau of Labor Statistics. (2006). *100 Years of U.S. Consumer Spending: Data for the Nation, New York City, and Boston, BLS Report 991*. Washington, DC: U.S. Government Printing Office.

United States Environmental Protection Agency. (2016). *Advancing Sustainable Materials Management: 2014 Fact Sheet: Assessing Trends in Materials Generation, Recycling, Composting, Combustion with Energy Recovery and Landfilling in the United States*. Washington, DC: U.S. Government Printing Office.

United States Environmental Protection Agency. (2017). *EPA report: Sustainable Management of Construction and Demolition Materials*. Washington, DC: U.S. Government Printing Office.

United States Green Building Council. (2009). *Construction and demolition waste management.* Retrieved from https://www.usgbc.org/credits/new-construction-core-and-shell-schools-new-construction-retail-new-construction-healthca-38

United States Green Building Council. (2016). *Recycled Content.* Retrieved from https://www.usgbc.org/credits/commercial-interiors-retail-ci/v2009/mrc4

United States Green Building Council. (2018). *About USGBC.* Retrieved from https://new.usgbc.org/about

Urban, F. (2018). China's rise: Challenging the north-south technology transfer paradigm for climate change mitigation and low carbon energy. *Energy Policy, 113*, 320–330. doi:10.1016/j.enpol.2017.11.007

Vallejos, M. E., Peresin, M. S., & Rojas, O. J. (2012). All-cellulose composite fibers obtained by electrospinning dispersions of cellulose acetate and cellulose nanocrystals. *Journal of Polymers and the Environment, 20*(4), 1075–1083. doi:10.100710924-012-0499-1

Van Lavieren, H., Burt, J., Feary, D., Cavalcante, G., Marquis, E., Benedetti, L., & Sale, P. (2011). *Managing the growing impacts of development on fragile coastal and marine ecosystems: Lessons from the gulf.* Hamilton, Ontario, Canada: United Nations University Institute for Water, Environment and Health.

Vatan, M. (2017). Evolution of construction systems: Cultural effects on traditional structures and their reflection on modern building construction. In G. Koc, M.-T. Claes, & B. Christiansen (Eds.), *Cultural influences on architecture* (pp. 35–57). Hershey, PA: IGI Global. doi:10.4018/978-1-5225-1744-3.ch002

Vigna, I., Bianco, L., Goia, F., & Serra, V. (2018). Phase Change Materials in Transparent Building Envelopes: A Strengths, Weaknesses, Opportunities and Threats (SWOT) Analysis. *Energies, 11*(1), 111. doi:10.3390/en11010111

Vijayalaxmi, J. (2010). Towards sustainable architecture – a case with Greentainer. *Local Environment: The International Journal of Justice and Sustainability, 15*(3), 245–259. doi:10.1080/13549830903575596

Vogel, W. (1994). Historical Development of Glass Chemistry. In *Glass Chemistry.* New York, NY: Springer-Verlag. doi:10.1007/978-3-642-78723-2_1

Wackernagel, M., & Rees, W. (1996). *Our Ecological Footprint: Reducing Human Impact on the Earth.* Gabriola Island, BC: New Society Publishers.

Walker, S. (2006). *Sustainable by Design: Explorations in theory and practice.* Oxford, UK: Earthscan.

Wallenberger, F. T., & Bingham, P. A. (Eds.). (2010). *Fiberglass and Glass Technology: Energy-Friendly Compositions and Applications*. New York, NY: Springer. doi:10.1007/978-1-4419-0736-3

Wambua, M. P. (2015). *Environmental and socio-economic impacts of sand harvesting on the community in river kivou catchment, Mwingi sub county, Kitui County, Kenya. Thesis*. Nairobi, Kenya: Kenyatta University. Retrieved from http://ir-library.ku.ac.ke/handle/123456789/14023

Wania, A., Bruse, M., Blond, N., & Weber, C. (2012). Analysing the influence of different street vegetation on traffic-induced particle dispersion using microscale simulations. *Journal of Environmental Management, 94*, 91-101.

Warerkar, T. (2018, January 16). Governors Island's 2018 City of Dreams Pavilion finalists announced. *Curbed NY*. Retrieved from https://ny.curbed.com/2018/1/16/16897862/

Wei, P., Rao, X., Yang, J., Guo, Y., Chen, H., Zhang, Y., & Wang, Z. (2016). Hot pressing of wood-based composites: A review. *Forest Products Journal, 66*(7), 419–427. doi:10.13073/FPJ-D-15-00047

Welland, M. (2009). *Sand: The never-ending story*. Berkeley, CA: University of California Press.

Wesley, D., Puffer, S. M., & Wade, B. (2018). *Mining and corporate social responsibility: Scotbar Proprietary Limited*. London, ON, Canada: Ivey Case Publishing.

White, N. (2017). Sydney's ghost homes: How 200,000 homes sit vacant in Australia's most expensive city - because foreign investors buy them and leave them to gather dust. *Associated Newspapers*. Retrieved from http://www.dailymail.co.uk/news/article-4270282/Sydney-200-000homes-foreign-investors-blamed.html

Williams, A. (2016, September 1). Multi-use community center makes the most of recycled materials. *New Atlas*. Retrieved from https://newatlas.com/re-ainbow-hp-architects/39175/

Williams, A., & Micallef, A. (2011). *Beach management: Principles and practice*. Earthscan. Retrieved from https://books.google.com/books?id=z55JHEwrlLMC

Winkler, G. (2010). *Recycling construction and demolition waste- A LEED based toolkit*. McGraw Hill.

Wooldridge, T., Henter, H. J., & Kohn, J. R. (2016). Effects of beach replenishment on intertidal invertebrates: A 15-month, eight beach study. *Estuarine, Coastal and Shelf Science*, *175*, 24–33. doi:10.1016/j.ecss.2016.03.018

World Atlas. (2017). Retrieved from https://www.worldatlas.com/articles/top-20-sand-exporting-countries.html

Worrell, E., Price, L., Martin, N., Hendriks, C., & Meida, O. (2001). Carbon dioxide emissions from the global cement industry. *Annual Review of Energy and the Environment*, *26*(1), 303–329. doi:10.1146/annurev.energy.26.1.303

Worrell, E., & Reuter, M. (Eds.). (2014). *Handbook of Recycling: State-of-the-Art for Practitioners, Analysts, and Scientists*. New York, NY: Elsevier.

Wu, Y., Connelly, K., Liu, Y., Gu, X., Gao, Y., & Chen, G. Z. (2016). Smart solar concentrators for building integrated photovoltaic facades. *Solar Energy*, *133*, 111–118. doi:10.1016/j.solener.2016.03.046

Yashiro, T., Kato, H., Yoshida, T., & Komatsu, Y. (Eds.). (1990). *Proceedings from CIB 90 W55/65 Symposium: Survey on real life span of office buildings in Japan*. Academic Press.

Yu, M. L. (2013). *Skins, envelopes, and enclosures: concepts for designing building exteriors*. New York, NY: Routledge.

Zainab, Z., Ismail, E., & Hashmi, A. (2008). *Use of waste plastic in concrete mixture as aggregate replacement*. Waste Management.

Zhang, F. (2008). Framework for building design recyclability. *ProQuest Dissertations & Theses Global,* 304617969.

Zhang, G., Song, J., Yang, J., & Liu, X. (2006). Performance of mortar and concrete made with a fine aggregate of desert sand. *Building and Environment*, *41*(11), 1478–1481. doi:10.1016/j.buildenv.2005.05.033

Related References

To continue our tradition of advancing research on topics in the field of engineering, we have compiled a list of recommended IGI Global readings. These references will provide additional information and guidance to further enrich your knowledge and assist you with your own research and future publications.

Abawajy, J. H., Pathan, M., Rahman, M., Pathan, A., & Deris, M. M. (2013). *Network and traffic engineering in emerging distributed computing applications*. Hershey, PA: IGI Global. doi:10.4018/978-1-4666-1888-6

Abu-Faraj, Z. O. (2012). Bioengineering/biomedical engineering education. In Z. Abu-Faraj (Ed.), *Handbook of research on biomedical engineering education and advanced bioengineering learning: Interdisciplinary concepts* (pp. 1–59). Hershey, PA: IGI Global. doi:10.4018/978-1-4666-0122-2.ch001

Abu-Nimeh, S., & Mead, N. R. (2012). Combining security and privacy in requirements engineering. In T. Chou (Ed.), *Information assurance and security technologies for risk assessment and threat management: Advances* (pp. 273–290). Hershey, PA: IGI Global. doi:10.4018/978-1-61350-507-6.ch011

Abu-Taieh, E., El Sheikh, A., & Jafari, M. (2012). *Technology engineering and management in aviation: Advancements and discoveries*. Hershey, PA: IGI Global. doi:10.4018/978-1-60960-887-3

Achumba, I. E., Azzi, D., & Stocker, J. (2010). Low-cost virtual laboratory workbench for electronic engineering. *International Journal of Virtual and Personal Learning Environments, 1*(4), 1–17. doi:10.4018/jvple.2010100101

Achumba, I. E., Azzi, D., & Stocker, J. (2012). Low-cost virtual laboratory workbench for electronic engineering. In M. Thomas (Ed.), *Design, implementation, and evaluation of virtual learning environments* (pp. 201–217). Hershey, PA: IGI Global. doi:10.4018/978-1-4666-1770-4.ch014

Addo-Tenkorang, R., & Eyob, E. (2013). Engineer-to-order: A maturity concurrent engineering best practice in improving supply chains. In Industrial engineering: Concepts, methodologies, tools, and applications (pp. 1780-1796). Hershey, PA: IGI Global. doi:10.4018/978-1-4666-1945-6.ch095

Aguilera, A., & Davim, J. (2014). *Research developments in wood engineering and technology*. Hershey, PA: IGI Global. doi:10.4018/978-1-4666-4554-7

Aharoni, A., & Reinhartz-Berger, I. (2013). Semi-automatic composition of situational methods. In K. Siau (Ed.), *Innovations in database design, web applications, and information systems management* (pp. 335–364). Hershey, PA: IGI Global. doi:10.4018/978-1-4666-2044-5.ch013

Ahmad, M., Jung, L. T., & Zaman, N. (2014). A comparative analysis of software engineering approaches for sequence analysis. In *Software design and development: Concepts, methodologies, tools, and applications* (pp. 1093–1102). Hershey, PA: IGI Global. doi:10.4018/978-1-4666-4301-7.ch053

Ahrens, A., Bassus, O., & Zaščerinska, J. (2014). Enterprise 2.0 in engineering curriculum. In M. Cruz-Cunha, F. Moreira, & J. Varajão (Eds.), *Handbook of research on enterprise 2.0: Technological, social, and organizational dimensions* (pp. 599–617). Hershey, PA: IGI Global. doi:10.4018/978-1-4666-4373-4.ch031

Akbar, D. (2012). Community engagement in engineering education: Needs and learning outcomes. In M. Rasul (Ed.), *Developments in engineering education standards: Advanced curriculum innovations* (pp. 301–317). Hershey, PA: IGI Global. doi:10.4018/978-1-4666-0951-8.ch017

Alam, F., Subic, A., Plumb, G., Shortis, M., & Chandra, R. P. (2012). An innovative offshore delivery of an undergraduate mechanical engineering program. In M. Rasul (Ed.), *Developments in engineering education standards: Advanced curriculum innovations* (pp. 233–245). Hershey, PA: IGI Global. doi:10.4018/978-1-4666-0951-8.ch013

Ali, D. F., Patil, A., & Nordin, M. S. (2012). Visualization skills in engineering education: Issues, developments, and enhancement. In A. Patil, H. Eijkman, & E. Bhattacharyya (Eds.), *New media communication skills for engineers and IT professionals: Trans-national and trans-cultural demands* (pp. 175–203). Hershey, PA: IGI Global. doi:10.4018/978-1-4666-0243-4.ch011

Aljawarneh, S. (2013). Cloud security engineering: Avoiding security threats the right way. In S. Aljawarneh (Ed.), *Cloud computing advancements in design, implementation, and technologies* (pp. 147–153). Hershey, PA: IGI Global. doi:10.4018/978-1-4666-1879-4.ch010

Alkhatib, G. (2012). *Models for capitalizing on web engineering advancements: Trends and discoveries*. Hershey, PA: IGI Global. doi:10.4018/978-1-4666-0023-2

Allee, T., Handorf, A., & Li, W. (2010). Electrospinning: Development and biomedical applications. In A. Shukla & R. Tiwari (Eds.), *Intelligent medical technologies and biomedical engineering: Tools and applications* (pp. 48–78). Hershey, PA: IGI Global. doi:10.4018/978-1-61520-977-4.ch003

Alsmadi, I. (2014). Website performance measurement: Process and product metrics. In *Software design and development: Concepts, methodologies, tools, and applications* (pp. 1801–1827). Hershey, PA: IGI Global. doi:10.4018/978-1-4666-4301-7.ch086

Altarawneh, H., Alamaro, S., & El Sheikh, A. (2012). Web engineering and business intelligence: Agile web engineering development and practice. In A. Rahman El Sheikh & M. Alnoukari (Eds.), *Business intelligence and agile methodologies for knowledge-based organizations: Cross-disciplinary applications* (pp. 313–344). Hershey, PA: IGI Global. doi:10.4018/978-1-61350-050-7.ch015

Altarawneh, H., & El-Shiekh, A. (2010). Web engineering in small Jordanian web development firms: An XP based process model. In A. Tatnall (Ed.), *Web technologies: Concepts, methodologies, tools, and applications* (pp. 1696–1707). Hershey, PA: IGI Global. doi:10.4018/978-1-60566-982-3.ch091

Alzoabi, Z. (2014). Agile software: Body of knowledge. In *Software design and development: Concepts, methodologies, tools, and applications* (pp. 96–116). Hershey, PA: IGI Global. doi:10.4018/978-1-4666-4301-7.ch006

Andrade-Campos, A. (2013). Development of an optimization framework for parameter identification and shape optimization problems in engineering. In J. Davim (Ed.), *Dynamic methods and process advancements in mechanical, manufacturing, and materials engineering* (pp. 1–24). Hershey, PA: IGI Global. doi:10.4018/978-1-4666-1867-1.ch001

Andreatos, A. (2012). Educating the 21st century's engineers and IT professionals. In A. Patil, H. Eijkman, & E. Bhattacharyya (Eds.), *New media communication skills for engineers and IT professionals: Trans-national and trans-cultural demands* (pp. 132–159). Hershey, PA: IGI Global. doi:10.4018/978-1-4666-0243-4.ch009

Annamalai, C., & Ramayah, T. (2013). Reengineering for enterprise resource planning (ERP) systems implementation: An empirical analysis of assessing critical success factors (CSFs) of manufacturing organizations. In Industrial engineering: Concepts, methodologies, tools, and applications (pp. 791-806). Hershey, PA: IGI Global. doi:10.4018/978-1-4666-1945-6.ch044

Antchev, M. (2010). Other applications of converters and systems of converters. In *Technologies for electrical power conversion, efficiency, and distribution: Methods and processes* (pp. 270–299). Hershey, PA: IGI Global. doi:10.4018/978-1-61520-647-6.ch011

Anzelotti, G., & Valizadeh, M. (2010). Purpose-oriented small software: A case study for some engineering subjects. In R. Luppicini & A. Haghi (Eds.), *Cases on digital technologies in higher education: Issues and challenges* (pp. 164–178). Hershey, PA: IGI Global. doi:10.4018/978-1-61520-869-2.ch012

Anzelotti, G., & Valizadeh, M. (2010). Sights inside the virtual engineering education. In D. Russell & A. Haghi (Eds.), *Web-based engineering education: Critical design and effective tools* (pp. 160–174). Hershey, PA: IGI Global. doi:10.4018/978-1-61520-659-9.ch012

Asadi, M., Mohabbati, B., Gašević, D., Bagheri, E., & Hatala, M. (2012). Developing semantically-enabled families of method-oriented architectures. *International Journal of Information System Modeling and Design, 3*(4), 1–26. doi:10.4018/jismd.2012100101

Augusti, G., & Feyo de Azevedo, S. (2011). Qualification frameworks and field-specific approaches to quality assurance: Initiatives in engineering and technical education. *International Journal of Quality Assurance in Engineering and Technology Education, 1*(1), 44–57. doi:10.4018/ijqaete.2011010104

Aung, Z., & Nyunt, K. K. (2014). Constructive knowledge management model and information retrieval methods for software engineering. In *Software design and development: Concepts, methodologies, tools, and applications* (pp. 253–269). Hershey, PA: IGI Global. doi:10.4018/978-1-4666-4301-7.ch014

Azad, A. K., Auer, M. E., & Harward, V. (2012). *Internet accessible remote laboratories: Scalable e-learning tools for engineering and science disciplines.* Hershey, PA: IGI Global. doi:10.4018/978-1-61350-186-3

Azar, A. T. (2013). Overview of biomedical engineering. In Bioinformatics: Concepts, methodologies, tools, and applications (pp. 1-28). Hershey, PA: IGI Global. doi:10.4018/978-1-4666-3604-0.ch001

Badr, K. B., Badr, A. B., & Ahmad, M. N. (2013). Phases in ontology building methodologies: A recent review. In M. Nazir Ahmad, R. Colomb, & M. Abdullah (Eds.), *Ontology-based applications for enterprise systems and knowledge management* (pp. 100–123). Hershey, PA: IGI Global. doi:10.4018/978-1-4666-1993-7.ch006

Baer, W., & Renfro, C. (2013). Information sources and collection planning for engineering. In S. Holder (Ed.), *Library collection development for professional programs: Trends and best practices* (pp. 128–144). Hershey, PA: IGI Global. doi:10.4018/978-1-4666-1897-8.ch008

Baghdadi, Y., & Kraiem, N. (2014). Business process modeling with services: reverse engineering databases. In R. Perez-Castillo & M. Piattini (Eds.), *Uncovering essential software artifacts through business process archeology* (pp. 177–200). Hershey, PA: IGI Global. doi:10.4018/978-1-4666-4667-4.ch007

Bakopoulou, A., Leyhausen, G., Geurtsen, W., & Koidis, P. (2013). Dental tissue engineering research and translational approaches towards clinical application. In A. Daskalaki (Ed.), *Medical advancements in aging and regenerative technologies: Clinical tools and applications* (pp. 279–312). Hershey, PA: IGI Global. doi:10.4018/978-1-4666-2506-8.ch013

Baporikar, N. (2012). Developing right graduate attributes through project-based teaching. In M. Rasul (Ed.), *Developments in engineering education standards: Advanced curriculum innovations* (pp. 64–79). Hershey, PA: IGI Global. doi:10.4018/978-1-4666-0951-8.ch004

Baraldi, E., & Nadin, G. (2012). "Network process re-engineering" in a home textile network: The importance of business relationships and actor bonds. In T. Choi (Ed.), *Fashion supply chain management: Industry and business analysis* (pp. 212–234). Hershey, PA: IGI Global. doi:10.4018/978-1-60960-756-2.ch012

Barbu, M. C., Hasener, J., & Bernardy, G. (2014). Modern testing of wood-based panels, process control, and modeling. In A. Aguilera & J. Davim (Eds.), *Research developments in wood engineering and technology* (pp. 90–130). Hershey, PA: IGI Global. doi:10.4018/978-1-4666-4554-7.ch003

Bas, T. G. (2013). Dual market(ing) in "bio-engineering high technology" new products: The risk of uncertainty and failure. *International Journal of Measurement Technologies and Instrumentation Engineering*, *3*(2), 63–74. doi:10.4018/ijmtie.2013040104

Bedi, P., Gandotra, V., & Singhal, A. (2014). Innovative strategies for secure software development. In *Software design and development: Concepts, methodologies, tools, and applications* (pp. 2099–2119). Hershey, PA: IGI Global. doi:10.4018/978-1-4666-4301-7.ch097

Bellatreche, L. (2010). *Data warehousing design and advanced engineering applications: Methods for complex construction*. Hershey, PA: IGI Global. doi:10.4018/978-1-60566-756-0

Bhattacharyya, S., & Dutta, P. (2013). *Handbook of research on computational intelligence for engineering, science, and business* (Vols. 1–2). Hershey, PA: IGI Global. doi:10.4018/978-1-4666-2518-1

Blicblau, A. S., & Richards, D. (2012). Development of "real world" project skills for engineering students. *International Journal of Quality Assurance in Engineering and Technology Education*, *2*(1), 1–13. doi:10.4018/ijqaete.2012010101

Boci, E. S., Sarkani, S., & Mazzuchi, T. A. (2013). Development of a complex geospatial/RF design model in support of service volume engineering design. In M. Bartolacci & S. Powell (Eds.), *Advancements and innovations in wireless communications and network technologies* (pp. 56–67). Hershey, PA: IGI Global. doi:10.4018/978-1-4666-2154-1.ch005

Boudreaux, A., & Primeaux, B. (2014). Modular game engine design. In *Software design and development: Concepts, methodologies, tools, and applications* (pp. 1179–1199). Hershey, PA: IGI Global. doi:10.4018/978-1-4666-4301-7.ch058

Boudriga, N., & Hamdi, M. (2014). *Security engineering techniques and solutions for information systems: Management and implementation*. Hershey, PA: IGI Global. doi:10.4018/978-1-61520-803-6

Brad, S. (2010). Competitive design of web-based courses in engineering education. In D. Russell & A. Haghi (Eds.), *Web-based engineering education: Critical design and effective tools* (pp. 119–148). Hershey, PA: IGI Global. doi:10.4018/978-1-61520-659-9.ch010

Brad, S. (2010). Designing effective web-based courses in engineering. In R. Luppicini & A. Haghi (Eds.), *Cases on digital technologies in higher education: Issues and challenges* (pp. 217–240). Hershey, PA: IGI Global. doi:10.4018/978-1-61520-869-2.ch016

Bradford, M., Gingras, R., & Hornby, J. (2010). Business process reengineering and ERP: Weapons for the global organization. In K. St.Amant (Ed.), *IT outsourcing: Concepts, methodologies, tools, and applications* (pp. 211–228). Hershey, PA: IGI Global. doi:10.4018/978-1-60566-770-6.ch012

Brennan, R. W., Hugo, R., & Rosehart, W. D. (2012). CDIO as an enabler for graduate attributes assessment: A Canadian case study. *International Journal of Quality Assurance in Engineering and Technology Education*, *2*(2), 45–54. doi:10.4018/ijqaete.2012040105

Burnett, M. (2012). End-user software engineering and why it matters. In A. Dwivedi & S. Clarke (Eds.), *End-user computing, development, and software engineering: New challenges* (pp. 185–201). Hershey, PA: IGI Global. doi:10.4018/978-1-4666-0140-6.ch009

Burns, G. U., & Chisohlm, C. (2011). Engineering professional development related to sustainability of quality. *International Journal of Quality Assurance in Engineering and Technology Education*, *1*(1), 15–29. doi:10.4018/ijqaete.2011010102

Byrne, D., Kelly, L., & Jones, G. J. (2014). Multiple multimodal mobile devices: Lessons learned from engineering lifelong solutions. In *Software design and development: Concepts, methodologies, tools, and applications* (pp. 2014–2032). Hershey, PA: IGI Global. doi:10.4018/978-1-4666-4301-7.ch093

Cervera, M., Albert, M., Torres, V., & Pelechano, V. (2012). A model-driven approach for the design and implementation of software development methods. *International Journal of Information System Modeling and Design*, *3*(4), 86–103. doi:10.4018/jismd.2012100105

Chang, L., Levy, M., & Powell, P. (2011). Small firm process re-engineering success. In M. Tavana (Ed.), *Managing adaptability, intervention, and people in enterprise information systems* (pp. 138–155). Hershey, PA: IGI Global. doi:10.4018/978-1-60960-529-2.ch007

Chinyemba, F. (2011). Mobility of engineering and technology professionals and its impact on the quality of engineering and technology education: The case of Chinhoyi University of Technology, Zimbabwe. *International Journal of Quality Assurance in Engineering and Technology Education*, *1*(2), 35–49. doi:10.4018/ijqaete.2011070104

Chiong, R. (2010). *Nature-inspired informatics for intelligent applications and knowledge discovery: Implications in business, science, and engineering*. Hershey, PA: IGI Global. doi:10.4018/978-1-60566-705-8

Chiprianov, V., Kermarrec, Y., & Rouvrais, S. (2014). Integrating DSLs into a software engineering process: Application to collaborative construction of telecom services. In *Software design and development: Concepts, methodologies, tools, and applications* (pp. 570–595). Hershey, PA: IGI Global. doi:10.4018/978-1-4666-4301-7.ch028

Chis, M. (2010). Introduction: A survey of the evolutionary computation techniques for software engineering. In M. Chis (Ed.), *Evolutionary computation and optimization algorithms in software engineering: Applications and techniques* (pp. 1–12). Hershey, PA: IGI Global. doi:10.4018/978-1-61520-809-8.ch001

Chiu, D. K. (2013). *Mobile and web innovations in systems and service-oriented engineering*. Hershey, PA: IGI Global. doi:10.4018/978-1-4666-2470-2

Chu, P. K., & Wu, S. (2012). Biomaterials. In Z. Abu-Faraj (Ed.), *Handbook of research on biomedical engineering education and advanced bioengineering learning: Interdisciplinary concepts* (pp. 238–283). Hershey, PA: IGI Global. doi:10.4018/978-1-4666-0122-2.ch006

Cimellaro, G. P. (2013). Optimal placement of controller for seismic structures. In N. Lagaros, V. Plevris, & C. Mitropoulou (Eds.), *Design optimization of active and passive structural control systems* (pp. 1–33). Hershey, PA: IGI Global. doi:10.4018/978-1-4666-2029-2.ch001

Clark, R., & Andrews, J. (2012). Engineering the future. In M. Rasul (Ed.), *Developments in engineering education standards: Advanced curriculum innovations* (pp. 143–155). Hershey, PA: IGI Global. doi:10.4018/978-1-4666-0951-8.ch008

Clark, T., & Willans, J. (2013). Software language engineering with XMF and XModeler. In M. Mernik (Ed.), *Formal and practical aspects of domain-specific languages: Recent developments* (pp. 311–340). Hershey, PA: IGI Global. doi:10.4018/978-1-4666-2092-6.ch011

Coll, R. K., & Zegwaard, K. E. (2012). Enculturation into engineering professional practice: Using legitimate peripheral participation to develop communication skills in engineering students. In A. Patil, H. Eijkman, & E. Bhattacharyya (Eds.), *New media communication skills for engineers and IT professionals: Trans-national and trans-cultural demands* (pp. 22–33). Hershey, PA: IGI Global. doi:10.4018/978-1-4666-0243-4.ch003

Coll, R. K., & Zegwaard, K. E. (2012). Enculturation into engineering professional practice: Using legitimate peripheral participation to develop communication skills in engineering students. In A. Patil, H. Eijkman, & E. Bhattacharyya (Eds.), *New media communication skills for engineers and IT professionals: Trans-national and trans-cultural demands* (pp. 22–33). Hershey, PA: IGI Global. doi:10.4018/978-1-4666-0243-4.ch003

Colomb, R. M. (2013). Representation of action is a primary requirement in ontologies for interoperating information systems. In M. Nazir Ahmad, R. Colomb, & M. Abdullah (Eds.), *Ontology-based applications for enterprise systems and knowledge management* (pp. 68–76). Hershey, PA: IGI Global. doi:10.4018/978-1-4666-1993-7.ch004

Cooklev, T. (2013). The role of standards in engineering education. In K. Jakobs (Ed.), *Innovations in organizational IT specification and standards development* (pp. 129–137). Hershey, PA: IGI Global. doi:10.4018/978-1-4666-2160-2.ch007

Costa, L., Loughran, N., & Grønmo, R. (2014). Model-driven engineering, services and interactive real-time applications. In *Software design and development: Concepts, methodologies, tools, and applications* (pp. 178–202). Hershey, PA: IGI Global. doi:10.4018/978-1-4666-4301-7.ch010

Dai, Y., Chakraborty, B., & Shi, M. (2011). *Kansei engineering and soft computing: Theory and practice*. Hershey, PA: IGI Global. doi:10.4018/978-1-61692-797-4

Daud, M. F., Taib, J. M., & Shariffudin, R. S. (2012). Assessing mechanical engineering undergraduates' conceptual knowledge in three dimensional computer aided design (3D CAD). In K. Yusof, N. Azli, A. Kosnin, S. Yusof, & Y. Yusof (Eds.), *Outcome-based science, technology, engineering, and mathematics education: Innovative practices* (pp. 350–363). Hershey, PA: IGI Global. doi:10.4018/978-1-4666-1809-1.ch017

Daugherty, A., Hires, W. E., & Braunstein, S. G. (2013). Collection development for the college of engineering at Louisiana State University libraries: Liaison responsibilities and duties. In S. Holder (Ed.), *Library collection development for professional programs: Trends and best practices* (pp. 291–305). Hershey, PA: IGI Global. doi:10.4018/978-1-4666-1897-8.ch017

Davim, J. (2013). *Dynamic methods and process advancements in mechanical, manufacturing, and materials engineering*. Hershey, PA: IGI Global. doi:10.4018/978-1-4666-1867-1 de Vere, I., & Melles, G. (2013). Integrating 'designerly' ways with engineering science: A catalyst for change within product design and development. In Industrial engineering: Concepts, methodologies, tools, and applications (pp. 56-78). Hershey, PA: IGI Global. doi:10.4018/978-1-4666-1945-6.ch005

del Sagrado Martinez, J., & del Aguila Cano, I. M. (2010). A Bayesian network for predicting the need for a requirements review. In F. Meziane & S. Vadera (Eds.), *Artificial intelligence applications for improved software engineering development: New prospects* (pp. 106–128). Hershey, PA: IGI Global. doi:10.4018/978-1-60566-758-4.ch006

Dhar-Bhattacharjee, S., & Takruri-Rizk, H. (2012). An Indo-British comparison. In C. Romm Livermore (Ed.), *Gender and social computing: Interactions, differences and relationships* (pp. 50–71). Hershey, PA: IGI Global. doi:10.4018/978-1-60960-759-3.ch004

Díaz, V. G., Lovelle, J. M., García-Bustelo, B. C., & Martinez, O. S. (2014). *Advances and applications in model-driven engineering*. Hershey, PA: IGI Global. doi:10.4018/978-1-4666-4494-6

Dogru, A., Senkul, P., & Kaya, O. (2012). Modern approaches to software engineering in the compositional era. In *Machine learning: Concepts, methodologies, tools and applications* (pp. 1903–1923). Hershey, PA: IGI Global. doi:10.4018/978-1-60960-818-7.ch803

Dogru, A. H., & Biçer, V. (2011). *Modern software engineering concepts and practices: Advanced approaches*. Hershey, PA: IGI Global. doi:10.4018/978-1-60960-215-4

Doll, W. J., & Deng, X. (2013). Antecedents of improvisation in IT-enabled engineering work. In A. Dwivedi & S. Clarke (Eds.), *Innovative strategies and approaches for end-user computing advancements* (pp. 242–264). Hershey, PA: IGI Global. doi:10.4018/978-1-4666-2059-9.ch013

Dominguez, U., & Magdaleno, J. (2011). Industrial training in engineering education in Spain. In P. Keleher, A. Patil, & R. Harreveld (Eds.), *Work-integrated learning in engineering, built environment and technology: Diversity of practice in practice* (pp. 72–84). Hershey, PA: IGI Global. doi:10.4018/978-1-60960-547-6.ch004

Dormido, S., Vargas, H., & Sánchez, J. (2013). AutomatL@bs consortium: A Spanish network of web-based labs for control engineering education. In *Industrial engineering: Concepts, methodologies, tools, and applications* (pp. 679–699). Hershey, PA: IGI Global. doi:10.4018/978-1-4666-1945-6.ch039

Dyro, J. F. (2012). Clinical engineering. In Z. Abu-Faraj (Ed.), *Handbook of research on biomedical engineering education and advanced bioengineering learning: Interdisciplinary concepts* (pp. 521–576). Hershey, PA: IGI Global. doi:10.4018/978-1-4666-0122-2.ch012

Easton, J. M., Davies, J. R., & Roberts, C. (2011). Ontology engineering the "what's", "why's", and "how's" of data exchange. *International Journal of Decision Support System Technology*, *3*(1), 40–53. doi:10.4018/jdsst.2011010103

Eijkman, H., & Kayali, O. (2011). Addressing the politics of accreditation in engineering education: The benefits of soft systems thinking. *International Journal of Quality Assurance in Engineering and Technology Education*, *1*(2), 1–10. doi:10.4018/ijqaete.2011070101

El-Khalili, N. H. (2013). Teaching agile software engineering using problem-based learning. *International Journal of Information and Communication Technology Education*, *9*(3), 1–12. doi:10.4018/jicte.2013070101

Favre, L. M. (2010). MDA-based object-oriented reverse engineering. In L. Favre (Ed.), *Model driven architecture for reverse engineering technologies: Strategic directions and system evolution* (pp. 199–229). Hershey, PA: IGI Global. doi:10.4018/978-1-61520-649-0.ch010

Favre, L. M. (2010). Reverse engineering and MDA: An introduction. In L. Favre (Ed.), *Model driven architecture for reverse engineering technologies: Strategic directions and system evolution* (pp. 1–14). Hershey, PA: IGI Global. doi:10.4018/978-1-61520-649-0.ch001

Fenton, N., Hearty, P., Neil, M., & Radlinski, L. (2010). Software project and quality modelling using Bayesian networks. In F. Meziane & S. Vadera (Eds.), *Artificial intelligence applications for improved software engineering development: New prospects* (pp. 1–25). Hershey, PA: IGI Global. doi:10.4018/978-1-60566-758-4.ch001

Ferreira, R., Brisolara, L., Mattos, J. C., Spech, E., & Cota, E. (2010). Engineering embedded software: From application modeling to software synthesis. In L. Gomes & J. Fernandes (Eds.), *Behavioral modeling for embedded systems and technologies: Applications for design and implementation* (pp. 245–270). Hershey, PA: IGI Global. doi:10.4018/978-1-60566-750-8.ch010

Ferreira da Silva Oliveira, M., Wehrmeister, M. A., Assis do Nascimento, F., & Pereira, C. E. (2010). High-level design space exploration of embedded systems using the model-driven engineering and aspect-oriented design approaches. In L. Gomes & J. Fernandes (Eds.), *Behavioral modeling for embedded systems and technologies: Applications for design and implementation* (pp. 114–146). Hershey, PA: IGI Global. doi:10.4018/978-1-60566-750-8.ch005

Ferris, T. L. (2013). Engineering design as research. In *Industrial engineering: Concepts, methodologies, tools, and applications* (pp. 1766–1779). Hershey, PA: IGI Global. doi:10.4018/978-1-4666-1945-6.ch094

Frank, M. (2010). Capacity for engineering systems thinking (CEST): Literature review, principles for assessing and the reliability and validity of an assessing tool. In M. Hunter (Ed.), *Strategic information systems: Concepts, methodologies, tools, and applications* (pp. 1171–1183). Hershey, PA: IGI Global. doi:10.4018/978-1-60566-677-8.ch076

Fraser, D. (2012). A case study of curriculum development in engineering: Insights gained over two decades. In K. Yusof, N. Azli, A. Kosnin, S. Yusof, & Y. Yusof (Eds.), *Outcome-based science, technology, engineering, and mathematics education: Innovative practices* (pp. 27–49). Hershey, PA: IGI Global. doi:10.4018/978-1-4666-1809-1.ch002

Fraser, D. (2012). Curriculum initiatives to help engineering students learn and develop. In K. Yusof, N. Azli, A. Kosnin, S. Yusof, & Y. Yusof (Eds.), *Outcome-based science, technology, engineering, and mathematics education: Innovative practices* (pp. 85–106). Hershey, PA: IGI Global. doi:10.4018/978-1-4666-1809-1.ch005

Fries, T. P. (2014). Reengineering structured legacy system documentation to UML object-oriented artifacts. In *Software design and development: Concepts, methodologies, tools, and applications* (pp. 749–771). Hershey, PA: IGI Global. doi:10.4018/978-1-4666-4301-7.ch036

Funabashi, M. (2014). Transdisciplinary science and technology and service systems. In M. Kosaka & K. Shirahada (Eds.), *Progressive trends in knowledge and system-based science for service innovation* (pp. 101–127). Hershey, PA: IGI Global. doi:10.4018/978-1-4666-4663-6.ch006

Gaetano, L., Puppato, D., & Balestra, G. (2012). Modeling clinical engineering activities to support healthcare technology management. In A. Kolker & P. Story (Eds.), *Management engineering for effective healthcare delivery: Principles and applications* (pp. 113–131). Hershey, PA: IGI Global. doi:10.4018/978-1-60960-872-9.ch005

Galvão, T. A., Neto, F., & Bonates, M. F. (2013). eRiskGame: A persistent browser-based game for supporting project-based learning in the risk management context. In Industrial engineering: Concepts, methodologies, tools, and applications (pp. 1243-1259). Hershey, PA: IGI Global. doi:10.4018/978-1-4666-1945-6.ch067

Garousi, V., Shahnewaz, S., & Krishnamurthy, D. (2013). UML-driven software performance engineering: A systematic mapping and trend analysis. In V. Díaz, J. Lovelle, B. García-Bustelo, & O. Martínez (Eds.), *Progressions and innovations in model-driven software engineering* (pp. 18–64). Hershey, PA: IGI Global. doi:10.4018/978-1-4666-4217-1.ch002

Ge, W., Yang, N., Wang, W., & Li, J. (2011). Interfacial interactions: Drag. In S. Pannala, M. Syamlal, & T. O'Brien (Eds.), *Computational gas-solids flows and reacting systems: Theory, methods and practice* (pp. 128–177). Hershey, PA: IGI Global. doi:10.4018/978-1-61520-651-3.ch004

Génova, G., Llorens, J., & Morato, J. (2014). Software engineering research: The need to strengthen and broaden the classical scientific method. In *Software design and development: Concepts, methodologies, tools, and applications* (pp. 1639–1658). Hershey, PA: IGI Global. doi:10.4018/978-1-4666-4301-7.ch079

Genvigir, E. C., & Vijaykumar, N. L. (2010). Requirements traceability. In M. Ramachandran & R. de Carvalho (Eds.), *Handbook of research on software engineering and productivity technologies: Implications of globalization* (pp. 102–120). Hershey, PA: IGI Global. doi:10.4018/978-1-60566-731-7.ch008

Ghosh, S. (2010). Online automated essay grading system as a web based learning (WBL) tool in engineering education. In D. Russell & A. Haghi (Eds.), *Web-based engineering education: Critical design and effective tools* (pp. 53–62). Hershey, PA: IGI Global. doi:10.4018/978-1-61520-659-9.ch005

Gill, A. Q., & Bunker, D. (2013). SaaS requirements engineering for agile development. In X. Wang, N. Ali, I. Ramos, & R. Vidgen (Eds.), *Agile and lean service-oriented development: Foundations, theory, and practice* (pp. 64–93). Hershey, PA: IGI Global. doi:10.4018/978-1-4666-2503-7.ch004

Giraldo, F. D., Villegas, M. L., & Collazos, C. A. (2014). The use of HCI approaches into distributed CSCL activities applied to software engineering courses. In *Software design and development: Concepts, methodologies, tools, and applications* (pp. 2033–2050). Hershey, PA: IGI Global. doi:10.4018/978-1-4666-4301-7.ch094

Girardi, R., & Leite, A. (2011). Knowledge engineering support for agent-oriented software reuse. In M. Ramachandran (Ed.), *Knowledge engineering for software development life cycles: Support technologies and applications* (pp. 177–195). Hershey, PA: IGI Global. doi:10.4018/978-1-60960-509-4.ch010

Goodhew, P. (2013). Why get your engineering programme accredited? In *Industrial engineering: Concepts, methodologies, tools, and applications* (pp. 18–20). Hershey, PA: IGI Global. doi:10.4018/978-1-4666-1945-6.ch002

Goossenaerts, J., Possel-Dölken, F., & Popplewell, K. (2011). Vision, trends, gaps and a broad roadmap for future engineering. In I. Management Association (Ed.), Global business: Concepts, methodologies, tools and applications (pp. 2229-2243). Hershey, PA: IGI Global. doi:10.4018/978-1-60960-587-2.ch802

Goyal, S. B., & Prakash, N. (2013). Functional method engineering. *International Journal of Information System Modeling and Design*, *4*(1), 79–103. doi:10.4018/jismd.2013010104

Gunasekaran, A., & Shea, T. (2010). Requirements management for ERP projects. In *Organizational advancements through enterprise information systems: Emerging applications and developments* (pp. 29–45). Hershey, PA: IGI Global. doi:10.4018/978-1-60566-968-7.ch003

Gustavsson, I., Claesson, L., Nilsson, K., Zackrisson, J., Zubia, J. G., & Jayo, U. H. … Claesson, I. (2012). The VISIR open lab platform. In A. Azad, M. Auer, & V. Harward (Eds.) Internet accessible remote laboratories: Scalable e-learning tools for engineering and science disciplines (pp. 294-317). Hershey, PA: IGI Global. doi:10.4018/978-1-61350-186-3.ch015

Habib, M. K., & Davim, J. (2013). *Engineering creative design in robotics and mechatronics*. Hershey, PA: IGI Global. doi:10.4018/978-1-4666-4225-6

Haghi, A. K. (2013). *Methodologies and applications for chemoinformatics and chemical engineering*. Hershey, PA: IGI Global. doi:10.4018/978-1-4666-4010-8

Haghi, A. K., & Noroozi, B. (2010). Adapting engineering education to the new century. In D. Russell & A. Haghi (Eds.), *Web-based engineering education: Critical design and effective tools* (pp. 30–41). Hershey, PA: IGI Global. doi:10.4018/978-1-61520-659-9.ch003

Haghpanahi, M., Nikkhoo, M., & Peirovi, H. A. (2010). Computer aided tissue engineering from modeling to manufacturing. In A. Lazakidou (Ed.), *Biocomputation and biomedical informatics: Case studies and applications* (pp. 75–88). Hershey, PA: IGI Global. doi:10.4018/978-1-60566-768-3.ch004

Hepsø, I. L., Rindal, A., & Waldal, K. (2013). The introduction of a hand-held platform in an engineering and fabrication company. In T. Rosendahl & V. Hepsø (Eds.), *Integrated operations in the oil and gas industry: Sustainability and capability development* (pp. 246–260). Hershey, PA: IGI Global. doi:10.4018/978-1-4666-2002-5.ch015

Hernández-López, A., Colomo-Palacios, R., García-Crespo, Á., & Cabezas-Isla, F. (2013). Software engineering productivity: Concepts, issues and challenges. In J. Wang (Ed.), *Perspectives and techniques for improving information technology project management* (pp. 69–79). Hershey, PA: IGI Global. doi:10.4018/978-1-4666-2800-7.ch006

Herrmann, A., & Morali, A. (2012). Interplay of security requirements engineering and reverse engineering in the maintenance of undocumented software. In J. Rech & C. Bunse (Eds.), *Emerging technologies for the evolution and maintenance of software models* (pp. 57–91). Hershey, PA: IGI Global. doi:10.4018/978-1-61350-438-3.ch003

Hochrainer, M., & Ziegler, F. (2013). Tuned liquid column gas damper in structural control: The salient features of a general purpose damping device and its application in buildings, bridges, and dams. In N. Lagaros, V. Plevris, & C. Mitropoulou (Eds.), *Design optimization of active and passive structural control systems* (pp. 150–179). Hershey, PA: IGI Global. doi:10.4018/978-1-4666-2029-2.ch007

Hochstein, L., Schott, B., & Graybill, R. B. (2013). Computational engineering in the cloud: Benefits and challenges. In A. Dwivedi & S. Clarke (Eds.), *Innovative strategies and approaches for end-user computing advancements* (pp. 314–332). Hershey, PA: IGI Global. doi:10.4018/978-1-4666-2059-9.ch017

Höhn, S., Lowis, L., Jürjens, J., & Accorsi, R. (2010). Identification of vulnerabilities in web services using model-based security. In C. Gutiérrez, E. Fernández-Medina, & M. Piattini (Eds.), *Web services security development and architecture: Theoretical and practical issues* (pp. 1–32). Hershey, PA: IGI Global. doi:10.4018/978-1-60566-950-2.ch001

Holton, D. L., & Verma, A. (2010). Designing animated simulations and web-based assessments to improve electrical engineering education. In D. Russell & A. Haghi (Eds.), *Web-based engineering education: Critical design and effective tools* (pp. 77–95). Hershey, PA: IGI Global. doi:10.4018/978-1-61520-659-9.ch007

Honnutagi, A. R., Sonar, R., & Babu, S. (2012). Quality accreditation system for Indian engineering education using knowledge management and system dynamics. *International Journal of Quality Assurance in Engineering and Technology Education*, *2*(3), 47–61. doi:10.4018/ijqaete.2012070105

Hua, G. B. (2013). Business process re-engineering. In *Implementing IT business strategy in the construction industry* (pp. 118–140). Hershey, PA: IGI Global. doi:10.4018/978-1-4666-4185-3.ch006

Hussein, B., Hage-Diab, A., Hammoud, M., Kawtharani, A., El-Hage, H., & Haj-Ali, A. (2014). Management response to improve the educational performance of engineering students: The case of the lebanese international university. In G. Khoury & M. Khoury (Eds.), *Cases on management and organizational behavior in an Arab context* (pp. 91–110). Hershey, PA: IGI Global. doi:10.4018/978-1-4666-5067-1.ch006

Hussey, M., Wu, B., & Xu, X. (2011). Co-operation models for industries and software education institutions. In *Software industry-oriented education practices and curriculum development: Experiences and lessons* (pp. 39–56). Hershey, PA: IGI Global. doi:10.4018/978-1-60960-797-5.ch003

Hussey, M., Wu, B., & Xu, X. (2011). Curriculum issues in industry oriented software engineering education. In *Software industry-oriented education practices and curriculum development: Experiences and lessons* (pp. 153–165). Hershey, PA: IGI Global. doi:10.4018/978-1-60960-797-5.ch010

Hussey, M., Wu, B., & Xu, X. (2011). Industry oriented curriculum and syllabus creation for software engineering series courses in the school of software. In *Software industry-oriented education practices and curriculum development: Experiences and lessons* (pp. 98–109). Hershey, PA: IGI Global. doi:10.4018/978-1-60960-797-5.ch006

Ilyas, Q. M. (2013). Ontology augmented software engineering. In K. Buragga & N. Zaman (Eds.), *Software development techniques for constructive information systems design* (pp. 406–413). Hershey, PA: IGI Global. doi:10.4018/978-1-4666-3679-8.ch023

Islam, S., Mouratidis, H., Kalloniatis, C., Hudic, A., & Zechner, L. (2013). Model based process to support security and privacy requirements engineering. *International Journal of Secure Software Engineering*, *3*(3), 1–22. doi:10.4018/jsse.2012070101

Jahan, K., Everett, J. W., Tang, G., Farrell, S., Zhang, H., Wenger, A., & Noori, M. (2010). Use of living systems to teach basic engineering concepts. In D. Russell & A. Haghi (Eds.), *Web-based engineering education: Critical design and effective tools* (pp. 96–107). Hershey, PA: IGI Global. doi:10.4018/978-1-61520-659-9.ch008

Jakovljevic, M. (2013). A conceptual model of creativity, invention, and innovation (MCII) for entrepreneurial engineers. In S. Buckley & M. Jakovljevic (Eds.), *Knowledge management innovations for interdisciplinary education: Organizational applications* (pp. 66–87). Hershey, PA: IGI Global. doi:10.4018/978-1-4666-1969-2.ch004

Jaroucheh, Z., Liu, X., & Smith, S. (2014). A software engineering framework for context-aware service-based processes in pervasive environments. In *Software design and development: Concepts, methodologies, tools, and applications* (pp. 71–95). Hershey, PA: IGI Global. doi:10.4018/978-1-4666-4301-7.ch005

Joiner, R., Iacovides, I., Darling, J., Diament, A., Drew, B., Duddley, J., ... Gavin, C. (2013). Racing academy: A case study of a digital game for supporting students learning of physics and engineering. In Y. Baek & N. Whitton (Eds.), *Cases on digital game-based learning: Methods, models, and strategies* (pp. 509–523). Hershey, PA: IGI Global. doi:10.4018/978-1-4666-2848-9.ch026

Jordan, M. E. (2014). Interweaving the digital and physical worlds in collaborative project-based learning experiences. In D. Loveless, B. Griffith, M. Bérci, E. Ortlieb, & P. Sullivan (Eds.), *Academic knowledge construction and multimodal curriculum development* (pp. 265–282). Hershey, PA: IGI Global. doi:10.4018/978-1-4666-4797-8.ch017

Juang, Y. (2010). WIRE: A highly interactive blended learning for engineering education. In D. Russell & A. Haghi (Eds.), *Web-based engineering education: Critical design and effective tools* (pp. 149–159). Hershey, PA: IGI Global. doi:10.4018/978-1-61520-659-9.ch011

Kadry, S. (2013). *Diagnostics and prognostics of engineering systems: Methods and techniques*. Hershey, PA: IGI Global. doi:10.4018/978-1-4666-2095-7

Kamboj, A., Kumar, S., & Singh, H. (2012). Design and development of hybrid stir casting process. *International Journal of Applied Industrial Engineering*, *1*(2), 1–6. doi:10.4018/ijaie.2012070101

Kamthan, P. (2010). A methodology for integrating the social web environment in software engineering education. In S. Dasgupta (Ed.), *Social computing: Concepts, methodologies, tools, and applications* (pp. 457–471). Hershey, PA: IGI Global. doi:10.4018/978-1-60566-984-7.ch031

Kamthan, P. (2010). A social web perspective of software engineering education. In S. Murugesan (Ed.), *Handbook of research on web 2.0, 3.0, and X.0: Technologies, business, and social applications* (pp. 472–495). Hershey, PA: IGI Global. doi:10.4018/978-1-60566-384-5.ch026

Kamthan, P. (2010). On the prospects and concerns of pattern-oriented web engineering. In G. Alkhatib & D. Rine (Eds.), *Web engineering advancements and trends: Building new dimensions of information technology* (pp. 97–128). Hershey, PA: IGI Global. doi:10.4018/978-1-60566-719-5.ch006

Kamthan, P. (2011). Using the social web environment for software engineering education. In L. Tomei (Ed.), *Online courses and ICT in education: Emerging practices and applications* (pp. 23–45). Hershey, PA: IGI Global. doi:10.4018/978-1-60960-150-8.ch003

Kamthan, P. (2013). An exploration of the social web environment for collaborative software engineering education. In N. Karacapilidis, M. Raisinghani, & E. Ng (Eds.), *Web-based and blended educational tools and innovations* (pp. 1–23). Hershey, PA: IGI Global. doi:10.4018/978-1-4666-2023-0.ch001

Kapos, G., Dalakas, V., Nikolaidou, M., & Anagnostopoulos, D. (2014). An integrated framework to simulate SysML models using DEVS simulators. In P. Fonseca i Casas (Ed.), *Formal languages for computer simulation: Transdisciplinary models and applications* (pp. 305–332). Hershey, PA: IGI Global. doi:10.4018/978-1-4666-4369-7.ch010

Karlsson, F., & Ågerfalk, P. J. (2011). Towards structured flexibility in information systems development: Devising a method for method configuration. In K. Siau (Ed.), *Theoretical and practical advances in information systems development: Emerging trends and approaches* (pp. 214–238). Hershey, PA: IGI Global. doi:10.4018/978-1-60960-521-6.ch010

Karpati, P., Sindre, G., & Matulevicius, R. (2012). Comparing misuse case and mal-activity diagrams for modelling social engineering attacks. *International Journal of Secure Software Engineering*, *3*(2), 54–73. doi:10.4018/jsse.2012040103

Keleher, P., & Patil, A. (2012). Conducting an effective residential school for an undergraduate materials science and engineering course. *International Journal of Quality Assurance in Engineering and Technology Education*, *2*(3), 41–46. doi:10.4018/ijqaete.2012070104

Ker, H. W. (2012). Engineering education and attitudes toward mathematics: A comparative study. *International Journal of Quality Assurance in Engineering and Technology Education*, *2*(1), 63–76. doi:10.4018/ijqaete.2012010105

Kinsner, W. (2012). Challenges in the design of adoptive, intelligent and cognitive systems. In Y. Wang (Ed.), *Software and intelligent sciences: New transdisciplinary findings* (pp. 47–67). Hershey, PA: IGI Global. doi:10.4018/978-1-4666-0261-8.ch004

Kirikova, M. (2011). Domain modeling approaches in IS engineering. In J. Osis & E. Asnina (Eds.), *Model-driven domain analysis and software development: Architectures and functions* (pp. 388–406). Hershey, PA: IGI Global. doi:10.4018/978-1-61692-874-2.ch018

Kljajic, M., & Farr, J. V. (2010). Importance of systems engineering in the development of information systems. In D. Paradice (Ed.), *Emerging systems approaches in information technologies: Concepts, theories, and applications* (pp. 51–66). Hershey, PA: IGI Global. doi:10.4018/978-1-60566-976-2.ch004

Kljajic, M., & Farr, J. V. (2010). The role of systems engineering in the development of information systems. In M. Hunter (Ed.), *Strategic information systems: Concepts, methodologies, tools, and applications* (pp. 369–381). Hershey, PA: IGI Global. doi:10.4018/978-1-60566-677-8.ch026

Knoell, H. D. (2010). User participation in the quality assurance of requirements specifications: An evaluation of traditional models and animated systems engineering techniques. In M. Hunter (Ed.), *Strategic information systems: Concepts, methodologies, tools, and applications* (pp. 1623–1638). Hershey, PA: IGI Global. doi:10.4018/978-1-60566-677-8.ch106

Kof, L. (2010). From textual scenarios to message sequence charts. In F. Meziane & S. Vadera (Eds.), *Artificial intelligence applications for improved software engineering development: New prospects* (pp. 83–105). Hershey, PA: IGI Global. doi:10.4018/978-1-60566-758-4.ch005

Köhler, B., Gluchow, M., & Brügge, B. (2014). Teaching basic software engineering to senior high school students. In *K-12 education: Concepts, methodologies, tools, and applications* (pp. 1634–1649). Hershey, PA: IGI Global. doi:10.4018/978-1-4666-4502-8.ch094

Kolker, A. (2012). Efficient managerial decision-making in healthcare settings: Examples and fundamental principles. In A. Kolker & P. Story (Eds.), *Management engineering for effective healthcare delivery: Principles and applications* (pp. 1–45). Hershey, PA: IGI Global. doi:10.4018/978-1-60960-872-9.ch001

Kolker, A., & Story, P. (2012). *Management engineering for effective healthcare delivery: Principles and applications*. Hershey, PA: IGI Global. doi:10.4018/978-1-60960-872-9

Kornyshova, E., Deneckère, R., & Claudepierre, B. (2013). Towards method component contextualization. In J. Krogstie (Ed.), *Frameworks for developing efficient information systems: Models, theory, and practice* (pp. 337–368). Hershey, PA: IGI Global. doi:10.4018/978-1-4666-4161-7.ch015

Krishnan, S. (2012). Problem-based learning curricula in engineering. In M. Rasul (Ed.), *Developments in engineering education standards: Advanced curriculum innovations* (pp. 23–40). Hershey, PA: IGI Global. doi:10.4018/978-1-4666-0951-8.ch002

Ku, H., & Thonglek, S. (2011). Running a successful practice school: Challenges and lessons learned. In P. Keleher, A. Patil, & R. Harreveld (Eds.), *Work-integrated learning in engineering, built environment and technology: Diversity of practice in practice* (pp. 131–163). Hershey, PA: IGI Global. doi:10.4018/978-1-60960-547-6.ch007

Lacuesta, R., Fernández-Sanz, L., & Romay, M. D. (2014). Requirements specification as basis for mobile software quality assurance. In *Software design and development: Concepts, methodologies, tools, and applications* (pp. 719–732). Hershey, PA: IGI Global. doi:10.4018/978-1-4666-4301-7.ch034

Lai, A., Zhang, C., & Busovaca, S. (2013). 2-SQUARE: A web-based enhancement of SQUARE privacy and security requirements engineering. *International Journal of Software Innovation, 1*(1), 41–53. doi:10.4018/ijsi.2013010104

Lane, J. A., Petkov, D., & Mora, M. (2010). Software engineering and the systems approach: A conversation with Barry Boehm. In M. Hunter (Ed.), *Strategic information systems: Concepts, methodologies, tools, and applications* (pp. 333–337). Hershey, PA: IGI Global. doi:10.4018/978-1-60566-677-8.ch024

Lansiquot, R. D. (2013). *Cases on interdisciplinary research trends in science, technology, engineering, and mathematics: Studies on urban classrooms*. Hershey, PA: IGI Global. doi:10.4018/978-1-4666-2214-2

Laribi, S., Le Bris, A., Huang, L. M., Olsson, P., & Guillemoles, J. F. (2013). Phononic engineering for hot carrier solar cells. In L. Fara & M. Yamaguchi (Eds.), *Advanced solar cell materials, technology, modeling, and simulation* (pp. 214–242). Hershey, PA: IGI Global. doi:10.4018/978-1-4666-1927-2.ch012

Lee, C. K., & Sidhu, M. S. (2013). Computer-aided engineering education: New learning approaches and technologies. In V. Wang (Ed.), *Handbook of research on teaching and learning in K-20 education* (pp. 317–340). Hershey, PA: IGI Global. doi:10.4018/978-1-4666-4249-2.ch019

Lee, J., Ma, S., Lee, S., Wu, C., & Lee, C. L. (2014). Towards a high-availability-driven service composition framework. In *Software design and development: Concepts, methodologies, tools, and applications* (pp. 1498–1520). Hershey, PA: IGI Global. doi:10.4018/978-1-4666-4301-7.ch073

Leng, J., Rhyne, T., & Sharrock, W. (2012). Visualization: Future technology and practices for computational science and engineering. In J. Leng & W. Sharrock (Eds.), *Handbook of research on computational science and engineering: Theory and practice* (pp. 381–413). Hershey, PA: IGI Global. doi:10.4018/978-1-61350-116-0.ch016

Leng, J., & Sharrock, W. (2012). *Handbook of research on computational science and engineering: Theory and practice* (Vols. 1–2). Hershey, PA: IGI Global. doi:10.4018/978-1-61350-116-0

Li, Z. J. (2013). Prototyping of robotic systems in surgical procedures and automated manufacturing processes. In *Industrial engineering: Concepts, methodologies, tools, and applications* (pp. 1969–1987). Hershey, PA: IGI Global. doi:10.4018/978-1-4666-1945-6.ch106

Lladó, C. M., Bonet, P., & Smith, C. U. (2014). Towards a multi-formalism multi-solution framework for model-driven performance engineering. In M. Gribaudo & M. Iacono (Eds.), *Theory and application of multi-formalism modeling* (pp. 34–55). Hershey, PA: IGI Global. doi:10.4018/978-1-4666-4659-9.ch003

Loo, A. (2013). *Distributed computing innovations for business, engineering, and science*. Hershey, PA: IGI Global. doi:10.4018/978-1-4666-2533-4

Ma, Z. M. (2011). Engineering design knowledge management. In D. Schwartz & D. Te'eni (Eds.), *Encyclopedia of knowledge management* (2nd ed.; pp. 263–269). Hershey, PA: IGI Global. doi:10.4018/978-1-59904-931-1.ch025

Maiti, C. K., & Maiti, A. (2013). Teaching technology computer aided design (TCAD) online. In Industrial engineering: Concepts, methodologies, tools, and applications (pp. 1043-1063). Hershey, PA: IGI Global. doi:10.4018/978-1-4666-1945-6.ch057

Male, S. A. (2012). Generic engineering competencies required by engineers graduating in Australia: The competencies of engineering graduates (CEG) project. In M. Rasul (Ed.), *Developments in engineering education standards: Advanced curriculum innovations* (pp. 41–63). Hershey, PA: IGI Global. doi:10.4018/978-1-4666-0951-8.ch003

Malmqvist, J. (2012). A comparison of the CDIO and EUR-ACE quality assurance systems. *International Journal of Quality Assurance in Engineering and Technology Education, 2*(2), 9–22. doi:10.4018/ijqaete.2012040102

Maree, D. J., & Maree, M. (2010). Factors contributing to the success of women working in science, engineering and technology (SET) careers. In A. Cater-Steel & E. Cater (Eds.), *Women in engineering, science and technology: Education and career challenges* (pp. 183–210). Hershey, PA: IGI Global. doi:10.4018/978-1-61520-657-5.ch009

Marichal, G. N., & González, E. J. (2012). Intelligent MAS in system engineering and robotics. In *Machine learning: Concepts, methodologies, tools and applications* (pp. 175–182). Hershey, PA: IGI Global. doi:10.4018/978-1-60960-818-7.ch204

Martinez, L., Favre, L., & Pereira, C. (2013). Architecture-driven modernization for software reverse engineering technologies. In V. Díaz, J. Lovelle, B. García-Bustelo, & O. Martínez (Eds.), *Progressions and innovations in model-driven software engineering* (pp. 288–307). Hershey, PA: IGI Global. doi:10.4018/978-1-4666-4217-1.ch012

Matanovic, D. (2014). The macondo 252 disaster: Causes and consequences. In D. Matanovic, N. Gaurina-Medjimurec, & K. Simon (Eds.), *Risk analysis for prevention of hazardous situations in petroleum and natural gas engineering* (pp. 115–131). Hershey, PA: IGI Global. doi:10.4018/978-1-4666-4777-0.ch006

Matanovic, D., Gaurina-Medjimurec, N., & Simon, K. (2014). *Risk analysis for prevention of hazardous situations in petroleum and natural gas engineering*. Hershey, PA: IGI Global. doi:10.4018/978-1-4666-4777-0

Matsuo, T., & Fujimoto, T. (2012). Analogical thinking based instruction method in IT professional education. In R. Colomo-Palacios (Ed.), *Professional advancements and management trends in the IT sector* (pp. 95–108). Hershey, PA: IGI Global. doi:10.4018/978-1-4666-0924-2.ch007

Mazo, R., Salinesi, C., Diaz, D., Djebbi, O., & Lora-Michiels, A. (2012). Constraints: The heart of domain and application engineering in the product lines engineering strategy. *International Journal of Information System Modeling and Design, 3*(2), 33–68. doi:10.4018/jismd.2012040102

Mead, N. R. (2010). Benefits and challenges in the use of case studies for security requirements engineering methods. *International Journal of Secure Software Engineering, 1*(1), 74–91. doi:10.4018/jsse.2010102005

Méausoone, P., & Aguilera, A. (2014). Inventory of experimental works on cutting tools' life for the wood industry. In A. Aguilera & J. Davim (Eds.), *Research developments in wood engineering and technology* (pp. 320–342). Hershey, PA: IGI Global. doi:10.4018/978-1-4666-4554-7.ch009

Medlin, B. D., & Cazier, J. (2011). Obtaining patient's information from hospital employees through social engineering techniques: An investigative study. In H. Nemati (Ed.), *Pervasive information security and privacy developments: Trends and advancements* (pp. 77–89). Hershey, PA: IGI Global. doi:10.4018/978-1-61692-000-5.ch006

Meziane, F., & Vadera, S. (2010). *Artificial intelligence applications for improved software engineering development: new prospects*. Hershey, PA: IGI Global. doi:10.4018/978-1-60566-758-4

Meziane, F., & Vadera, S. (2012). Artificial intelligence in software engineering: Current developments and future prospects. In *Machine learning: Concepts, methodologies, tools and applications* (pp. 1215–1236). Hershey, PA: IGI Global. doi:10.4018/978-1-60960-818-7.ch504

Milanovic, N. (2011). *Engineering reliable service oriented architecture: Managing complexity and service level agreements*. Hershey, PA: IGI Global. doi:10.4018/978-1-60960-493-6

Miller, W. S., & Summers, J. D. (2013). Tool and information centric design process modeling: Three case studies. In *Industrial engineering: Concepts, methodologies, tools, and applications* (pp. 1613–1637). Hershey, PA: IGI Global. doi:10.4018/978-1-4666-1945-6.ch086

Mishra, B., & Shukla, K. K. (2014). Data mining techniques for software quality prediction. In *Software design and development: Concepts, methodologies, tools, and applications* (pp. 401–428). Hershey, PA: IGI Global. doi:10.4018/978-1-4666-4301-7.ch021

Mishra, S. (2011). Social and ethical concerns of biomedical engineering research and practice. In A. Shukla & R. Tiwari (Eds.), *Biomedical engineering and information systems: Technologies, tools and applications* (pp. 54–80). Hershey, PA: IGI Global. doi:10.4018/978-1-61692-004-3.ch003

Moeller, D. P., & Sitzmann, D. (2012). Online computer engineering: combining blended e-learning in engineering with lifelong learning. In M. Rasul (Ed.), *Developments in engineering education standards: Advanced curriculum innovations* (pp. 194–215). Hershey, PA: IGI Global. doi:10.4018/978-1-4666-0951-8.ch011

Moeller, D. P., & Vakilzadian, H. (2012). Technology-enhanced learning standard through integration of modeling and simulation into engineering study programs. In M. Rasul (Ed.), *Developments in engineering education standards: Advanced curriculum innovations* (pp. 157–177). Hershey, PA: IGI Global. doi:10.4018/978-1-4666-0951-8.ch009

Mohan Baral, L., Kifor, C. V., Bondrea, I., & Oprean, C. (2012). Introducing problem based learning (PBL) in textile engineering education and assessing its influence on six sigma project implementation. *International Journal of Quality Assurance in Engineering and Technology Education*, *2*(4), 38–48. doi:10.4018/ijqaete.2012100104

Moore, S., May, D., & Wold, K. (2012). Developing cultural competency in engineering through transnational distance learning. In R. Hogan (Ed.), *Transnational distance learning and building new markets for universities* (pp. 210–228). Hershey, PA: IGI Global. doi:10.4018/978-1-4666-0206-9.ch013

Mora, M., Gelman, O., Frank, M., Paradice, D. B., & Cervantes, F. (2010). Toward an interdisciplinary engineering and management of complex IT-intensive organizational systems: A systems view. In D. Paradice (Ed.), *Emerging systems approaches in information technologies: Concepts, theories, and applications* (pp. 1–24). Hershey, PA: IGI Global. doi:10.4018/978-1-60566-976-2.ch001

Mora, M., O'Connor, R., Raisinghani, M. S., Macías-Luévano, J., & Gelman, O. (2013). An IT service engineering and management framework (ITS-EMF). In P. Ordóñez de Pablos & R. Tennyson (Eds.), *Best practices and new perspectives in service science and management* (pp. 76–91). Hershey, PA: IGI Global. doi:10.4018/978-1-4666-3894-5.ch005

Moser, T., Biffl, S., Sunindyo, W. D., & Winkler, D. (2013). Integrating production automation expert knowledge across engineering domains. In N. Bessis (Ed.), *Development of distributed systems from design to application and maintenance* (pp. 152–167). Hershey, PA: IGI Global. doi:10.4018/978-1-4666-2647-8.ch009

Mouratidis, H. (2011). *Software engineering for secure systems: Industrial and research perspectives*. Hershey, PA: IGI Global. doi:10.4018/978-1-61520-837-1

Mourtos, N. J. (2012). Defining, teaching, and assessing engineering design skills. *International Journal of Quality Assurance in Engineering and Technology Education*, *2*(1), 14–30. doi:10.4018/ijqaete.2012010102

Moustafa, A. (2012). Damage assessment of inelastic structures under simulated critical earthquakes. In V. Plevris, C. Mitropoulou, & N. Lagaros (Eds.), *Structural seismic design optimization and earthquake engineering: Formulations and applications* (pp. 128–151). Hershey, PA: IGI Global. doi:10.4018/978-1-4666-1640-0.ch006

Nasir, M. H., Alias, N. A., Fauzi, S. S., & Massatu, M. H. (2012). Implementation of the personal software process in academic settings and current support tools. In S. Fauzi, M. Nasir, N. Ramli, & S. Sahibuddin (Eds.), *Software process improvement and management: Approaches and tools for practical development* (pp. 117–148). Hershey, PA: IGI Global. doi:10.4018/978-1-61350-141-2.ch007

Nhlabatsi, A., Nuseibeh, B., & Yu, Y. (2012). Security requirements engineering for evolving software systems: A survey. In K. Khan (Ed.), *Security-aware systems applications and software development methods* (pp. 108–128). Hershey, PA: IGI Global. doi:10.4018/978-1-4666-1580-9.ch007

Noroozi, B., Valizadeh, M., & Sorial, G. A. (2010). Designing of e-learning for engineering education in developing countries: Key issues and success factors. In D. Russell & A. Haghi (Eds.), *Web-based engineering education: Critical design and effective tools* (pp. 1–19). Hershey, PA: IGI Global. doi:10.4018/978-1-61520-659-9.ch001

Oh Navarro, E. (2011). On the role of learning theories in furthering software engineering education. In *Instructional design: Concepts, methodologies, tools and applications* (pp. 1645–1666). Hershey, PA: IGI Global. doi:10.4018/978-1-60960-503-2.ch709

Onwubiko, C. (2014). Modelling situation awareness information and system requirements for the mission using goal-oriented task analysis approach. In *Software design and development: Concepts, methodologies, tools, and applications* (pp. 460–478). Hershey, PA: IGI Global. doi:10.4018/978-1-4666-4301-7.ch023

Osis, J., & Asnina, E. (2011). Is modeling a treatment for the weakness of software engineering? In J. Osis & E. Asnina (Eds.), *Model-driven domain analysis and software development: Architectures and functions* (pp. 1–14). Hershey, PA: IGI Global. doi:10.4018/978-1-61692-874-2.ch001

Othman, R., & Awang, Z. (2012). Using multiple methods in assessing oral communication skills in the final year project design course of an undergraduate engineering program. In K. Yusof, N. Azli, A. Kosnin, S. Yusof, & Y. Yusof (Eds.), *Outcome-based science, technology, engineering, and mathematics education: Innovative practices* (pp. 263–287). Hershey, PA: IGI Global. doi:10.4018/978-1-4666-1809-1.ch013

Ozcelik, Y. (2010). IT-enabled reengineering: Productivity impacts. In K. St.Amant (Ed.), *IT outsourcing: Concepts, methodologies, tools, and applications* (pp. 371–376). Hershey, PA: IGI Global. doi:10.4018/978-1-60566-770-6.ch022

Ožvoldová, M., & Schauer, F. (2012). Remote experiments in freshman engineering education by integrated e-learning. In A. Azad, M. Auer, & V. Harward (Eds.), *Internet accessible remote laboratories: Scalable e-learning tools for engineering and science disciplines* (pp. 60–83). Hershey, PA: IGI Global. doi:10.4018/978-1-61350-186-3.ch004

Paay, J., Pedell, S., Sterling, L., Vetere, F., & Howard, S. (2011). The benefit of ambiguity in understanding goals in requirements modelling. *International Journal of People-Oriented Programming, 1*(2), 24–49. doi:10.4018/ijpop.2011070102

Palmer, S., & Holt, D. (2010). Online discussion in engineering education: Student responses and learning outcomes. In L. Shedletsky & J. Aitken (Eds.), *Cases on online discussion and interaction: Experiences and outcomes* (pp. 105–122). Hershey, PA: IGI Global. doi:10.4018/978-1-61520-863-0.ch005

Pandian, A., Ismail, S. A., & Abdullah, A. S. (2012). Communication framework to empower 21st century engineers and IT professionals. In A. Patil, H. Eijkman, & E. Bhattacharyya (Eds.), *New media communication skills for engineers and IT professionals: Trans-national and trans-cultural demands* (pp. 34–54). Hershey, PA: IGI Global. doi:10.4018/978-1-4666-0243-4.ch004

Pasupathy, K. S. (2010). Systems engineering and health informatics: Context, content, and implementation. In S. Kabene (Ed.), *Healthcare and the effect of technology: Developments, challenges and advancements* (pp. 123–144). Hershey, PA: IGI Global. doi:10.4018/978-1-61520-733-6.ch009

Pasupathy, K. S. (2010). Transforming healthcare: Leveraging the complementarities of health informatics and systems engineering. *International Journal of Healthcare Delivery Reform Initiatives*, *2*(2), 35–55. doi:10.4018/jhdri.2010040103

Pasupathy, K. S. (2011). Systems engineering and health informatics. In *Clinical technologies: Concepts, methodologies, tools and applications* (pp. 1684–1705). Hershey, PA: IGI Global. doi:10.4018/978-1-60960-561-2.ch606

Patel, C., & Ramachandran, M. (2010). Story card process improvement framework for agile requirements. In M. Ramachandran & R. de Carvalho (Eds.), *Handbook of research on software engineering and productivity technologies: Implications of globalization* (pp. 61–54). Hershey, PA: IGI Global. doi:10.4018/978-1-60566-731-7.ch006

Patro, C. S. (2012). A study on adaptability of total quality management in engineering education sector. *International Journal of Quality Assurance in Engineering and Technology Education*, *2*(4), 25–37. doi:10.4018/ijqaete.2012100103

Peña de Carrillo, C. I., Choquet, C., Després, C., Iksal, S., Jacoboni, P., Lekira, A., … Thi-Ngoc, D. P. (2014). Engineering and reengineering of technology enhanced learning scenarios using context awareness processes. In Software design and development: Concepts, methodologies, tools, and applications (pp. 1289-1313). Hershey, PA: IGI Global. doi:10.4018/978-1-4666-4301-7.ch063

Peng, F., & Li, H. (2011). A steganalysis method for 2D engineering graphics based on the statistic of geometric features. *International Journal of Digital Crime and Forensics*, *3*(2), 35–40. doi:10.4018/jdcf.2011040103

Pérez, J. L., Rabuñal, J., & Abella, F. M. (2012). Soft computing techniques in civil engineering: Time series prediction. In *Computer engineering: Concepts, methodologies, tools and applications* (pp. 1982–1997). Hershey, PA: IGI Global. doi:10.4018/978-1-61350-456-7.ch811

Pérez-Castillo, R., Rodríguez de Guzmán, I. G., & Piattini, M. (2011). Architecture-driven modernization. In A. Dogru & V. Biçer (Eds.), *Modern software engineering concepts and practices: Advanced approaches* (pp. 75–103). Hershey, PA: IGI Global. doi:10.4018/978-1-60960-215-4.ch004

Pérez-Castillo, R., Rodríguez de Guzmán, I. G., & Piattini, M. (2012). Model-driven reengineering. In J. Rech & C. Bunse (Eds.), *Emerging technologies for the evolution and maintenance of software models* (pp. 200–229). Hershey, PA: IGI Global. doi:10.4018/978-1-61350-438-3.ch008

Petkov, D., Edgar-Nevill, D., Madachy, R., & O'Connor, R. (2010). Information systems, software engineering, and systems thinking: Challenges and opportunities. In M. Hunter (Ed.), *Strategic information systems: Concepts, methodologies, tools, and applications* (pp. 315–332). Hershey, PA: IGI Global. doi:10.4018/978-1-60566-677-8.ch023

Petkov, D., Edgar-Nevill, D., Madachy, R., & O'Connor, R. (2012). Towards a wider application of the systems approach in information systems and software engineering. In *Computer engineering: Concepts, methodologies, tools and applications* (pp. 1627–1645). Hershey, PA: IGI Global. doi:10.4018/978-1-61350-456-7.ch701

Petre, L., Sere, K., & Troubitsyna, E. (2012). *Dependability and computer engineering: Concepts for software-intensive systems*. Hershey, PA: IGI Global. doi:10.4018/978-1-60960-747-0

Poels, G., Decreus, K., Roelens, B., & Snoeck, M. (2013). Investigating goal-oriented requirements engineering for business processes. *Journal of Database Management, 24*(2), 35–71. doi:10.4018/jdm.2013040103

Polgar, J., & Adamson, G. (2011). *New generation of portal software and engineering: Emerging technologies*. Hershey, PA: IGI Global. doi:10.4018/978-1-60960-571-1

Power, C., Freire, A. P., & Petrie, H. (2011). Integrating accessibility evaluation into web engineering processes. In G. Alkhatib (Ed.), *Web engineered applications for evolving organizations: Emerging knowledge* (pp. 315–339). Hershey, PA: IGI Global. doi:10.4018/978-1-60960-523-0.ch018

Praeg, C. (2011). Framework for IT service value engineering: Managing value and IT service quality. In C. Praeg & D. Spath (Eds.), *Quality management for IT services: Perspectives on business and process performance* (pp. 274–297). Hershey, PA: IGI Global. doi:10.4018/978-1-61692-889-6.ch016

Prescott, J., & Bogg, J. (2013). *Gendered occupational differences in science, engineering, and technology careers*. Hershey, PA: IGI Global. doi:10.4018/978-1-4666-2107-7

Prpic, J. K., & Moore, G. (2012). E-portfolios as a quantitative and qualitative means of demonstrating learning outcomes and competencies in engineering. In K. Yusof, N. Azli, A. Kosnin, S. Yusof, & Y. Yusof (Eds.), *Outcome-based science, technology, engineering, and mathematics education: Innovative practices* (pp. 124–154). Hershey, PA: IGI Global. doi:10.4018/978-1-4666-1809-1.ch007

Puteh, M. M., & Ismail, K. (2012). Quality assurance through innovation policy: The pedagogical implications on engineering education. In *Human resources management: Concepts, methodologies, tools, and applications* (pp. 40–49). Hershey, PA: IGI Global. doi:10.4018/978-1-4666-1601-1.ch004

Puthanpurayil, A. M., Dhakal, R. P., & Carr, A. J. (2013). Optimal passive damper positioning techniques: State-of-the-art. In N. Lagaros, V. Plevris, & C. Mitropoulou (Eds.), *Design optimization of active and passive structural control systems* (pp. 85–111). Hershey, PA: IGI Global. doi:10.4018/978-1-4666-2029-2.ch004

Ramachandran, M. (2010). Knowledge engineering support for software requirements, architectures and components. In F. Meziane & S. Vadera (Eds.), *Artificial intelligence applications for improved software engineering development: New prospects* (pp. 129–145). Hershey, PA: IGI Global. doi:10.4018/978-1-60566-758-4.ch007

Ramachandran, M. (2011). *Knowledge engineering for software development life cycles: Support technologies and applications*. Hershey, PA: IGI Global. doi:10.4018/978-1-60960-509-4

Ramachandran, M., & de Carvalho, R. (2010). *Handbook of research on software engineering and productivity technologies: Implications of globalization*. Hershey, PA: IGI Global. doi:10.4018/978-1-60566-731-7

Rasul, M., Nouwens, F., Swift, R., Martin, F., & Greensill, C. V. (2012). Assessment of final year engineering projects: A pilot investigation on issues and best practice. In M. Rasul (Ed.), *Developments in engineering education standards: Advanced curriculum innovations* (pp. 80–104). Hershey, PA: IGI Global. doi:10.4018/978-1-4666-0951-8.ch005

Razali, Z. B., & Trevelyan, J. (2012). An evaluation of students' practical intelligence and ability to diagnose equipment faults. In K. Yusof, N. Azli, A. Kosnin, S. Yusof, & Y. Yusof (Eds.), *Outcome-based science, technology, engineering, and mathematics education: Innovative practices* (pp. 328–349). Hershey, PA: IGI Global. doi:10.4018/978-1-4666-1809-1.ch016

Reimann, D. (2011). Shaping interactive media with the sewing machine: Smart textile as an artistic context to engage girls in technology and engineering education. *International Journal of Art, Culture and Design Technologies, 1*(1), 12–21. doi:10.4018/ijacdt.2011010102

Rikkilä, J. (2014). Agile, lean, and service-oriented development, continuum, or chasm. In *Software design and development: concepts, methodologies, tools, and applications* (pp. 132–163). Hershey, PA: IGI Global. doi:10.4018/978-1-4666-4301-7.ch008

Rocci, L. (2010). Engineering and environmental technoethics. In R. Luppicini (Ed.), *Technoethics and the evolving knowledge society: Ethical issues in technological design, research, development, and innovation* (pp. 146–162). Hershey, PA: IGI Global. doi:10.4018/978-1-60566-952-6.ch008

Rodrigues, D., Estrella, J. C., Monaco, F. J., Branco, K. R., Antunes, N., & Vieira, M. (2012). Engineering secure web services. In V. Cardellini, E. Casalicchio, K. Castelo Branco, J. Estrella, & F. Monaco (Eds.), *Performance and dependability in service computing: Concepts, techniques and research directions* (pp. 360–380). Hershey, PA: IGI Global. doi:10.4018/978-1-60960-794-4.ch016

Rodríguez, J., Fernández-Medina, E., Piattini, M., & Mellado, D. (2011). A security requirements engineering tool for domain engineering in software product lines. In N. Milanovic (Ed.), *Non-functional properties in service oriented architecture: Requirements, models and methods* (pp. 73–92). Hershey, PA: IGI Global. doi:10.4018/978-1-60566-794-2.ch004

Rojko, A., Zürcher, T., Hercog, D., & Stebler, R. (2012). Implementation of remote laboratories for industrial education. In A. Azad, M. Auer, & V. Harward (Eds.), *Internet accessible remote laboratories: Scalable e-learning tools for engineering and science disciplines* (pp. 84–107). Hershey, PA: IGI Global. doi:10.4018/978-1-61350-186-3.ch005

Rosado, D. G., Mellado, D., Fernandez-Medina, E., & Piattini, M. G. (2013). *Security engineering for cloud computing: Approaches and tools.* Hershey, PA: IGI Global. doi:10.4018/978-1-4666-2125-1

Rose, S., Lauder, M., Schlereth, M., & Schürr, A. (2011). A multidimensional approach for concurrent model-driven automation engineering. In J. Osis & E. Asnina (Eds.), *Model-driven domain analysis and software development: Architectures and functions* (pp. 90–113). Hershey, PA: IGI Global. doi:10.4018/978-1-61692-874-2.ch005

Russell, C. (2012). Conceptual mapping, visualisation, and systems thinking in engineering. In A. Patil, H. Eijkman, & E. Bhattacharyya (Eds.), *New media communication skills for engineers and IT professionals: Trans-national and trans-cultural demands* (pp. 72–93). Hershey, PA: IGI Global. doi:10.4018/978-1-4666-0243-4.ch006

Russell, D., & Haghi, A. (2010). *Web-based engineering education: Critical design and effective tools*. Hershey, PA: IGI Global. doi:10.4018/978-1-61520-659-9

Russo, B., Scotto, M., Sillitti, A., & Succi, G. (2010). Improving agile methods. In B. Russo, M. Scotto, A. Sillitti, & G. Succi (Eds.), *Agile technologies in open source development* (pp. 189–231). Hershey, PA: IGI Global. doi:10.4018/978-1-59904-681-5.ch012

Russo, B., Scotto, M., Sillitti, A., & Succi, G. (2010). Requirements management. In B. Russo, M. Scotto, A. Sillitti, & G. Succi (Eds.), *Agile technologies in open source development* (pp. 268–286). Hershey, PA: IGI Global. doi:10.4018/978-1-59904-681-5.ch015

Russo, M. T. (2012). Humanities in engineering education. In M. Rasul (Ed.), *Developments in engineering education standards: Advanced curriculum innovations* (pp. 285–300). Hershey, PA: IGI Global. doi:10.4018/978-1-4666-0951-8.ch016

Saidane, A., & Guelfi, N. (2014). Towards test-driven and architecture model-based security and resilience engineering. In *Software design and development: Concepts, methodologies, tools, and applications* (pp. 2072–2098). Hershey, PA: IGI Global. doi:10.4018/978-1-4666-4301-7.ch096

Sala, N. (2012). Virtual reality in architecture, in engineering and beyond. In E. Abu-Taieh, A. El Sheikh, & M. Jafari (Eds.), *Technology engineering and management in aviation: Advancements and discoveries* (pp. 336–345). Hershey, PA: IGI Global. doi:10.4018/978-1-60960-887-3.ch020

Salzmann, C., Gillet, D., Esquembre, F., Vargas, H., Sánchez, J., & Dormido, S. (2012). Web 2.0 open remote and virtual laboratories in engineering education. In A. Okada, T. Connolly, & P. Scott (Eds.), *Collaborative learning 2.0: Open educational resources* (pp. 369–390). Hershey, PA: IGI Global. doi:10.4018/978-1-4666-0300-4.ch020

Samaras, G. (2012). Human-centered systems engineering: Managing stakeholder dissonance in healthcare delivery. In A. Kolker & P. Story (Eds.), *Management engineering for effective healthcare delivery: Principles and applications* (pp. 148–171). Hershey, PA: IGI Global. doi:10.4018/978-1-60960-872-9.ch007

Sampaio, A. Z., Henriques, P. G., Cruz, C. O., & Martins, O. P. (2011). Interactive models based on virtual reality technology used in civil engineering education. In G. Vincenti & J. Braman (Eds.), *Teaching through multi-user virtual environments: Applying dynamic elements to the modern classroom* (pp. 387–413). Hershey, PA: IGI Global. doi:10.4018/978-1-61692-822-3.ch021

Sanabria, S. J., Furrer, R., Neuenschwander, J., Niemz, P., & Sennhauser, U. (2014). Bonding defect imaging in glulam with novel air-coupled ultrasound testing. In A. Aguilera & J. Davim (Eds.), *Research developments in wood engineering and technology* (pp. 221–246). Hershey, PA: IGI Global. doi:10.4018/978-1-4666-4554-7.ch006

Sánchez de la Rosa, J. L., Miranda, S. A., & González, C. S. (2013). Evaluation of transversal competences of the engineering students and their relation to the enterprise requirements. In K. Patel & S. Vij (Eds.), *Enterprise resource planning models for the education sector: Applications and methodologies* (pp. 1–17). Hershey, PA: IGI Global. doi:10.4018/978-1-4666-2193-0.ch001

Sarirete, A., & Chikh, A. (2012). A knowledge management process in communities of practice of engineering based on the SECI model for knowledge. In E. Ng, N. Karacapilidis, & M. Raisinghani (Eds.), *Evaluating the impact of technology on learning, teaching, and designing curriculum: Emerging trends* (pp. 134–149). Hershey, PA: IGI Global. doi:10.4018/978-1-4666-0032-4.ch009

Sass, L. (2013). Direct building manufacturing of homes with digital fabrication. In *Industrial engineering: Concepts, methodologies, tools, and applications* (pp. 1231–1242). Hershey, PA: IGI Global. doi:10.4018/978-1-4666-1945-6.ch066

Sauvet, B., Boukhicha, M., Balan, A., Hwang, G., Taverna, D., Shukla, A., & Régnier, S. (2012). Selective pick-and-place of thin film by robotic micromanipulation. *International Journal of Intelligent Mechatronics and Robotics, 2*(3), 24–37. doi:10.4018/ijimr.2012070103

Schnabel, M. A. (2013). Learning parametric designing. In *Industrial engineering: Concepts, methodologies, tools, and applications* (pp. 197–210). Hershey, PA: IGI Global. doi:10.4018/978-1-4666-1945-6.ch013

Schrödl, H., & Wind, S. (2013). Requirements engineering for cloud application development. In A. Bento & A. Aggarwal (Eds.), *Cloud computing service and deployment models: Layers and management* (pp. 137–150). Hershey, PA: IGI Global. doi:10.4018/978-1-4666-2187-9.ch007

Schulz, T., Radlinski, L., Gorges, T., & Rosenstiel, W. (2011). Software process model using dynamic bayesian networks. In M. Ramachandran (Ed.), *Knowledge engineering for software development life cycles: Support technologies and applications* (pp. 289–310). Hershey, PA: IGI Global. doi:10.4018/978-1-60960-509-4.ch016

Sevillian, D. B. (2013). Aircraft development and design: Enhancing product safety through effective human factors engineering design solutions. In *Industrial engineering: Concepts, methodologies, tools, and applications* (pp. 858–886). Hershey, PA: IGI Global. doi:10.4018/978-1-4666-1945-6.ch048

Shan, T. C., & Hua, W. W. (2010). Strategic technology engineering planning. In M. Hunter (Ed.), *Strategic information systems: Concepts, methodologies, tools, and applications* (pp. 414–434). Hershey, PA: IGI Global. doi:10.4018/978-1-60566-677-8.ch029

Sharma, A., & Maurer, F. (2014). A roadmap for software engineering for the cloud: Results of a systematic review. In *Software design and development: Concepts, methodologies, tools, and applications* (pp. 1–16). Hershey, PA: IGI Global. doi:10.4018/978-1-4666-4301-7.ch001

Sharma, N., Singh, K., & Goyal, D. (2014). Software engineering, process improvement, and experience management: Is the nexus productive? Clues from the Indian giants. In *Software design and development: Concepts, methodologies, tools, and applications* (pp. 1401–1414). Hershey, PA: IGI Global. doi:10.4018/978-1-4666-4301-7.ch068

Shirk, S., Arreola, V., Wobig, C., & Russell, K. (2012). Girls' e-mentoring in science, engineering, and technology based at the University of Illinois at Chicago women in science and engineering (WISE) program. In Computer engineering: Concepts, Methodologies, tools and applications (pp. 1144-1163). Hershey, PA: IGI Global. doi:10.4018/978-1-61350-456-7.ch505

Shukla, A., & Tiwari, R. (2011). *Biomedical engineering and information systems: Technologies, tools and applications*. Hershey, PA: IGI Global. doi:10.4018/978-1-61692-004-3

Sidhu, M. S. (2010). Challenges and trends of TAPS packages in enhancing engineering education. In M. Sidhu (Ed.), *Technology-assisted problem solving for engineering education: Interactive multimedia applications* (pp. 158–166). Hershey, PA: IGI Global. doi:10.4018/978-1-60566-764-5.ch011

Sidhu, M. S. (2010). Evaluation of TAPS packages. In M. Sidhu (Ed.), *Technology-assisted problem solving for engineering education: Interactive multimedia applications* (pp. 128–147). Hershey, PA: IGI Global. doi:10.4018/978-1-60566-764-5.ch009

Sidhu, M. S. (2010). Technology assisted problem solving packages: A new approach to learning, visualizing, and problem solving in engineering. In M. Sidhu (Ed.), *Technology-assisted problem solving for engineering education: Interactive multimedia applications* (pp. 69–90). Hershey, PA: IGI Global. doi:10.4018/978-1-60566-764-5.ch006

Sidhu, M. S. (2010). *Technology-assisted problem solving for engineering education: Interactive multimedia applications*. Hershey, PA: IGI Global. doi:10.4018/978-1-60566-764-5

Sidhu, M. S., & Kang, L. C. (2012). Emerging trends and technologies for enhancing engineering education: An overview. In L. Tomei (Ed.), *Advancing education with information communication technologies: Facilitating new trends* (pp. 320–330). Hershey, PA: IGI Global. doi:10.4018/978-1-61350-468-0.ch026

Sidhu, M. S., & Kang, L. C. (2012). New trends and futuristic information communication technologies for engineering education. In S. Chhabra (Ed.), *ICTs for advancing rural communities and human development: Addressing the digital divide* (pp. 251–262). Hershey, PA: IGI Global. doi:10.4018/978-1-4666-0047-8.ch017

Simonette, M. J., & Spina, E. (2014). Enabling IT innovation through soft systems engineering. In M. Pańkowska (Ed.), *Frameworks of IT prosumption for business development* (pp. 64–72). Hershey, PA: IGI Global. doi:10.4018/978-1-4666-4313-0.ch005

Singh, T., Verma, M. K., & Singh, R. (2014). Role of emotional intelligence in academic achievement: An empirical study on engineering students. In P. Ordóñez de Pablos & R. Tennyson (Eds.), *Strategic approaches for human capital management and development in a turbulent economy* (pp. 255–263). Hershey, PA: IGI Global. doi:10.4018/978-1-4666-4530-1.ch016

Smolnik, S., Teuteberg, F., & Thomas, O. (2012). *Semantic technologies for business and information systems engineering: Concepts and applications*. Hershey, PA: IGI Global. doi:10.4018/978-1-60960-126-3

Solemon, B., Sahibuddin, S., & Ghani, A. A. (2014). Requirements engineering process improvement and related models. In *Software design and development: Concepts, methodologies, tools, and applications* (pp. 203–218). Hershey, PA: IGI Global. doi:10.4018/978-1-4666-4301-7.ch011

Sorial, G. A., & Noroozi, B. (2012). Improvement of engineering students education by e-learning. In *Virtual learning environments: Concepts, methodologies, tools and applications* (pp. 870–883). Hershey, PA: IGI Global. doi:10.4018/978-1-4666-0011-9.ch414

Srinivasan, A., López-Ribot, J. L., & Ramasubramanian, A. K. (2011). Microfluidic applications in vascular bioengineering. In A. Shukla & R. Tiwari (Eds.), *Biomedical engineering and information systems: Technologies, tools and applications* (pp. 1–30). Hershey, PA: IGI Global. doi:10.4018/978-1-61692-004-3.ch001

Steinbuch, R. (2013). Evolutionary optimization of passive compensators to improve earthquake resistance. In N. Lagaros, V. Plevris, & C. Mitropoulou (Eds.), *Design optimization of active and passive structural control systems* (pp. 250–273). Hershey, PA: IGI Global. doi:10.4018/978-1-4666-2029-2.ch011

Stephenson, S. V., & Sage, A. P. (2010). Information and knowledge perspectives in systems engineering and management for innovation and productivity through enterprise resource planning. In M. Hunter (Ed.), *Strategic information systems: Concepts, methodologies, tools, and applications* (pp. 338–368). Hershey, PA: IGI Global. doi:10.4018/978-1-60566-677-8.ch025

Strasser, T., Zoitl, A., & Rooker, M. (2013). Zero-downtime reconfiguration of distributed control logic in industrial automation and control. In *Industrial engineering: Concepts, methodologies, tools, and applications* (pp. 2024–2051). Hershey, PA: IGI Global. doi:10.4018/978-1-4666-1945-6.ch109

Sudeikat, J., & Renz, W. (2010). Building complex adaptive systems: On engineering self-organizing multi-agent systems. In M. Hunter (Ed.), *Strategic information systems: Concepts, methodologies, tools, and applications* (pp. 767–787). Hershey, PA: IGI Global. doi:10.4018/978-1-60566-677-8.ch050

Sugiyama, S., & Burgess, L. (2012). Principle for engineering service based system by swirl computing. In X. Liu & Y. Li (Eds.), *Advanced design approaches to emerging software systems: Principles, methodologies and tools* (pp. 48–60). Hershey, PA: IGI Global. doi:10.4018/978-1-60960-735-7.ch003

Sun, Z., Han, J., Dong, D., & Zhao, S. (2010). Engineering of experience based trust for e-commerce. In M. Wang & Z. Sun (Eds.), *Handbook of research on complex dynamic process management: Techniques for adaptability in turbulent environments* (pp. 342–367). Hershey, PA: IGI Global. doi:10.4018/978-1-60566-669-3.ch014

Sunindyo, W. D., Moser, T., Winkler, D., Mordinyi, R., & Biffl, S. (2013). Workflow validation framework in collaborative engineering environments. In A. Loo (Ed.), *Distributed computing innovations for business, engineering, and science* (pp. 285–299). Hershey, PA: IGI Global. doi:10.4018/978-1-4666-2533-4.ch015

Szeto, A. Y. (2014). Assistive technology and rehabilitation engineering. In *Assistive technologies: Concepts, methodologies, tools, and applications* (pp. 277–331). Hershey, PA: IGI Global. doi:10.4018/978-1-4666-4422-9.ch015

Takruri-Rizk, H., Sappleton, N., & Dhar-Bhattacharjee, S. (2010). Progression of UK women engineers: Aids and hurdles. In A. Cater-Steel & E. Cater (Eds.), *Women in engineering, science and technology: Education and career challenges* (pp. 280–300). Hershey, PA: IGI Global. doi:10.4018/978-1-61520-657-5.ch013

Tekinerdogan, B., & Aksit, M. (2011). A comparative analysis of software engineering with mature engineering disciplines using a problem-solving perspective. In A. Dogru & V. Biçer (Eds.), *Modern software engineering concepts and practices: Advanced approaches* (pp. 1–18). Hershey, PA: IGI Global. doi:10.4018/978-1-60960-215-4.ch001

Tellioglu, H. (2012). About representational artifacts and their role in engineering. In G. Viscusi, G. Campagnolo, & Y. Curzi (Eds.), *Phenomenology, organizational politics, and IT design: The social study of information systems* (pp. 111–130). Hershey, PA: IGI Global. doi:10.4018/978-1-4666-0303-5.ch007

Terawaki, Y., Takahashi, Y., Kodama, Y., & Yana, K. (2013). The development of educational environment suited to the Japan-specific educational service using requirements engineering techniques: Case study of running Sakai with PostgreSQL. In V. Kumar & F. Lin (Eds.), *System and technology advancements in distance learning* (pp. 261–270). Hershey, PA: IGI Global. doi:10.4018/978-1-4666-2032-2.ch019

Thomas, K. D., & Muga, H. E. (2012). Sustainability: The new 21st century general education requirement for engineers. In M. Rasul (Ed.), *Developments in engineering education standards: Advanced curriculum innovations* (pp. 263–284). Hershey, PA: IGI Global. doi:10.4018/978-1-4666-0951-8.ch015

Trigo, B., Olguin, G., & Matai, P. (2010). The use of applets in an engineering chemistry course: Advantages and new ideas. In D. Russell & A. Haghi (Eds.), *Web-based engineering education: Critical design and effective tools* (pp. 108–118). Hershey, PA: IGI Global. doi:10.4018/978-1-61520-659-9.ch009

Tsadimas, A., Nikolaidou, M., & Anagnostopoulos, D. (2014). Model-based system design using SysML: The role of the evaluation diagram. In P. Fonseca i Casas (Ed.), *Formal languages for computer simulation: Transdisciplinary models and applications* (pp. 236–266). Hershey, PA: IGI Global. doi:10.4018/978-1-4666-4369-7.ch008

Unhelkar, B., Ghanbary, A., & Younessi, H. (2010). *Collaborative business process engineering and global organizations: Frameworks for service integration*. Hershey, PA: IGI Global. doi:10.4018/978-1-60566-689-1

Uziak, J., Oladiran, M. T., & Kommula, V. P. (2012). Integrating general education courses into engineering curriculum: Students' perspective. In M. Rasul (Ed.), *Developments in engineering education standards: Advanced curriculum innovations* (pp. 247–262). Hershey, PA: IGI Global. doi:10.4018/978-1-4666-0951-8.ch014

Valizadeh, M., Anzelotti, G., & Salehi, A. S. (2010). Web-based approaches in engineering education. In R. Luppicini & A. Haghi (Eds.), *Cases on digital technologies in higher education: Issues and challenges* (pp. 241–256). Hershey, PA: IGI Global. doi:10.4018/978-1-61520-869-2.ch017

Valizadeh, M., Anzelotti, G., & Salehi, S. (2010). Web-based training: An applicable tool for engineering education. In D. Russell & A. Haghi (Eds.), *Web-based engineering education: Critical design and effective tools* (pp. 186–198). Hershey, PA: IGI Global. doi:10.4018/978-1-61520-659-9.ch014

Valverde, R., Toleman, M., & Cater-Steel, A. (2010). Design science: A case study in information systems re-engineering. In M. Hunter (Ed.), *Strategic information systems: Concepts, methodologies, tools, and applications* (pp. 490–503). Hershey, PA: IGI Global. doi:10.4018/978-1-60566-677-8.ch033

Vaquero, L. M., Rodero-Merino, L., Cáceres, J., Chapman, C., Lindner, M., & Galán, F. (2012). Principles, methodology and tools for engineering cloud computing systems. In X. Liu & Y. Li (Eds.), *Advanced design approaches to emerging software systems: Principles, methodologies and tools* (pp. 250–273). Hershey, PA: IGI Global. doi:10.4018/978-1-60960-735-7.ch012

Vargas, E. P. (2014). Quality, improvement and measurements in high risk software. In *Software design and development: Concepts, methodologies, tools, and applications* (pp. 733–748). Hershey, PA: IGI Global. doi:10.4018/978-1-4666-4301-7.ch035

Vasant, P. (2013). *Meta-heuristics optimization algorithms in engineering, business, economics, and finance*. Hershey, PA: IGI Global. doi:10.4018/978-1-4666-2086-5

Venkatraman, S. (2013). Software engineering research gaps in the cloud. *Journal of Information Technology Research, 6*(1), 1–19. doi:10.4018/jitr.2013010101

Villazón-Terrazas, B. C., Suárez-Figueroa, M., & Gómez-Pérez, A. (2012). A pattern-based method for re-engineering non-ontological resources into ontologies. In A. Sheth (Ed.), *Semantic-enabled advancements on the web: Applications across industries* (pp. 17–54). Hershey, PA: IGI Global. doi:10.4018/978-1-4666-0185-7.ch002

Vinod, J. S. (2010). Dem simulations in geotechnical earthquake engineering education. *International Journal of Geotechnical Earthquake Engineering, 1*(1), 61–69. doi:10.4018/jgee.2010090804

Virdi, J. (2014). Business risk analysis: Obsolescence management in requirements engineering. In *Software design and development: Concepts, methodologies, tools, and applications* (pp. 1736–1763). Hershey, PA: IGI Global. doi:10.4018/978-1-4666-4301-7.ch083

Walk, S., Pöschko, J., Strohmaier, M., Andrews, K., Tudorache, T., Noy, N. F., ... Musen, M. A. (2013). PragmatiX: An interactive tool for visualizing the creation process behind collaboratively engineered ontologies. *International Journal on Semantic Web and Information Systems, 9*(1), 45–78. doi:10.4018/jswis.2013010103 PMID:24465189

Wang, J. (2013). Architects and engineers. In *Challenging ICT applications in architecture, engineering, and industrial design education* (pp. 48–64). Hershey, PA: IGI Global. doi:10.4018/978-1-4666-1999-9.ch003

Wang, J. (2013). *Challenging ICT applications in architecture, engineering, and industrial design education*. Hershey, PA: IGI Global. doi:10.4018/978-1-4666-1999-9

Wang, J. (2013). Pedagogy and curriculum in architecture and engineering. In *Challenging ICT applications in architecture, engineering, and industrial design education* (pp. 65–92). Hershey, PA: IGI Global. doi:10.4018/978-1-4666-1999-9.ch004

Wang, J. (2013). Professionalism in architecture and engineering. In *Challenging ICT applications in architecture, engineering, and industrial design education* (pp. 137–156). Hershey, PA: IGI Global. doi:10.4018/978-1-4666-1999-9.ch007

Wang, J. (2013). Reviewing engineers and introducing industrial designers. In *Challenging ICT applications in architecture, engineering, and industrial design education* (pp. 111–136). Hershey, PA: IGI Global. doi:10.4018/978-1-4666-1999-9.ch006

Wang, Y. (2010). On cognitive properties of human factors and error models in engineering and socialization. In Y. Wang (Ed.), *Discoveries and breakthroughs in cognitive informatics and natural intelligence* (pp. 93–109). Hershey, PA: IGI Global. doi:10.4018/978-1-60566-902-1.ch006

Wang, Y. (2012). Convergence of software science and computational intelligence: A new transdisciplinary research field. In Y. Wang (Ed.), *Software and intelligent sciences: New transdisciplinary findings* (pp. 1–13). Hershey, PA: IGI Global. doi:10.4018/978-1-4666-0261-8.ch001

Wang, Y., & Patel, S. (2012). Exploring the cognitive foundations of software engineering. In Y. Wang (Ed.), *Software and intelligent sciences: New transdisciplinary findings* (pp. 232–251). Hershey, PA: IGI Global. doi:10.4018/978-1-4666-0261-8.ch014

Wei, J., Chen, J., & Zhu, Q. (2010). Service science, management and engineering education: A unified model for university. *International Journal of Service Science, Management, Engineering, and Technology, 1*(2), 51–69. doi:10.4018/jssmet.2010040104

Weigang, L., Barros, A. D., & Romani de Oliveira, I. (2010). *Computational models, software engineering, and advanced technologies in air transportation: Next generation applications*. Hershey, PA: IGI Global. doi:10.4018/978-1-60566-800-0

Westh Nicolajsen, H. (2010). Limitations and perspectives on use of e-services in engineering consulting. In *Electronic services: Concepts, methodologies, tools and applications* (pp. 1280–1295). Hershey, PA: IGI Global. doi:10.4018/978-1-61520-967-5.ch078

Wilson, R., & Younis, H. (2013). *Business strategies for electrical infrastructure engineering: Capital project implementation*. Hershey, PA: IGI Global. doi:10.4018/978-1-4666-2839-7

Winter, A. (2010). The smart women – smart state strategy: A policy on women's participation in science, engineering and technology in Queensland, Australia. In A. Cater-Steel & E. Cater (Eds.), *Women in engineering, science and technology: Education and career challenges* (pp. 1–20). Hershey, PA: IGI Global. doi:10.4018/978-1-61520-657-5.ch001

Winter, R., & van Beijnum, I. (2014). Inter-domain traffic engineering using the origin preference attribute. In M. Boucadair & D. Binet (Eds.), *Solutions for sustaining scalability in internet growth* (pp. 18–38). Hershey, PA: IGI Global. doi:10.4018/978-1-4666-4305-5.ch002

Worth, D., Greenough, C., & Chin, S. (2014). Pragmatic software engineering for computational science. In *Software design and development: Concepts, methodologies, tools, and applications* (pp. 663–694). Hershey, PA: IGI Global. doi:10.4018/978-1-4666-4301-7.ch032

Wu, J. (2013). *Biomedical engineering and cognitive neuroscience for healthcare: Interdisciplinary applications*. Hershey, PA: IGI Global. doi:10.4018/978-1-4666-2113-8

Wu, Y., & Koszalka, T. A. (2013). Instructional design of an advanced interactive discovery environment: Exploring team communication and technology use in virtual collaborative engineering problem solving. In *Industrial engineering: Concepts, methodologies, tools, and applications* (pp. 117–136). Hershey, PA: IGI Global. doi:10.4018/978-1-4666-1945-6.ch009

Yadav, V. (2011). Research review: Globally distributed requirements engineering and agility. *International Journal of Innovation in the Digital Economy*, *2*(1), 1–11. doi:10.4018/jide.2011010101

Yang, J. J., Liu, J. F., Kurokawa, T., Kitamura, N., Yasuda, K., & Gong, J. P. (2013). Tough double-network hydrogels as scaffolds for tissue engineering: Cell behavior in vitro and in vivo test. In J. Wu (Ed.), *Technological advancements in biomedicine for healthcare applications* (pp. 213–222). Hershey, PA: IGI Global. doi:10.4018/978-1-4666-2196-1.ch023

Yeoh, W., Gao, J., & Koronios, A. (2010). Empirical investigation of critical success factors for implementing business intelligence systems in multiple engineering asset management organisations. In M. Hunter (Ed.), *Strategic information systems: Concepts, methodologies, tools, and applications* (pp. 2039–2063). Hershey, PA: IGI Global. doi:10.4018/978-1-60566-677-8.ch129

Yerrick, R., Lund, C., & Lee, Y. (2013). Online simulator use in the preparing chemical engineers. *International Journal of Online Pedagogy and Course Design, 3*(2), 1–24. doi:10.4018/ijopcd.2013040101

Yilmaz, A. E., & Yilmaz, I. B. (2012). Natural language processing techniques in requirements engineering. In *Computer engineering: Concepts, methodologies, tools and applications* (pp. 533–545). Hershey, PA: IGI Global. doi:10.4018/978-1-61350-456-7.ch303

Yong, E. (2012). Literature review skills for undergraduate engineering students in large classes. In K. Yusof, N. Azli, A. Kosnin, S. Yusof, & Y. Yusof (Eds.), *Outcome-based science, technology, engineering, and mathematics education: Innovative practices* (pp. 240–261). Hershey, PA: IGI Global. doi:10.4018/978-1-4666-1809-1.ch012

Yun, H., Xu, J., Xiong, J., & Wei, M. (2013). A knowledge engineering approach to develop domain ontology. In V. Kumar & F. Lin (Eds.), *System and technology advancements in distance learning* (pp. 55–70). Hershey, PA: IGI Global. doi:10.4018/978-1-4666-2032-2.ch004

Zardari, S., Faniyi, F., & Bahsoon, R. (2013). Using obstacles for systematically modeling, analysing, and mitigating risks in cloud adoption. In I. Mistrik, A. Tang, R. Bahsoon, & J. Stafford (Eds.), *Aligning enterprise, system, and software architectures* (pp. 275–296). Hershey, PA: IGI Global. doi:10.4018/978-1-4666-2199-2.ch014

Zaščerinska, J., Ahrens, A., & Bassus, O. (2012). Enterprise 2.0 and 3.0 in education: Engineering and business students' view. In M. Cruz-Cunha, P. Gonçalves, N. Lopes, E. Miranda, & G. Putnik (Eds.), *Handbook of research on business social networking: Organizational, managerial, and technological dimensions* (pp. 472–494). Hershey, PA: IGI Global. doi:10.4018/978-1-61350-168-9.ch025

Zašcerinska, J., Ahrens, A., & Bassus, O. (2012). Enterprise 2.0 and 3.0 in education: Engineering and business students' view. In M. Cruz-Cunha, P. Gonçalves, N. Lopes, E. Miranda, & G. Putnik (Eds.), *Handbook of research on business social networking: Organizational, managerial, and technological dimensions* (pp. 472–494). Hershey, PA: IGI Global. doi:10.4018/978-1-61350-168-9.ch025

Zhang, D. (2012). Machine learning and value-based software engineering. In Y. Wang (Ed.), *Software and intelligent sciences: New transdisciplinary findings* (pp. 287–301). Hershey, PA: IGI Global. doi:10.4018/978-1-4666-0261-8.ch017

Zhang, H., Xu, C., Su, W., & Luo, H. (2014). Routing optimization for inter-domain traffic engineering under identifier network. In M. Boucadair & D. Binet (Eds.), *Solutions for sustaining scalability in internet growth* (pp. 127–147). Hershey, PA: IGI Global. doi:10.4018/978-1-4666-4305-5.ch007

Zhong, Y. (2013). Processing of 3D unstructured measurement data for reverse engineering. In S. Sirouspour (Ed.), *Advanced engineering and computational methodologies for intelligent mechatronics and robotics* (pp. 118–127). Hershey, PA: IGI Global. doi:10.4018/978-1-4666-3634-7.ch008

Zhou, Z., Wang, H., & Lou, P. (2010). *Manufacturing intelligence for industrial engineering: Methods for system self-organization, learning, and adaptation.* Hershey, PA: IGI Global. doi:10.4018/978-1-60566-864-2

Zhu, L., Jayaram, U., & Kim, O. (2011). Online semantic knowledge management for product design based on product engineering ontologies. *International Journal on Semantic Web and Information Systems*, *7*(4), 36–61. doi:10.4018/jswis.2011100102

About the Contributors

Gülşah Koç is earning her Master's degree in Architecture from Yildiz Technical University in Istanbul, Turkey.

Bryan Christiansen is the Chief Executive Officer of Global Research Society, LLC in Michigan, USA. A former business lecturer at universities in Turkey, Russia, and the USA, he has traveled to 41 countries where he has conducted international business since 1985 in multiple languages and various industries with Global 500 firms and smaller. Christiansen received his Bachelor of Science degree in Marketing at the University of the State of New York in 1996 and his MBA degree at Capella University in 2003. The author of 25 Reference books on business, economics, and psychology, he is currently working on his Doctor of Business Administration degree at Middlesex University in London, England and is expected to graduate in 2020.

* * *

Graciela Affranchino is Chemical Engineer Director of the Water, Soils and Tributaries Laboratory.

Namdeo N. Bhujbal is working as Head and Associate Professor in Department of Chemistry, PDEA's Annasaheb Magar Mahavidyalaya, Hadapsar, Pune, India. He has 31 years of teaching experience. His research areas of expertise are identification of the molecules and chemical compounds; chromatography, etc. He has 10 publications to his credit.

Blaine Brownell is an architect and former Fulbright scholar to Japan with a focus on emergent materials and applications. He is an associate professor and director of graduate studies at the University of Minnesota School of Architecture and principal of the design and research practice Transstudio. Brownell authored the three-volume Transmaterial series as well as Matter in the Floating World,

Material Strategies, and Hypernatural (co-authored with Marc Swackhamer) with Princeton Architectural Press. He also writes the Mind & Matter column for Architect magazine. Considered a leading scholar on advanced materials for architecture and design, Brownell has been published in over 70 architecture, design, science, and news journals including The New York Times, The London Times, The Wall Street Journal, The Boston Globe, New Scientist, and Discover, and he has lectured widely in the Americas, Europe, and Asia. Blaine's latest book is Transmaterial Next: A Catalog of Materials that Redefine Our Future.

Andrea Ciaramella is senior lecture at the ABC Department (Architecture, Built Environment and Construction engineering) at the Politecnico di Milano, with which he has collaborated since 1993. He does teaching, research and consulting work, specializing in the area of technological innovation and in the relationship between the construction process, real estate development and property management. He teaches degree courses at the Politecnico di Milano: Integrated course on "The Technical and Financial Management of Organizations", a degree course in the Architecture and Building Production Program, School of Architecture and Society; Laboratory of "Facility Management", a Master of Science in the Construction Management Program, 6th School of Engineering. He currently teaches and coordinates diverse third-tier educational initiatives (Specialising Masters, Post-Graduate Courses and Permanent Education Courses). Since the beginning of the program (1997), he has been a professor of the First-Level Master's Degree course "Real Estate Management" at BEST Department, Politecnico di Milano, which he has coordinated for several program periods. Since 2001 he has been a professor of the refresher course "Methods, techniques and professional instruments for the real-estate sector" at BEST Department, Politecnico di Milano. Since 2004 he has been the co-director of the Executive Course "Assets, Property and Facility Management - FPM" at MIP, the Business School of Politecnico di Milano. Since 2005 he has been the co-director of the Master in Real Estate, promoted by SDA Bocconi and MIP Politecnico di Milano, and provided by the two schools in joint venture. Since 2005 he has been a member of the Centre of Competence in Construction Management at MIP, Politecnico di Milano. He has been Chair of Royal Institution of Chartered Surveyor's Italian board for 4 years. He has been a member of the Continental European Standard Board Committee (CESB), Royal Institution of Chartered Surveyors for 6 years. Currently he's involved in the CESB as "academic expert". He is one of the independent members, nominated by The Ministry of Economy, in the Advisory Board of the Fondo Immobili Pubblici (the most important real estate investment fund with public owned asset in Italy). He's member of the Advisory Board within different rear estate investment funds managed by Generali Real Estate (Sammartini, Mascagni, Haydin). He has also been called onto projects as consultant with regards to management and organisational problematics arising from the restructuring and the management of buildings.

Nadeem Faisal, B.Tech (Mechanical Engineering, ITM University, Gwalior, India), did his M.E. (Design of Mechanical Equipment, Birla Institute of Technology, Mesra, India). He has over 1 year of Industrial experience. His areas of interests are Optimization, Material Science, Product and Process Design, CAD/CAM/CAE and Rapid Prototyping. He has 2 books, 12 Book Chapters, 2 SCI Indexed and 2 SCOPUS international publication to his credit.

Aparna B. Gunjal is the Director of Asian Agri Food Consultancy Services, Pune, India. She has 5 years of teaching experience. She has 32 publications to her credit (which includes research papers, review and book chapter). She has also received many national and international awards as well as travel grants for the conferences. She has presented many research papers in National and International Conferences. She is life member of many organisations and societies.

Aletheia Ida, Ph.D., is an architect, designer, and philosopher developing interdisciplinary design theory for robust frameworks to inform applied research in emergent environmental building technologies. With over twelve years of experience in professional architecture practice and fluency with building performance analytics, she is developing contextual and practical applications for air-to-water metabolic systems through parallel explorations of socio-environmental criteria, material inventions, and innovative digital and physical design methods. Her research is published in Metropolis "Source Materials," ACADIA, World Sustainable Building Conference Proceedings, Society for Optics and Photonics, Materials Research Society Programmable Matter, MRS Advances Cambridge University Press journal, the Building Research Information Knowledgebase sponsored by the AIA and NIST, and ACSA-AIA Intersections. She is currently an Assistant Professor in the School of Architecture at the University of Arizona, where she is Chair of the Master of Science in Architecture (MS.Arch) Emerging Building Technology (EBT) program, overseeing international research projects around topics of resilient futures, environmental performance, and emerging materials for architectural design. Aletheia recently received a National Science Foundation EAGER award for her research on building integrated dehumidification material innovations.

Kaushik Kumar, B.Tech (Mechanical Engineering, REC (Now NIT), Warangal), MBA (Marketing, IGNOU), and Ph.D (Engineering, Jadavpur University), is presently an Associate Professor in the Department of Mechanical Engineering, Birla Institute of Technology, Mesra, Ranchi, India. He has 14 years of Teaching & Research and over 11 years of industrial experience in a manufacturing unit of Global repute. His areas of teaching and research interest are Quality Management Systems, Optimization, Non-conventional machining, CAD / CAM, Rapid Prototyping and Composites. He has 9 Patents,15 Book, 6 Edited Book 35 Book Chapters, 120 international Journal publications, 18 International and 8 National Conference publications to his credit. He is on the editorial board and review panel of 7 International and 1 National Journals of repute. He has been felicitated with many awards and honours.

Jennifer Loy is Professor of Integrated Product Design at the University of Technology Sydney. She has a PhD in Industrial Design. Loy's background is in design for manufacturing, with a focus on sustainability and advanced manufacturing, particularly in relation to the opportunities being provided by digital technologies. Her current research interests are on the impact of additive manufacturing (3D printing) and associated technologies on design and human development across disciplines, from product service systems to humanitarian logistics, construction products to fashion design and medical devices.

Neelu N. Nawani is working as Professor in Dr. D.Y. Patil's Vidyapeeth, Dr. D.Y. Patil Biotechnology and Bioinformatics Institute, Tathawade, Pune, India. She has 20 years of teaching experience. To her credit, she has 75 publications (which includes research papers, book chapters and reviews). She has six patents to her credit. She has presented many research papers in National and International Conferences. Dr. Nawani is life member of many organisations and scientific societies.

Caroline O'Donnell is principal of CODA, best known perhaps for winning the MoMA PS1 Young Architects Program with "Party Wall" which was built at PS1 in 2013, and for two prize winning Europan housing schemes. CODA's work ranges in scale from the object and installation to the urban. CODA's work focuses on responsive, dynamic architectures, the misuse of materials, and the manipulation of perceptions. O'Donnell is also a writer and educator. She is O'Donnell is the Edgar A. Tafel assistant professor and director of the M.Arch at program Cornell University. She has a B.Arch from Manchester University, England (2000) with a specialization in Bioclimatics and an M.Arch from Princeton University (2004).

She is the editor-in-chief of the Cornell Journal of Architecture, and co-founder of Pidgin magazine. Her book "Niche Tactics: Generative Relationships between Architecture and Site" was published with Routledge in 2016. This book describes a design agenda that looks toward natural and local resources to produce meaningful environment.

Neha N. Patil is working as Head and Associate Professor in Department of Microbiology, PDEA's Annasaheb Magar Mahavidyalaya, Hadapsar, Pune, India. She has 30 years of teaching experience. To her credit, she has 13 publications (which includes research papers, book chapter and review). She has presented many research papers in National and International Conferences.

Dillon Pranger is a designer, fabricator, and academic. His current research involves material limitations along with formal hierarchies at the moment of the detail. Specifically, he is interested in the representational misreading of formal elements and how the intrinsic value of found objects can manifest themselves within architecture. These explorations have been realized through his latest focus of design, at the pavilion scale, as the project lead and manager for the 2017 MoMA PS1 Young Architects Program, Lumen. His works have been featured in Cornell University's John Hartell Gallery, Association vol. 8, PLAT Journal, and Architect Magazine, among others. Pranger has served as a Teaching Associate in the Department of Architecture at Cornell University from 2016-2018. He holds a B.S. in Architecture from the University of Cincinnati (2012) and an M.Arch from Cornell University (2015).

Sheila M. Puffer is University Distinguished Professor at Northeastern University, Boston, USA, where she is a professor of international business and Strategy at the D'Amore-McKim School of Business. She is also a fellow at the Davis Center for Russian and Eurasian Studies at Harvard University, and has served as program director of the Gorbachev Foundation of North America. In 2015 she was a visiting research professor at the Graduate School of Business at Stanford University. Her coauthored book, Hammer & Silicon: The Soviet Diaspora in the US Innovation Economy, was published by Cambridge University Press in 2018. Dr. Puffer has more than 160 publications, including over 80 refereed articles and 11 books. She earned a diploma from the executive management program at the Plekhanov Institute of the National Economy in Moscow, and holds BA (Slavic Studies) and MBA degrees from the University of Ottawa, Canada, and a PhD in business administration from the University of California, Berkeley.

Valentina Puglisi, architect, Assistant Professor, and PhD in "Building Technology and Environmental Design" at the ABC Department where she worked since 2008 to research, training and consulting. She does teaching, research and consulting work, specializing in the area of technological innovation and in the relationship between the construction process, real estate development and management. She teaches Architecture Technology in the Course "Laboratory Design of Architecture". Since 2009 She is Coordinator of worksite safety and from 2009 to 2015 She is coordinator of the Master "MRE - Master of Real Estate" organized by MIP Politecnico di Milano and SDA Bocconi. Since 2015 She is a reviewer of several international Journal of Civil Engineering and Architecture. Since 2015 She is a member of scientific committee of international conference on Advanced Building Skins "Energy Forum". She is chair and speaker in different national and international conferences. She is the author of the book "The rating of the building " published in 2012 by Sole24Ore and other books . She collaborates with several magazines specialized in the construction industry.

Tim Schork is an Associate Professor in the School of Architecture at the University of Technology Sydney, where he co-leads the Transformative Technologies & Data Poetics Research Group. He received his PhD from RMIT University in 2013 for his research on the transformative effects and contributory role of integrative computational design strategies on the practices of architecture. His research investigates the progressive integration of computational design and simulation techniques with contemporary digital fabrication processes and material technology in terms of their architectonic, constructive, economic, ecological and social potentials. His research focuses primarily on novel design, construction and building processes, materials as well as customised production and digital fabrication. Internationally renowned for his design excellence and explorative, creative and innovative work, his research links diverse disciplines, researchers, institutions and places and combines a sophisticated design philosophy with advanced technology in order to create novel design solutions that address contemporary social and cultural agendas.

Meghmala S. Waghmode is working as Assistant Professor in Department of Microbiology, at PDEA's Annasaheb Magar Mahavidyalaya, Hadapsar, Pune, India. She has 10 years of teaching experience. Her research areas of expertise are waste management; bioremediation; nanotechnology; plant growth promoting rhizobacteria; etc. She has 13 publications to her credit (which includes research papers, book chapter and review).

David Wesley is case research manager at Northeastern University's D'Amore McKim School of Business in Boston, Massachusetts. His research encompasses a range of strategic management topics, including international strategy, cultural diversity, intellectual property, and new product development. His award-winning cases have appeared in 30 management textbooks in multiple editions. Dr. Wesley is co-author of a leading book on video game marketing and innovation. He teaches graduate level courses in global strategy and culture. He has a doctorate from Northeastern University.

Index

S

T

U

V

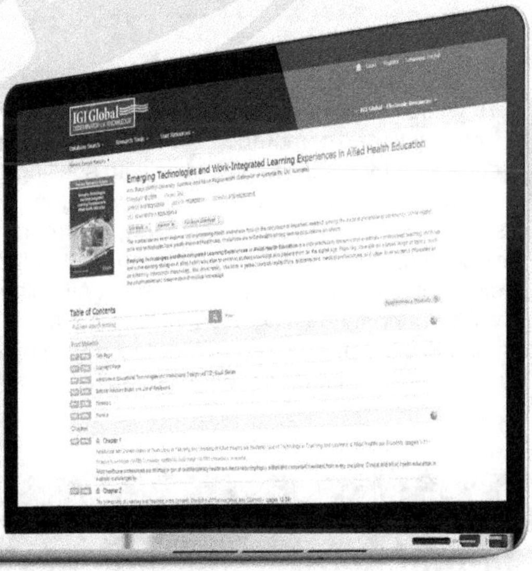

www.ingramcontent.com/pod-product-compliance
Lightning Source LLC
LaVergne TN
LVHW061543290926
849162LV00004B/891